Proceedings of the 29th International Geological Congress Part D

Also available from VSP

PROCEEDINGS OF THE 29TH INTERNATIONAL GEOLOGICAL CONGRESS - PART A:
Metamorphic Reaction: Kinetics and Mass Transfer
Edited by T. Nishiyama and G.W. Fisher
Sandstone Petrology in Relation to Tectonics
Edited by F. Kumon and K.M. Yu
Evaporite and Desert Environment
Edited by Y. Watanabe and A. Motamed

PROCEEDINGS OF THE 29TH INTERNATIONAL GEOLOGICAL CONGRESS - PART B:
Reconstruction of the Paleo-Asian Ocean
Edited by R.G. Coleman
Quaternary Environmental Changes
Edited by E.H. Juvigné

PROCEEDINGS OF THE 29TH INTERNATIONAL GEOLOGICAL CONGRESS - PART C:
Siliceous, Phosphatic and Glauconitic Sediments of the Tertiary and Mesozoic
Edited by A. Iijima, A.M. Abed and R.E. Garrison

Related titles

Facies Models in Exploration and Development of Hydrocarbon and Ore Deposits
Edited by A.H. Bouma and R.M. Carter

Regional Metamorphism of Ore Deposits and Genetic Implications
Edited by P.G. Spry and L.T. Bryndzia

Tectonics of Circum-Pacific Continental Margins
Edited by J. Aubouin and J. Bourgois

PROCEEDINGS OF THE 29TH INTERNATIONAL GEOLOGICAL CONGRESS PART D

Kyoto, Japan, 24 August - 3 September 1992

Circum-Pacific Ophiolites

Editors: A. Ishiwatari, J. Malpas and H. Ishizuka

Utrecht, The Netherlands, 1994

VSP BV
P.O. Box 346
3700 AH Zeist
The Netherlands

© VSP BV 1994

First published in 1994

ISBN 90-6764-176-6

All rights reserved. No part of this publication may be reproduced, stored in a retrieval system, or transmitted in any form or by any means, electronic, mechanical, photocopying, recording or otherwise, without the prior permission of the copyright owner.

CIP-DATA KONINKLIJKE BIBLIOTHEEK, DEN HAAG

Proceedings

Proceedings of the 29th International Geological Congress.
- Utrecht : VSP
Pt. D / ed.: A. Ishiwatari ... [et al.].
ISBN 90-6764-176-6 bound
NUGI 816
Subject headings: geology.

Printed in The Netherlands by A-D Druk, Zeist.

CONTENTS

Preface   1

CIRCUM-PACIFIC OPHIOLITES

Circum-Pacific Phanerozoic multiple ophiolite belts
A. Ishiwatari   7

Comparative genesis and tectonic setting of ophiolitic rocks of the South and North Islands of New Zealand
J. Malpas, I.E.M. Smith, and D. Williams   29

The Cambrian Trinity ophiolite and related rocks of the lower Paleozoic Trinity complex, northern California, U.S.A.
N. Lindsley-Griffin   47

The Shulaps ophiolite complex of British Columbia, Canada: a Paleozoic/Mesozoic arc-related microterrane.
J. Malpas, T.J. Calon, R.W.J. MacDonald   69

The diversity of accreted oceanic lithosphere in the Brooks Range, Alaska
K.R. Wirth, J.M. Bird, and J.N. Wessels   89

Tectonic position of the Mesozoic ophiolitic and island arc formations in the Koryak region (northeastern Russia)
N. Filatova and V. Vishnevskaya   109

Island-arc mafic-ultramafic plutonic complexes of North Kamchatka
V.G. Batanova and O.V. Astrakhantsev   129

High and low pressure cumulates of Paleozoic ophiolites in Primorye, eastern Russia
S.V. Vysotskiy   145

Geology and petrology of the Shimokawa ophiolite (Hokkaido, Japan): ophiolite possibly generated near R-T-T triple junction
S. Miyashita and A. Yoshida   163

OTHER OPHIOLITES

Time-space distribution and tectonic types of ophiolites in China
X. Wang and Z. Hao   183

The Hongguleleng island-arc ophiolite, a sequence of multiple intrusions in the Paleozoic ophiolites of Xinjiang, China
R. Laurent, X. Wang, and P. Bao   205

Petrology and tectonic settings of the Neyriz ophiolite, southeastern Iran
K. Sarkarinejad   221

Polychronous ophiolite belts of central Kazakhstan and their evolution
A.S. Yakubchuk   235

Geological and structural conditions localizing ornamental stone occurrences in the ophiolites of Itmurunda zone, Kazakhstan
I. Kovalenko, G. Aerov, and Z. Bagrova   255

Riphean ophiolites of the northern Baikal region (East Siberia)
M.I. Grudinin and I.A. Demin   263

Late Proterozoic ophiolite pulse
A.S. Yakubchuk, A.M. Nikishin, and A. Ishiwatari   273

Preface

Ophiolite complexes occur in every orogenic belt on the face of the Earth as crudely stratified series of rocks consisting of peridotite at the base, gabbro in the middle, and basalt at the upper levels. The size of a typical complex is greater than 100 km in length, 10 km wide but most are only a few km thick. In the early 1970's, these ophiolite complexes were identified as fragments of ancient oceanic lithosphere (i.e. oceanic crust and parts of the underlying upper mantle) incorporated as part of the orogenic belts by plate tectonic processes. It is not surprising then, that since that time, ophiolite studies have contributed much to the understanding of plate tectonics, the origin of orogenic belts and the nature of construction and alteration of oceanic lithosphere itself.

There is now a number of well-studied typical ophiolite complexes including the Samail Complex of Oman, the Troodos Complex of Cyprus, and the Bay of Islands Complex of Newfoundland, Canada. These occur in orogenic belts formed during continental collision and now rest upon one of the continental blocks, marking the suture zone as remnants of the otherwise subducted oceanic lithosphere. Other ophiolites occur in the circum-Pacific orogenic belts which have been interpreted as assemblages of juxtaposed micro-continents (or accreted terranes). These ophiolites are often quoted as dismembered and metamorphosed examples, marking sutures between micro-continents. However, recent studies of accretionary complexes, together with the rapid development of radiolarian biostratigraphy, have shown that most circum-Pacific orogenic belts consist of sedimentary sequences deposited directly onto ocean floor and trench slopes, and accreted into subduction zones over a very long time span (>200 m.y.). The circum-Pacific ophiolites do not therefore occur between micro-continents but amongst these accretionary complexes of oceanic origin. They are characterized by temporal diversity and a multiplicity of igneous and tectonic processes occurring during a long-lasting, accretionary process. Such characteristics of the circum-Pacific ophiolites were actually pointed out from the Klamath Mountains in California in the late 1970's, but their pan-Pacific geotectonic importance has not been recognized until recently.

This volume focuses in the multiplicity and diversity of the circum-Pacific Phanerozoic ophiolites and their intra-continental analogues. After an introductory overview, papers deal with particular segments of the ophiolite belts in an order passing counter-clockwise around the Pacific Ocean from New Zealand to Japan with stops in California, British Columbia, Alaska, Koryakia, Kamchatka, and Sikhote Alin. These papers are followed by contributions on multiple ophiolite belts within the Asian continent (in China and Kazakhstan), as well as a paper on a Tethyan ophiolite in Iran. The last portion of this volume includes a report from Baikal and a general review of Late Proterozoic ophiolites.

This volume comprises the proceedings of the ophiolite symposium (No. I-3-26) of the 29th International Geological Congress (IGC) held in Kyoto, Japan, on the 26th and 27th of August, 1992. The editors wish to thank the participants of the symposium whose contributions are presented in the following pages. Abstracts of these contributions are printed on Pages 132-140 of the Abstracts Volume 1 of the 29th IGC. In addition, the editors wish to express their sincere thanks to the following people who have kindly reviewed the papers included in this volume; N. Lindsley-Griffin, K. Wirth, A.H.F. Robertson, T. Falloon, C. Xenophontos and a number of anonymous reviewers. We also acknowledge the financial support for this publication offered by the 29th IGC Organizing Committee.

A. Ishiwatari, J. Malpas and H. Ishizuka
Kyoto, January 1994

29th International Geological Congress, Kyoto 1992
Symposium I-3-26 "Ophiolites"
Executed Program

Oral Session: 26 Aug. 1992 (2nd day; Wednesday)
Poster Session: 27 Aug. 1992 (3rd day; Thursday)

Conveners: A. Ishiwatari (Kanazawa Univ., Kanazawa 920, Japan)
J. Malpas (Memorial Univ., Newfoundland A1B 3X5, Canada)
H. Ishizuka (Kochi Univ., Kochi 780, Japan)

ORAL SESSION (The numbers correspond to those in the Abstract Volume)

Convener: A. Ishiwatari
O-1. 8:30-8:48 Ishiwatari, A. (Dept. Earth Sci., Kanazawa Univ., Japan)
Circum-Pacific ophiolites: Time-space distribution and petrologic diversity.
O-3. 8:48-9:06 Malpas, J. and Moore, P. (Memorial Univ. of Newfoundland, Canada)
Ophiolites of the Dun Mountain Belt, Nelson, New Zealand.
O-4. 9:06-9:24 Smith, I.E.M., *Malpas, J., Black, M. and Sporli, K.B. (Dept. Geol., Univ. Auckland, N.Z., *Dept. Earth Sci., Memorial Univ. Newfoundland, Canada)
The Northland ophiolite, New Zealand.
O-5. 9:24-9:42 Spadea, P. (Earth Sci. Inst., Univ. of Udine, 33100 Italy)
Petrology and geochemistry of Mesozoic ophiolites from the southwestern Colombian Andes, South America.
O-2. 9:42-10:00 Watanabe, T. (Dept.Geol.Min., Hokkaido Univ., Sapporo, Japan), Leitch, E.C. (Univ.Tech. Sydney, N.S.W., Australia), Fukui, S. (Okayama Univ.Sci., Okayama, Japan), Ikeda, Y. (D.G.M., Hokkaido Univ.), Allan, A.D (North Broken Hill-Peko, N.S.W., Australia) and Itaya, T. (Okayama Univ.Sci., Japan)
The origin and tectonic significance of meta-ophiolites in the Woodsreef terrane, New England Fold Belt, eastern Australia.

Coffee Break: 10:00-10:12

Convener: J. Malpas
O-6. 10:12-10:30 Lindsley-Griffin, N. (Dept. Geol., Univ. of Nebraska, Lincoln, NE 68588-0340, U.S.A.)
The ophiolitic lower Paleozoic Trinity Complex of northern California, U.S.A.
O-7. 10:30-10:48 Wirth, K.R. (Geol. Dept., Macalester Coll., St. Paul, MN 55105, U.S.A.) and Bird, J.M. (Dept. Geol. Sci., Snee Hall, Cornell Univ., Ithaca, NY 14853, U.S.A.)
The diversity of accreted oceanic lithosphere in the Brooks Range, Alaska.
O-8. 10:48-11:06 Batanova, V.G. and Astrakhantsev, O.V. (Geol. Inst., Acad. Sci., Moscow, Russia)
Tectonic setting and genesis of island arc ultramafic-mafic plutonic complexes from North Kamchatka.
O-9. 11:06-11:24 Filatova, N. and Vishnevskaya, V. (Inst. Lithosphere, Academy of Sciences, Moscow, Russia)
Position of the Circum-Pacific ophiolitic and island-arc formations in the Mesozoic framework of the NW Pacific continental margin.

O-10. 11:24–11:42 Vysotsky, S.V. (Far East Geol. Inst., Vladivostok, Russia)
Mineralogy and petrology of ophiolites from the North Primorye: evidences of high-pressure crystallization.
O-11. 11:42–12:00 Miyashita, S. (Dept. Geol. Min., Niigata Univ., Japan)
The Shimokawa ophiolite was probably generated at a RTT triple junction.

Lunch: 12:00–13:30

Convener: H. Ishizuka
O-12. 13:30–13:48 Yumul, G.P.Jr. (National Inst. Geol. Sci., Univ. Philippines, Quezon, Philippines)
REE Geochemistry of a Marginal Basin Ophiolite: Zambales ophiolite complex, Luzon, Philippines.
O-13. 13:48–14:06 Ishii, T. (Ocean Res.Inst., Univ.Tokyo, Tokyo 164, Japan), Robinson, P.T. (Centre for Marine Geol., Dalhousie Univ., Halifax, B3H 3J5 Canada), Maekawa, H. (Dept.Earth Sci., Kobe Univ., Kobe 657, Japan) and Fiske, R. (Dept.Min.Sci., NHB-119, Smithsonian, Wash.D.C., 20560 U.S.A.)
Proto-ophiolite; peridotites from diapiric serpentinite seamounts in the Izu–Ogasawara–Mariana forearc, ODP Leg 125.
O-14. 14:06–14:24 Robertson, A.H.F. (Dept. Geol. Geophys., Univ. of Edinburgh, U.K.) and Jones, G. (Dept.Earth Sci., Univ. Leeds, Leeds, U.K.)
Tethyan versus circum-Pacific ophiolite terrains: comparison and overview.
O-15. 14:24–14:42 Nicolas, A. (Sci. de la Terre, Univ. Montpellier II, France)
Dynamic magma chambers at fast spreading ridges: evidence from the Oman ophiolite.
O-16. 14:42–15:00 Pedersen, R.B. (Geol. Inst., Univ. Bergen, Norway)
Boninitic magmatism within Caledonian ophiolites: age relations, isotope geochemistry, petrology and petrogenetic constraints.

Coffee Break: 15:00–15:12

Convener: A. Ishiwatari
O-17. 15:12–15:30 Bédard, J.H. (GSC, Centre. Géosci. Québec, Canada) and Hébert, R. (Dépt. Géol., Univ. Laval, Québec, Canada)
The lower crust of the Bay of Islands ophiolite, Canada; evidence for multiple syn-kinematic intrusion, hybridization and assimilation.
O-18. 15:30–15:48 Jedrysek, M.O. (Inst. Geol. Sci., Wroclaw Univ., Poland)
An inhomogeneity of mesostructural features of Sleza Ophiolite SW Poland.
O-19. 15:48–16:06 Laurent, R. (Dept. Geol., Univ. Laval, Quebec, Canada), Wang, X. and Bao, P. (Inst. Geol., Chinese Acad. Geol. Sci., Beijing, China)
The Hongguleleng island-arc ophiolite, a sequence of multiple intrusions in the Paleozoic ophiolites of Xinjiang, China.
O-20. 16:06–16:24 Zhu, X. (Lab. Ore Depo., Geochem. Inst., Academia Sinica, Guiyang, China)
Geochemistry of boron in Dalabut ophiolites, Northwest China.
O-21. 16:24–16:42 Sabzehei, M. (Geological Survey, Kerman, Iran)
New reflections on the textures and structures of ophiolitic layered gabbros: evidences from Southeast Iranian ophiolites.

POSTER SESSION (In alphabetical order of the first author)

P-10. Hebert, R., Varfalvy, V. (Dept.Geol., Univ.Laval, Quebec, Canada) and Bedard, J.H. (Centre Geosci. de Quebec, Canada)
Petrogenesis of trapped melts in upper mantle peridotites exposed in North Arm Massif, Bay of Islands ophiolite, Newfoundland, Canada.

P-11. Kovalenko, I.V. (VNIISIMS, Russia), Akrov, G.D. (Balkhash, Kazakh Repub.) and Bagrova, Z.A. (Russia)
Location peculiarities of coloured stones in the Itmurundinsky ophiolite zone, Kazakhstan.

P-1. Malpas, J. and Calon, T. (Dept. Earth Sci., Memorial Univ. Newfoundland, Canada)
Shulaps ophiolite in the Canadian Cordillera.

P-16. Osozawa, S. (Inst. Geol. Paleont., Fac. Sci., Tohoku Univ., Japan) and Okamura, M. (Dept. Geol., Fac. Sci., Kochi Univ., Japan)
The Troodos ophiolite was probably formed through subduction of an active mid-ocean ridge.

P-2. Sano, S. (Geol. Labo., Fac. Educ., Ehime Univ., Japan)
Petrogenesis of the Red Hills peridotite in the Dun Mountain ophiolite belt, New Zealand.

P-3. Sarkarinejad, K. (Dept. Geol., Shiraz Univ., Shiraz, Iran)
Petrology and tectonic settings of the Neyris ophiolite, southeastern Iran.

P-4. Shahpasandi, I. and Shiva Kumar, B.S. (Geol.Dept., Bangl.Univ., India)
Chromite bearing ultramafics as "ophiolitic assemblage" in the Nuggehalli schist belt of Hassan district, Karnataka, India.

P-6. Vuollo, J.I., Piirainen, T.A., Nykaenen, V. (Dept. of Geology, Univ. of Oulu, Finland) and E. Kontas (Geol. Surv., Rovaniemi, Finland)
The Outokumpu-Jormua ophiolite belt in Eastern Finland and preliminary data on its PGE distribution

P-26. Yakubchuk, A.S. (Geol. Fac., Moscow Univ., Russia)
The ophiolites of the Central Kazakhstan.

P-27. Yakubchuk, A.S. and Nikishin, A.M. (Geol. Fac., Moscow Univ., Russia)
The ophiolite and oceanic crust pulses during Phanerozoic.

ABSTRACTS (In alphabetical order of the first author)

O-22. Anand, R. (Dept. Earth Sci., Manipur Univ., Imphal, India)
Boninites from Dras island arc setting of Indus Ophiolite, Ladakh, India.

P-8. Deng, W. (Lab. Lithosph. Tect. Evol., Inst. Geol. China)
A preliminary investigation of the ophiolite and tectonic evolution in Northwest Tibetan Plateau, China.

P-9. Fedorchuk, A.V. (Inst. of Lithosphere, Acad. Sci., Moscow, Russia)
MORB-like and BABB-like remnants in accretionary complexes of the eastern Kamchatka.

P-7. Grudinin, M.I. and Demin, I.A. (Earth Crust Inst., Irkutsk, Russia)
Riphean ophiolites of the northern Baikal Region (East Siberia)

P-12. Krylov, K. and Grigoriev, V. (Geol.Inst., Moscow, Acad.Sci., Russia)
Kuyul ophiolite complex, northern Kamchatka, Russia (age, composition, and structure).

P-13. Laz'ko, E.E. (Inst. Ore Depo. Geol., Acad. Sci., Moscow, Russia)
The layered mafic-ultramafic series of Khabarny ophiolite, South Ural, Russia: an unusual example of island arc related magma mixing.

P-14. Li, Y., Lu, G. and Hao, J. (Inst. Geol., Academia Sinica, Beijing, China)
Banxi Group and Banxi ophiolite melange. (South China)

P-15. Liu, C. (Chengdu Inst. Geol. Min. Res., Chengdu, China)
Paired basic-ultrabasic rock belts and relations to plate tectonics in the Qinghai-Xizang (Tibet) plateau.

P-17. Oxman, V.S., Parfenov, L.M., Prokopiev, A.V., Timofeev, V.F. and Tretyakov, F.F. (Yakutian Geol.Inst., Yakutsk, Russia)
The Chersky Range ophiolite (North-East Asia)

P-18. Palandzhyan, S.A. (Northeast Interdisc. Res. Inst., Magadan, Russia)
Chemical geodynamics of the peridotites: petrochemical characteristics of the ophiolite's mantle sequence in various geodynamic environments (GDE).

P-19. Rai, H. (Wadia Inst. of Himalayan Geol., Dehradun, India)
Volcanism in the Shyok ophiolite melange zone, northern Ladakh, Himalaya, India.

P-20. Satian, M. (Lab. Lith., I.G.S., Armen. Ac. Sc., Yerevan, Russia)
"Atypical" ophiolites (Lesser Caucasus)

P-21. Schira, W. (Residenzplatz 10a, W-8390 Passau, Germany)
Geochemical signatures of the Coast Range ophiolite, south-central Chile.

P-22. Shteynberg, D. and Chaschchukhin, I. (Inst. Geol. & Geochem., Urals Branch, Academy of Sciences, Russia)
The Alpine-type ultramafites: origin and geological position

P-23. Varma, O.P. (Indian Geological Congress, Univ. Roorkee, India)
Proterozoic ophiolitic mafic and ultramafic complexes of Orissa (India): a study of their tectonic features, petrogeneses and metallogeneses for a genetic correlation.

P-24. Vashchilov, Yu.Ya. (NE Interdisc. Res. Inst., Magadan, Russia)
Ophiolites of Anadyr-Koryak region; deep structure and dynamics.

P-25. Vedernikov, N.N. (VNIIgeolnerud, Zinin, Kazan 420097, Russia)
Ophiolites in structural evolution of the earth crust and their mineragenetic meaning.

P-28. Wang, X. (Inst. Geol., Chinese Acad. Geol. Sci., Beijing, China)
Tectonic types of ophiolites in the orogenic belts of China.

Note: Abstracts of the papers listed above are printed in the Abstract Volume 1 (pp. 132-140) of the 29th IGC. However, the last two abstracts (P-25 and P-28) are mistakenly lacking in the Abstract Volume, and are printed in the following page.

Additional Abstracts of the 29th IGC Ophiolite Symposium (I-3-26):

P-25

Ophiolites in structural evolution of the earth crust and their mineragenetic meaning

N.N. VEDERNIKOV (VNII geolnerud, Kazan 420097, Russia)

We can distinguish two groups among ophiolite complexes of the earth crust. Ophiolites of Group A, by character of their geologic section, are the analogues of oceanic crust. Those of Group B, including also ultramafites, mafites, and volcanogenic sediments, have different rock composition and thickness. The appearance of the first group in the crust took place due exclusively to tectonic-magmatic processes. The ratio of these two ophiolites groups in fold systems have changed through the history of the earth crust evolution. Ophiolites of Group A dominated the initial stages, and ophiolites of Group B did the later stages.

According to the conditions of localization, the following types of ophiolites complexes are distinguished. They formed:
1) near the edge of the platforms, microcontinents, and island arcs on the sialic (continental) base;
2) in the margin of the platforms or in microcontinent, in connection with the first type;
3) as a result of clustered and arched uplift of the crust near island arcs on the simatic base;
4) in zones of main deep faults as a result of extrusion and following migration of large ophiolite plates.

The first and second types dominate the ophiolites of the A group, and the third and fourth types do the B group. The evolution of the two ophiolite types reflects geotectonic and mineragenetic special features of folded zones.

P-28

Tectonic types of ophiolites in the orogenic belts of China

X. WANG and P. BAO (Institute of Geology, Chinese Academy of Geological Sciences, Beijing, China)

Ophiolites of China extensively occur in the large orogenic belts of different geological age, constituting 14 large ophiolite belts. The types of ophiolites are complicated and multiple. They can be divided into three main tectonic types on the basis of the formation stage of ophiolite:

The first is of initial oceanic basin type. It is formed in initial oceanic basin dominated by subsidence prior to spreading. It is characterized by absence of sheeted dyke swarms and occurrence of interlayered volcanics and abyssal sediments with large thickness. This type of ophiolite belongs to the transitional crust and can be called a transitional type of ophiolite.

The second is of mature oceanic basin type. It is formed in mature oceanic basin at the spreading stage. It is characterized by development of sheeted dyke swarms and a thinner oceanic crust. This type belongs to typical oceanic crust and can be called a normal ophiolite.

The third is of closed oceanic basin type. It is formed in remnant oceanic basin at closing stage. The occurrence of pillow lava in the upper crust and of basic pluton enriched in iron in the lower crust and absence of cumulates and mantle peridotites are characteristic of this type. This "ophiolite" is a part of continental crust and may be characterized as "non-ophiolite".

The ophiolites in the orogenic belts of China are composed mainly of the first type and secondly of the second type.

Circum-Pacific Phanerozoic Multiple Ophiolite Belts

A. ISHIWATARI

Department of Earth Sciences, Faculty of Science, Kanazawa University, Kanazawa 920-11, JAPAN

Abstract. Ophiolites in most of the circum-Pacific orogenic belts have very diverse ages, which covers full-Phanerozoic time span even within a single orogenic belt. Tectonic superposition of ophiolite nappes and melanges of widely varying ages, which are generally younger in structurally lower nappes, as well as extreme petrologic diversity and highly dismembered occurrence, are characteristic of the circum-Pacific ophiolites in Japan, eastern Russia, and western U.S.A.
On the other hand, ophiolites in the southwestern Pacific margin from the Philippines to New Caledonia are mostly young and uniform in age (Late Cretaceous to Eocene). The uniform age of the ophiolites, the contemporary voluminous production of oceanic lithosphere in the adjacent marginal basins, the presence of currently spreading back-arc basins, and a well-developed upper-mantle low-velocity zone suggest that the ophiolites in the southwestern Pacific regions formed in oceanic rift zones, which were as extensive as other mid-oceanic ridges but which subsequently decomposed into many island arc-marginal basin systems.
The ophiolites of the circum-Pacific regions mostly represent portions of fore-arc lithosphere thrust onto the subducting oceanic plate, as demonstrated by some subaqueous ophiolite outcrops in the present trench walls. The circum-Pacific ophiolites are found mostly on or among the accreted oceanic and trench sediments characterizing active continental margins. This is in clear contrast to the Tethyan ophiolite occurrences, where ophiolites are mostly emplaced on passive continental margins through continental collision events.
The temporal distribution of Phanerozoic ophiolites, both in circum-Pacific and other areas, show distinct peaks in Jurassic-Cretaceous and Ordovician times with a smaller peak in the Permian. These pulses of ophiolite generation correspond to the long periods without geomagnetic reversals, high eustatic sea levels, major oil and coal production, and voluminous oceanic crust production, suggesting the explosions and subsequent surges of superplumes possibly originating at the core/mantle boundary. Although formation and emplacement of an ophiolite may be a superficial plate-tectonic process, episodic contemporaneous generation and subsequent emplacement of ophiolites on a global scale may be better explained by the earth's thermal pulses with the plume-surge tectonics.

Keywords: ophiolite pulses, superplumes, surge tectonics

INTRODUCTION

Circum-Pacific ophiolites occur primarily among accretionary complexes, which consist of sequences of nappes, each of which are characterized by a unique rock sequence called the "oceanic plate stratigraphy" [37]. The rock sequence starts with thin basal chert, grades upward into siliceous and then terrigenous shale, and ends in thick sandstone and olistostrome (Fig. 1). This sequence represents a long period of pelagic deep-sea sedimentation (chert) followed by a short period of terrigenous trench-fill sedimentation (sandstone) on a portion of oceanic crust that traveled from mid-oceanic ridge to subduction zone. The sequence also includes occasional fragments of seamount-limestone complexes and island-arc volcanics or tuffs among the terrigenous sediments. Older continental basement is generally lacking in the underlying accretionary complex, which is generally younger than the overlying ophiolite. Such occurrences of circum-Pacific ophiolites are clearly different from those of the Alpine-Himalayan ophiolites emplaced upon old crystalline basement with the sedimentary cover which constituted a passive continental margin. The sedimentary lithologies in a nappe become gradually younger in structurally lower levels in the nappe sequence (generally oceanward in plan view) (Fig. 1), suggesting long-lasting (e.g. from middle Paleozoic to the present) subduction-accretion processes such as successive underplating of younger sediments beneath an older accretionary wedge.

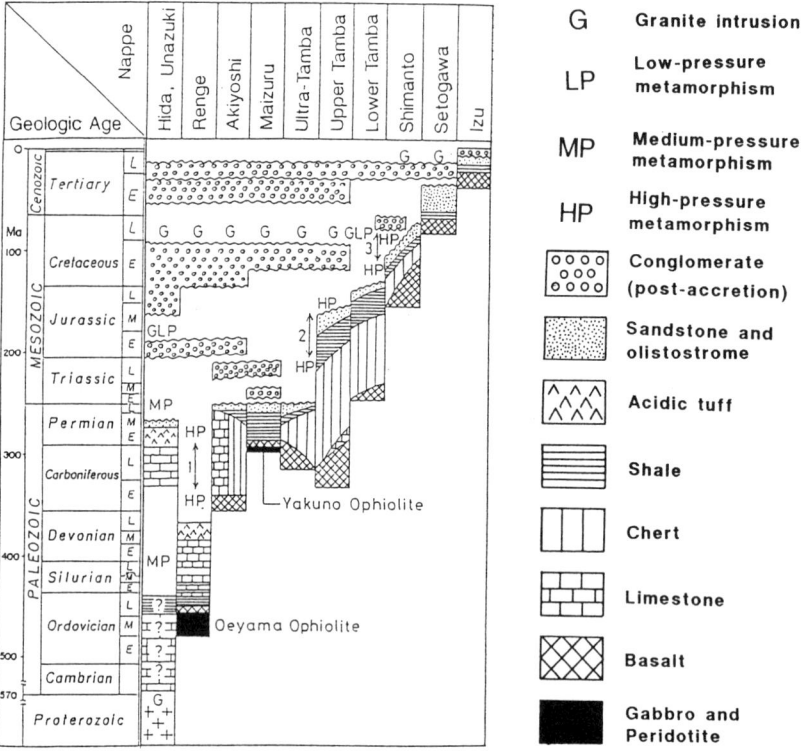

Figure 1. Time-lithology diagram of each nappe (or nappe group) in southwestern Japan [34, 63]. The columns are arranged from right to left as going downward (generally oceanward in plan) through the nappe pile. Major high-pressure metamorphic events are numbered (1: Sangun-Renge, 2: Sangun-Hatto and -Suo, 3: Sambagawa), E, M, and L in the time scale stand for early, middle, and late, respectively. Note that the length of the column represents time span, but does not correspond to actual thickness.

Ophiolitic nappes and melanges in the accretionary orogenic belts also show the same downward-younging trend [32, 33]. The ophiolites occupying structurally higher positions may be of Paleozoic age, while those in the lower part may be of Mesozoic or even Cenozoic age. The multiplicity of the circum-Pacific ophiolites both in space (esp. vertical superposition) and time is in clear contrast to the temporal singularity among the orogenic belts in the continental hemisphere. For example, the Alpine-Himalayan ophiolite belt includes only Jurassic (Alps to Greece) and Cretaceous (Greece to Himalaya) ophiolites [1], while the Appalachian-Caledonian-Uralian ophiolite belts are dominated by Ordovician ophiolites [20].

The first purpose of this paper is to establish the concept of multiple ophiolite belts using typical examples in Japan, eastern Russia, and the western U.S.A., and compare less typical examples in Alaska, eastern Australia, New Zealand, etc.

Most ophiolites in the world are younger than 1 Ga, and their formation ages show distinct peaks in the Jurassic-Cretaceous, Cambro-Ordovician, and Late Proterozoic (Late Riphean) times (Fig. 2) [1, 64, 32, 33]. The Jurassic-Cretaceous and Cambro-Ordovician pulses are apparent also in the circum-Pacific areas, and are not consistent with models of ophiolite being continuously generated by magmatic processes along mid-ocean ridges and randomly emplaced along convergent plate boundaries. The youngest and most voluminous ophiolite belts in the Western Pacific area are of almost uniform in age (Late Cretaceous to Eocene) [68], and do not fit the North-Pacific multiple accretionary model. This Late Cretaceous-Cenozoic ophiolite belt may represent the latest, small-scale ophiolite pulse.

The second purpose of this paper is to discuss origin of the ophiolites in terms of "superplumes" [47, 48] and related new ideas, and to consider their relationship to steady-state subduction-accretion processes.

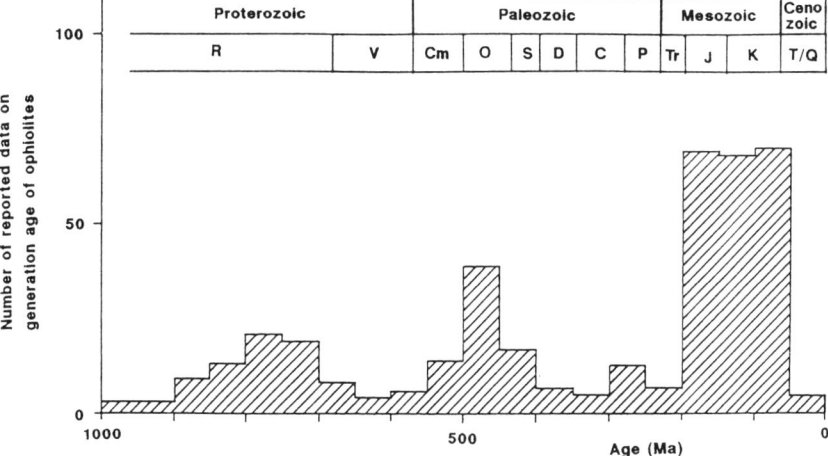

Figure 2. Histogram of the formation ages of ophiolites on the earth for every 50 m.y. since 1 Ga based on the compilation of Abbate et al. [1] and Yakubchuk et al. [85]. Geologic ages are shown in the upper part. The 750, 450 and 150 Ma peaks are strongly biased with Arabian (Pan-African), Appalachian-Caledonian, and Alpine-Himalayan data, respectively.

TYPICAL MULTIPLE OPHIOLITE BELTS

1. Japan and Sikhote Alin
The Japanese Islands are typical of the circum-Pacific Phanerozoic multiple ophiolite belts [32, 33]. The formation ages of the ophiolites here range from early Paleozoic (Ordovician) to Cenozoic. In southwestern Japan (Fig. 3), the Oeyama ophiolite (Ordovician) is thrust over the early Permian Yakuno ophiolite, which in turn is thrust over the Permian Ultra-Tamba accretionary complex. These Paleozoic ophiolites and accretionary complexes are underlain by the Jurassic Tamba (Mino) accretionary complex, whose outlier, distributed along the Pacific coast (Chichibu complex), is further underlain by Cretaceous (Shimanto) and then by Tertiary (Mineoka-Setogawa) accretionary complexes. The Chichibu complex includes the Mikabu ophiolitic melange in its basal part, and the Shimanto and Mineoka complexes also

include some ophiolitic melanges. Thus, the full Phanerozoic temporal range and the downward (oceanward) younging trend of the accretionary complexes and ophiolites are well demonstrated by this example. On the other hand, the Oeyama and Yakuno ophiolites in the Hida marginal belt are tectonically overlain by the Hida gneiss complex, in which Paleozoic clastic, basic volcanic, and calcareous protoliths were by polymetamorphosed during the Paleozoic and then cut by Triassic granite. A recent geologic review of southwestern Japan is presented in more detail by Nakajima et al. [63].

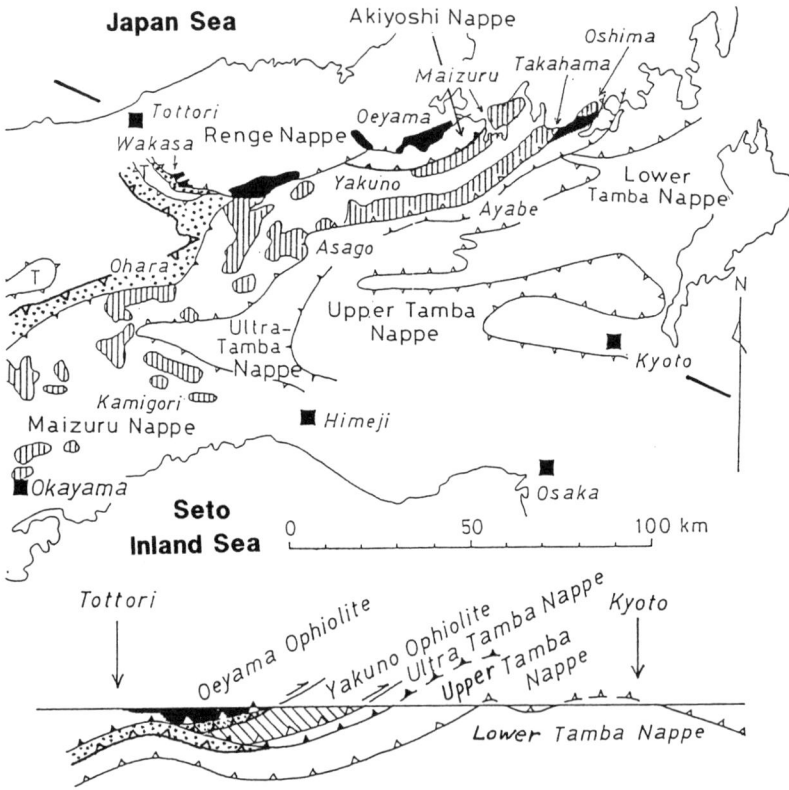

Figure 3. Geologic structure of the Inner Zone of southwestern Japan [34]. Black area: ophiolitic ultramafic rocks, lined area: ophiolitic mafic rocks, stippled area: high-pressure metamorphic rocks, sawtoothed line: thrust fault. Square represent large cities. See Figure 1 for lithostratigraphy of each nappe.

The structure of the accretionary complex is essentially the same in northeastern Japan [32, 33]. The Ordovician Miyamori ophiolite is underlain by the Jurassic accretionary complex of the northern Kitakami Mountains and southern Hokkaido. The Jurassic Horokanai ophiolite in central Hokkaido is underlain by the Kamuikotan melange, which in turn is underlain by the Cretaceous Hidaka accretionary complex that contains some ophiolitic fragments. However, the tectonic relationships in south-central Hokkaido have been complicated by westward backthrusting during the Miocene Hidaka orogeny.

The Japanese ophiolites also show wide petrological and geochemical diversity. The Ordovician Oeyama and Miyamori ophiolites are dominated by mantle peridotite with minor ultramafic and mafic cumulates and very rare volcanic rocks. The mantle peridotite of the Oeyama ophiolite is relatively fertile, consisting of clinopyroxene-rich harzburgite that closely resembles abyssal peridotite. Peridotite in the Miyamori ophiolite is relatively depleted harzburgite that is phlogopite- and amphibole-bearing, and is interpreted to represent the hydrated and metasomatised wedge mantle over a subduction zone [65]. More pronounced diversity is exhibited among younger, structurally complete ophiolites. The Jurassic Horokanai ophiolite in Hokkaido [36] is characterized by highly depleted harzburgite and orthopyroxene-type cumulates, the Permian Yakuno ophiolite in southwestern Japan [31] includes moderately depleted, clinopyroxene-bearing harzburgite associated with clinopyroxene-type cumulates, and the Cretaceous(?) Poroshiri ophiolite [61] in Hokkaido is characterized by fertile mantle lherzolite and plagioclase-type cumulates. This variation results from widely different degrees of partial melting in the mantle [31, 33].

The ophiolites also exhibits diversity in the metamorphic sequences of their crustal sections. The Yakuno ophiolite exhibits "ocean-floor metamorphism" whose grade increases with depth from basalt through gabbro. Granulite-facies is attained at the Moho level, where green spinel-Al opx-Al cpx-plagioclase assemblages prevail. The olivine-plagioclase reaction forming spinel and two pyroxenes indicates pressures between 5 and 10 kb, suggesting a 15 to 30 km thick crust [30]. This is unusually thick for oceanic crust (normally 5 km thick). Well developed ocean-floor metamorphism down to the Moho and relatively high recrystallization pressure at the Moho is characteristic of some circum-Pacific ophiolites.

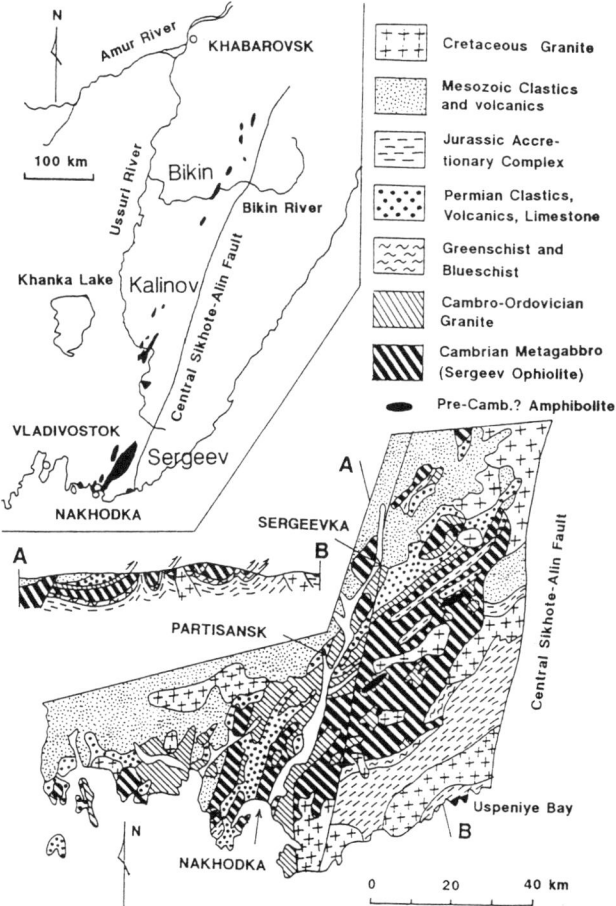

Figure 4. Distribution of ophiolites in Sikhote Alin (inset) and geologic map of the Sergeev ophiolite (simplified after Khanchuk's map in Zakharov *et al.* [86]).

The Sikhote-Alin Mountains may have been a northern continuation of southwestern Japan before the Miocene opening of the Japan Sea; the Proterozoic-early Paleozoic Khanka massif may continue to the Hida gneiss [87], the ophiolite belt on the east may continue to the Hida marginal belt, which includes the Oeyama and Yakuno ophiolites, and the Jurassic accretionary complex further to the east may continue to the Tamba (=Mino) accretionary complex [45]. The Sikhote-Alin ophiolite belt extends from Nakhodka to the northeast of Bikin over 400 km (Fig. 3a). The allochthonous nature of these ophiolites was first noted by Mazarovich [54]. Vysotskiy [79] divided this belt into the Sergeev, Kalinov, and Bikin ophiolites with different lithologies (and ages?).

The Sergeev ophiolite consists mainly of hornblende metagabbros with amphibolite inclusions, and is frequently cut by granite with 491 Ma muscovite Ar-Ar ages. This ophiolite is unconformably overlain by deformed Devonian clastics at the seaside outcrop to the west of Nakhodka. The ophiolite occurs as a nappe tectonically underlain by the Jurassic accretionary complex, which also appears in some tectonic windows (Fig. 3b). The Sergeev ophiolite may be of Cambro-Ordovician age, and is a possible equivalent of the Oeyama ophiolite, though its lithology is completely different. The Kalinov ophiolite is characterized by cpx-hornblende metagabbro, and the olivine-plagioclase reaction to form spinel and two pyroxenes takes place in the associated troctolite. The Bikin ophiolite is dominated by gneissose spinel-two pyroxene metagabbro [79] identical to that of the Moho level of the Yakuno ophiolite. The Kalinov and Bikin ophiolites may represent lower parts of relatively thick oceanic crust. Indeed, Khanchuk and Panchenko [44] recently discovered garnet-bearing metagabbro indicating high pressure in the Kalinov massif.

2. Koryak Mountains

Ophiolites ranging from early Paleozoic to late Mesozoic occur in the Koryak Mountains, which extend over 1000 km from the northern part of Kamchatka to Anadyr along the Bering Sea coast (Fig. 4a) [66, 76]. This accretionary orogenic belt is characterized by a nappe-pile structure similar to that in Japan (Fig. 4b) [71].

The oldest ophiolites of Ordovician to Devonian ages, appears along the Ust'-Belaya zone with some glaucophane schists and eclogites. The younger ophiolites of Permian to Jurassic ages, appear in the central zone of the orogenic belt (Koryak nappe system) mostly as klippen and melanges. The most conspicuous example is the Krasnaya Mountain, an isolated klippe of depleted harzburgite resting upon a Cretaceous accretionary complex. These klippen and melanges also include sediments ranging from Late Proterozoic(?) to Early Cretaceous [21]. The youngest ophiolites (Late Cretaceous to Paleocene) appear in the Olyutor zone along the Bering Sea coast of north Kamchatka with contemporary Alaskan-type mafic-ultramafic plutons of island-arc origin [5].

The Koryak-Kamchatka accretionary orogenic belt is structurally very similar to the Japanese analogue, but is very different in the constituent materials of the accretionary complexes [76]. Clastic material derived from Proterozoic continental crust is dominant among Japanese (and Sikhote-Alin) accretionary complexes, while volcaniclastic material is dominant in the Koryakian example, though both include common oceanic material (pillow basalt, limestone, and radiolarian chert) as tectonic blocks.

The Koryakian ophiolites are very diverse petrologically. Highly depleted harzburgite of the Krasnaya Mountain is found only a few km away from the Yagel serpentine melange, which includes mantle lherzolite. It should be noted that diamond and pyrope garnet were reported from some ophiolitic peridotites in the Koryak nappe system [74, 49]. Recently, coesite and in some cases diamond indicating more than 2.5 GPa pressure have been found from some ultra-high-pressure metamorphic terranes such as in the Alps, Norway, Urals,

Tien Shan, Dabie Shan, etc. From Shandong Province in eastern China, Wang et al. [81] reported coesite relics that survived recrystallization in granulite facies conditions (800°C, 1 GPa). It is possible that the diamond in ophiolitic peridotite is a relict mineral of deep mantle origin that survived partial melting in the upper mantle.

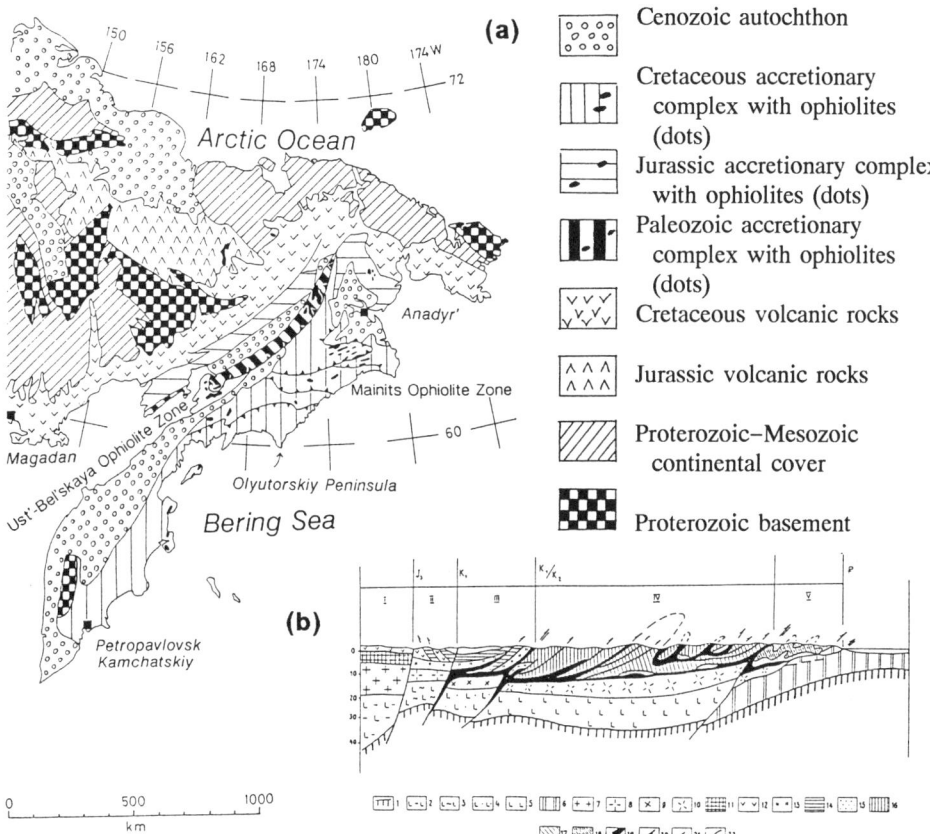

Figure 5. (a) Accretionary complexes and ophiolites in the Koryak Mountains (Russia) compiled after various sources. (b) Cross section of the Koryak accretionary complex along the 62°N latitude [71]. 1:Moho, 2-5:Basaltic layer of continental crust, 6:Oceanic crust, 7-10:Granitic layer of continental crust, 11:Proterozoic-Mesozoic cover, 12:Jurassic volcanics, 13:Paleozoic deep sea sediments, 14:Mesozoic autochthone, 15:Upper Paleozoic and lower Mesozoic with Tethyan fauna, 16-18:Allochthonous upper Jurassic to Cretaceous, 19:Ophiolites, 20-22:Tectonic boundaries. I:Omolon block, II:Jurassic zone, III:Ust'-Bel'skaya ophiolite zone, IV:Koryak nappe system, V:Ekonai nappe system.

3. Western U.S.A.

A beautiful example of a circum-Pacific Phanerozoic multiple ophiolite belt was first described in the Klamath Mountains by Irwin [27, 28, 29] and Coleman [15], though its pan-Pacific geologic significance has been only recently recognized. Here, the Cambro-Ordovician Trinity ophiolite [51] occupies the highest position of the nappe pile. It is underlain by a Devonian blueschist belt, which in turn is underlain by accretionary complexes (Permian to early Jurassic) with many ophiolitic fragments. These accretionary complexes are further underlain by the Josephine ophiolite of late Jurassic age, which in turn is underlain by the Franciscan accretionary complex of Cretaceous age.

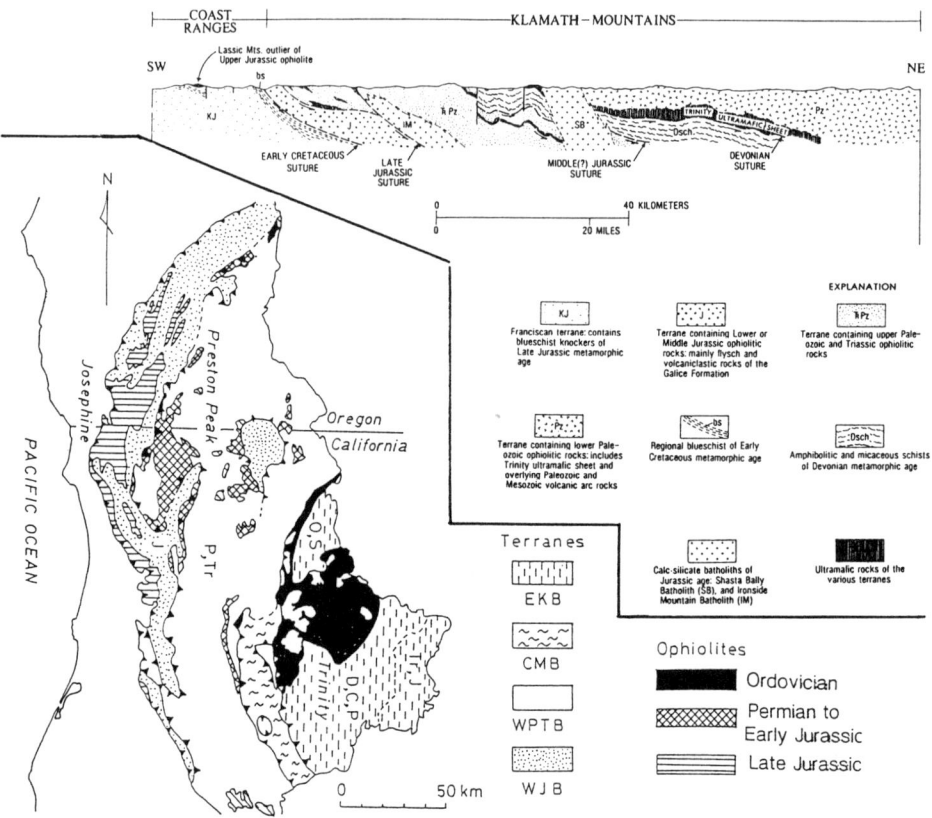

Figure 6. Geologic structure of the Klamath Mountains [28, 29]. EKB:Eastern Klamath belt, CMB:Central metamorphic belt, WPTB:Western Paleozoic-Triassic belt (Jurassic complex was also found recently), WJB:Western Jurassic belt. Franciscan accretionary complex of Cretaceous age occupies the coastal areas.

LESS TYPICAL EXAMPLES OF MULTIPLE OPHIOLITE BELTS

1. Alaska

An ophiolite belt is present along the Border Ranges Fault in southern coastal Alaska. The Peninsular terrane, with an ophiolitic basement, was thrust to south over the Chugach terrane. The Tonsina complex in this belt is interpreted as the basal part of the Jurassic island arc crust of the Peninsular terrane, which was thrust onto a Jurassic-Cretaceous subduction complex and accretionary wedge with some blueschists (Chugach terrane). This ophiolite belt may be a segment of the circum-Pacific ophiolite belt in view of the oceanward thrusting on a younger terrane. The deep seated nature of the Tonsina ophiolitic complex is well evidenced by the garnet granulite mineral assemblage at its Moho level, and is interpreted to represent the basal part of a thick island-arc crust [17].

Another ophiolite belt is present in the Brooks Range and around the Yukon-Koyukuk basin, northwestern Alaska [82]. The ages of these ophiolitic fragments range from

Carboniferous to Jurassic. They occur as nappes occupying the highest tectonic positions. These ophiolites seldom form a coherent succession, but are often composed of upper gabbro-peridotite nappes and lower basalt nappes separated by clear thrust boundaries. The gabbro-peridotite sequences may be mostly of Jurassic age, but the basaltic sequences are of Carboniferous-Permian (partly Devonian) ages. The underlying tectonic unit (Ruby "Geanticline") consists of Paleozoic and Precambrian(?) rocks (pelitic, carbonate, quartzose schists) metamorphosed in the greenschist and partly blueschist facies. This unit is frequently intruded by Cretaceous granite. The structurally lowest unit may be unmetamorphosed "Paleozoic" rocks including carbonate rocks, quartzite, conglomerate, shale, and chert. These ophiolites are believed to have been thrust to the north [69], the opposite sense to the oceanward thrusting pervasive in circum-Pacific areas. The nappe pile is also unusual in its upward-younging polarity. This is a peculiar example of the circum-Pacific ophiolite belts.

2. Chile
The Cordilleran orogenic belt along the Pacific coast includes Ordovician ophiolitic rocks, a late Paleozoic subduction complex containing ophiolitic fragments, and Early Cretaceous marginal-basin ophiolites (e.g. the Sarmiento and Tortuga complexes) [23].

3. Eastern Australia and New Zealand
Eastern Australia is underlain by a huge accretionary orogenic belt formed throughout the full-Paleozoic time span. The accretionary complexes in this orogenic belt show an eastward younging trend from the Cambrian Kanmantoo belt to the Permo-Triassic New England belt [50]. Fragmental ophiolite bodies, as represented by the Great Serpentine Belt marking the western edge of the New England belt, are widespread throughout the orogenic belt. The age of these ophiolites, however, seems to be uniformly Cambrian everywhere [2].

New Zealand may represent a younger accretionary orogenic belt originally attached to the eastern coast of Australia. The age of the accretionary complex shows an eastward-younging trend [53]. The "terranes" to the west of the Haast Schist, including the Permian (280 Ma) Dun Mountain ophiolite, may represent a Permian accretionary complex, which is in tectonic contact with a Cretaceous accretionary complex that contains various oceanic fragments as old as Carboniferous (Torlesse Terrane). On the North Island, these older terranes are discordantly cut by the Tertiary Waipapa terrane with Cretaceous or younger Northland ophiolite, which was emplaced during the Oligocene [53]. The protolith ages of Haast Schist are not yet known, but this unit may represent a Jurassic accretionary complex by analogy with Japan.

Although this Australian-New Zealand segment is not well interpreted in the context of accretionary tectonics, it is very probable that this segment has formed through a Paleozoic-Mesozoic accretionary history similar to Japan and the Klamath Mountains.

4. Indochina and Tibet
Paleozoic to Mesozoic ophiolites are present within Precambrian continental blocks in Indochina and the Malay peninsula, while Mesozoic ophiolites are present in Sumatra, and Cenozoic ophiolites in the Andaman and Mentawai Islands are along the convergent boundary between the Indian and Eurasian Plates [25]. These ophiolitic fragments form a belt that is analogous with circum-Pacific-type multiple ophiolite belts. These belts may extend to the north through Myanmar to the Qinghai-Tibet plateau, where six or more ophiolite belts of early Paleozoic to late Mesozoic ages are distributed between the Tarim basin and the Himalayas, with a southward younging polarity [80].

CENOZOIC OPHIOLITE BELTS IN SOUTHWEST PACIFIC

Ophiolites are of latest Cretaceous to Eocene ages in the Western Pacific area extending from the Philippine Sea to the Coral Sea. This segment of the circum-Pacific orogenic belts is clearly different from the other segments in the relatively young and uniform ages of the ophiolites.

1. Japan and Taiwan

The Japanese Islands are a typical segment of the circum-Pacific Phanerozoic multiple ophiolite belts as described above, and bear Cenozoic ophiolite fragments in Oligocene and Miocene melanges around the Izu peninsula [3], the collision point of the Izu-Mariana arc against the Japanese mainland (Honshu). These ophiolites include lithologies ranging from plagioclase lherzolite to highly depleted dunite (Fo_{94} olivine), as well as amphibolite, high-magnesian andesite, picrite, MORB, and plagiogranite. Subaqueous fore-arc ophiolite exposures reported from the Izu-Mariana trench walls will be mentioned later. The Liji melange of Taiwan bears the youngest on-land ophiolite (16-18 Ma) [38].

2. Philippines

The Zambales ophiolite extends north-south along the South China Sea coast for 150 km with 40 km width. This ophiolite bears economically important chromite deposits, and is well studied. Hawkins and Evans [22] concluded that the ophiolite is a fragment of an Eocene island arc-marginal basin lithosphere, consisting of the Acoje island arc segment and the Coto marginal basin segment which are in fault contact with each other. The Acoje block originated from a mantle source far more depleted than that of the Coto block. In the Acoje and coto segments, olivine is $Fo_{91.6}$ and $Fo_{89.9}$, spinel is Cr#73 and Cr#47, plagioclase is An_{95} and An_{85}, orthopyroxene is abundant and scarce, respectively. This ophiolite was thrust from east to west before Miocene time, and may not be a fragment of the Miocene South China Sea lithosphere to the east.

Palawan Island bridges Luzon and Borneo, and separates the South China Sea from the Sulu Sea. Ophiolitic rocks occupy more than one half of the island's surface, and are aligned in a northeast direction. Rashka et al. [72] noted that this ophiolite lacks a sheeted dike complex, though all other stratigraphic members are present. Ophiolitic basalts are covered by Middle Eocene flysch, and were thrust onto deformed flysch of the same age. The age of formation may be latest Cretaceous to early Eocene, and the age of emplacement may be Oligocene or Miocene. Amphibolite beneath the ophiolite nappe is dated as 40 Ma (K-Ar age of hornblende and mica) [72].

3. Eastern Indonesia

The Halmahera ophiolite was first thoroughly mapped by Bessho [7], who concluded that the peridotite bodies of Cretaceous age were thrust to the west as thirteen or more independent nappes in Miocene time. His idea is supported by recent British studies; Ballantyne [4] gave detailed petrological data of this ophiolite, and concludes that the rocks closely resemble those dredged and drilled from the Mariana fore-arc area.

The Celebes (Sulawesi) ophiolite is one of the largest on-land exposures of oceanic lithosphere on the earth. This ophiolite is of late Cretaceous or Eocene age, and consists of multiple thrust sheets emplaced from the north in middle Miocene time [46]. De Roever [18] first proposed that an alpine-type peridotite (basal part of an ophiolite) represents a tectonically transported fragment of the earth's mantle on the basis of his observation of this ophiolite. Silver & McCaffrey [75] give geophysical evidence indicating that the ophiolite is continuous with oceanic lithosphere to the north.

The Darvel Bay ophiolite in northern Borneo is also of the same age, and is characterized by amphibolite- and granulite-facies metagabbros constituting its lower crustal portion [26].

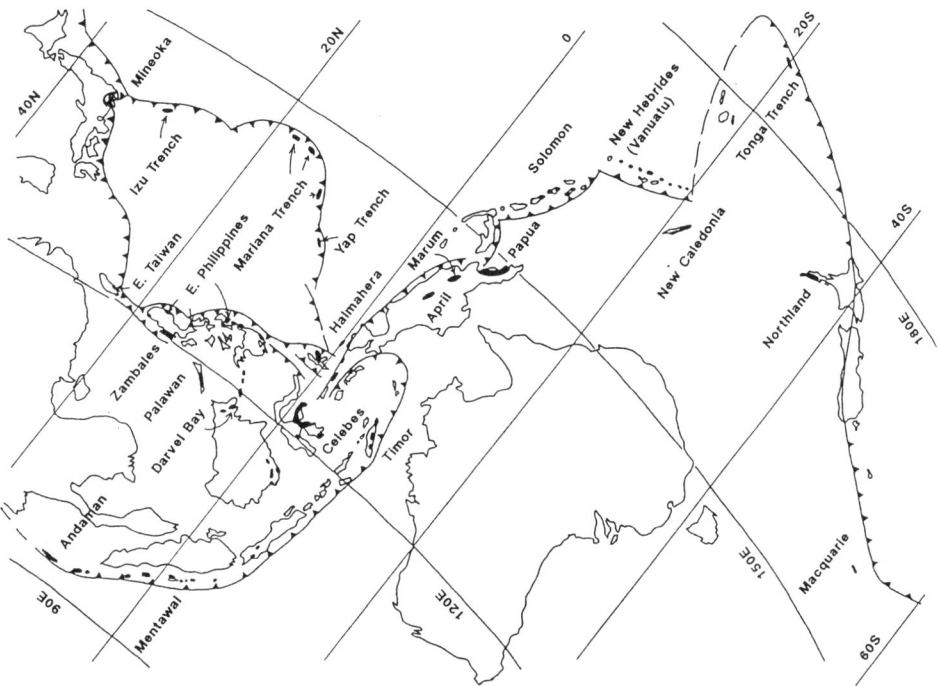

Figure 7. Cenozoic ophiolites in the Western Pacific areas.

4. Papua New Guinea

The Papuan ophiolite may be the largest in the circum-Pacific region, and is exposed over an area 250 km long and 50 km wide. This ophiolite formed in the late Jurassic to Cretaceous time as indicated by 150 and 147 Ma K-Ar ages for gabbros and 116 Ma K-Ar age for a basalt [16], was intruded by Eocene tonalite (50-55 Ma), and was covered by Eocene andesite and dacite. The Papuan ophiolite is characterized by very depleted mantle harzburgite, orthopyroxene-rich cumulate (including "harzburgite cumulate"), and basaltic volcanics with island arc signatures [73]. A sheeted dike complex is only locally present [40].

Geophysical data indicate that the Papuan ophiolite may be continuous with the present oceanic lithosphere of the Solomon Sea to the north [58]. This provides the basis for the idea of ophiolite obduction [13, 14]. Nicolas [64] classifies this ophiolite as a "passive continental margin" type. However, outcrops of "continental crust" underlying the ophiolite are mostly thick Cretaceous felsic clastic rocks with tuffs, limestones, and basalts metamorphosed in the greenschist and blueschist facies in the Eocene [16]. These lithologies rather suggest an active continental margin. It is possible that the Cretaceous sediments represent an accretionary complex developed in a fore-arc setting. Jaques [39] considers that the neighboring Marum ophiolite represents fore-arc lithosphere of the New Britain arc. The Papuan ophiolite itself is also located on the western extension of the Woodlark arc or

Pocklington arc (see Karig's map [43]), and may also therefore represent a late Cretaceous or Eocene fore-arc.

5. New Caledonia

The New Caledonian ophiolite occurs as vast nappes overlying Eocene flysch (partly including ophiolitic detritus) and older strata. A Nd-Sm age of gabbro is reported to be 131 Ma (Early Cretaceous) and that of a basalt is reported to be 58 Ma (Paleocene) [67]. The nappes are unconformably covered by early Miocene molasse. This ophiolite appears to have formed during the Cretaceous and have been emplaced in the Eocene. Geophysical data suggest that the ophiolite is continuous with the oceanic lithosphere to the north as in the case of Papua.

The pre-Permian(?) rocks underlying the ophiolite are very fine-grained, finely alternated pyroclastic and clastic sediments without fossils. Some alkali basaltic pillow lava bodies are intercalated with these sediments. The metamorphic grade is greenschist to blueschist facies. Permian and Triassic shallow-water sediments resembling those in the Maizuru Zone of Japan are also present. It is likely that the New Caledonian basement rocks comprise a circum-Pacific type accretionary complex as found in Japan and the Klamath Mountains, and hence represent an active continental margin.

6. New Zealand and Macquarie Island

Malpas et al. [53] described the Northland ophiolite, which may have been formed in Cretaceous to Eocene time and was emplaced in the Oligocene. Most basaltic rocks in the Northland ophiolite chemically resemble MORBs, but some alkali basalt with hornblende is also present. This ophiolite is a part of the Cenozoic ophiolite belt of the Southwest Pacific.

The termination of this belt may be Macquarie Island located far to the south of New Zealand. The 30 km-long island is made up wholly of ophiolitic rocks such as harzburgite, gabbro, basaltic sheeted dikes, and pillow lavas [78]. Deep-sea oozes alternating with pillow lavas bear early to late Miocene foraminifera, indicating uplift of 2000 to 4000 m since Miocene time.

7. Izu-Mariana-Yap and Tonga Trenches: Submarine Fore-Arc Ophiolites

Dredge and drill studies of the inner (arc-side) wall of the Izu-Mariana-Yap and Tonga trenches have revealed submarine exposures of "ophiolitic" rocks, which in some cases comprise an ophiolite stratigraphy [24, 10, 9]. The ophiolitic rocks are almost identical to those of the on-land ophiolites (e.g. [4]). They represent fore-arc oceanic crust tectonically emplaced on subducting oceanic crust, and are hence called "fore-arc ophiolites". Maekawa et al. [52] reported the occurrence of high-pressure, low-temperature metamorphic rocks in association with a fore-arc ophiolite in the Izu arc. This suggests that active thrusting of the fore-arc ophiolite dragged the high-pressure metamorphic rocks from a deeper part of the subduction zone, and continued tectonic erosion of the fore-arc thrust front by the subducting plate exposed the deep-seated rocks (fore-arc mantle rocks and blueschist) on the trench slope.

8. Marginal basins

Twenty eight out of thirty-two marginal basins on the earth are present in the Western Pacific area, and nearly all were formed (or "trapped") in Eocene or younger times [77]. Some marginal basins such as Okinawa, Mariana, Andaman, Manus, Woodlark, North Fiji, and Lau-Havre are currently forming through magmatism and rifting. Several marginal basin floors are 600 to 800 m deeper than the floors of major ocean of the same age [41, 77], suggesting a faster cooling rate in the lithosphere underlying marginal basins. It is important to note that the formation of marginal basin lithosphere occurred synchronously with the emplacement of ophiolites (Eocene to Miocene) in the Western Pacific region. Most ophiolites in this region may be the fragments of island arc-marginal basin lithosphere, which were emplaced onto accretionary wedges over subducting oceanic lithosphere.

DISCUSSION

1. Ophiolites: foot wall or hanging wall of subduction zone?

In the plate tectonic hypotheses of the early 1970's, ophiolites formed at mid-oceanic ridges, traveled long distances under the ocean, detached from the subducting oceanic lithosphere at trenches, and were emplaced onto continental margins. In this model, ophiolites represent the fragments of foot walls of subduction zones. However, geologic evidence such as the short time span represented by the overlying sedimentary sequences and the high-temperature thermal aureoles beneath some ophiolite nappes supports short travel histories. This difficulty might be overcome if we assume "obduction of a spreading ridge" (Fig. 8a), though this type of ophiolite emplacement does not occur along the Pacific coast of the Americas, where oceanic ridges are currently subducting. Recent finds of ophiolite sequences and blueschists exposed on the inner (arc-side) trench walls, and the geochemical identification of island arc signatures in most of the ophiolitic rocks, strongly supports the idea that ophiolites represent island arc-marginal basin lithosphere (i.e. hanging walls of subduction zones) (Fig. 8b).

Ophiolites are frequently associated with blueschists, which generally underlie the ophiolite nappe [13]. The blueschists are believed to form in the subduction zone itself where geothermal gradient (dT/dP) is small and deformation is intense, and the tectonically overlying ophiolite nappe may naturally represent fore-arc lithosphere. Recent discoveries of blueschist with lawsonite, aragonite, alkali pyroxene (chloromelanite), and blue amphibole (winchite) in a sea-floor outcrop of serpentine melange in the Izu-Mariana fore-arc (ODP Leg 125 [52]) is evidence in support of a fore-arc origin of the ophiolite-blueschist assemblage.

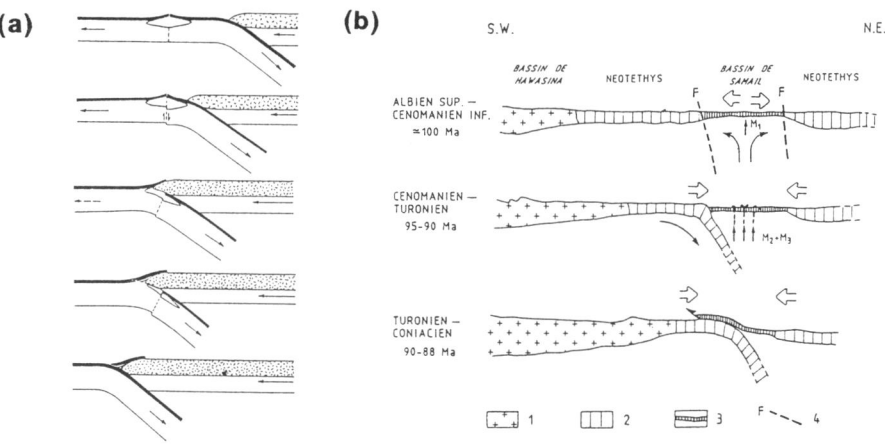

Figure 8. Two models of ophiolite emplacement. (a) Obduction of subducting ridge after Christensen and Salisbury [17]. (b) Thrusting of island arc-marginal basin (supra-subduction zone) lithosphere that originally formed as normal oceanic crust between rifted continental blocks after Beurrier et al. [8]. 1:continental crust, 2: old oceanic crust, 3: new Upper Cretaceous oceanic crust, 4: major faults.

Petrologic and geochemical studies have also revealed supra-subduction zone origin (fore-arc, volcanic arc, or back-arc) of most ophiolites after the landmark study of Miyashiro [59]. The signatures include; (1) Boninites and high-magnesian andesites indicating secondary hydrous melting of a previously depleted mantle source [11, 41]. (2) Ophiolitic mantle peridotites that are more depleted than the mantle peridotites obtained from the ocean floor [19]. Extremely depleted (Papua-type) mantle peridotite in some ophiolites may form as a residual product of a secondary melting event [31]. (3) Ophiolitic cumulates showing co-precipitation of highly calcic plagioclase and relatively iron-rich mafic minerals, which is characteristic of island-arc tholeiite [6, 35]. (4) Ophiolitic cumulates with abundant orthopyroxene which crystallized in an early stage of magmatic evolution. Orthopyroxene never appears in normal MORBs, but is common in island-arc volcanics [31, 4]. (5) Ophiolitic volcanic rocks with low concentrations of high field strength elements (HFSEs) such as Ti, Zr, Nb, and Ta, which are well expressed in their low ratios with the other incompatible elements such as Ti/V or Zr/Y. These chemical signatures are characteristic of island-arc volcanics [e.g. 73].

These geological, petrological, and geochemical relationships indicate that most ophiolites are of island arc (supra-subduction zone) origin. The ophiolites may represent fragments of newly formed oceanic lithosphere, generated either in a supra-subduction zone environment or a mid-ocean ridge environment quickly changed into that environment (possibly *in situ*), and were then promptly thrust onto a subducting plates in association with high-pressure metamorphic rocks which were previously buried deep in the subduction zone.

2. Tectonic setting of ophiolite emplacement

Nicolas [64] classified the tectonic settings of ophiolite emplacement into (1) passive continental margins, (2) active continental margins, and (3) continental collision zones. He classified most ophiolites as "collision type", some fragmental ophiolites (e.g. those in the Franciscan melange) as "active margin type", and some huge ophiolite nappes which are still continuous in nearby oceanic crust (e.g. Papua and New Caledonia) as "passive margin type". This classification, however, is not suitable in view of the recent findings, which reveal that the "continental crust" tectonically underlying ophiolites is mostly younger than the ophiolites themselves, and has formed through accretionary processes and magmatism *in situ*. Major continental collision has never taken place in the circum-Pacific orogenic belts.

The tectonic setting of ophiolite emplacement, whether along passive or active margins, must be judged from the stratigraphic and lithologic sequence of the unit underlying the ophiolite. For example, Alpine ophiolites were emplaced onto the European continental crust covered with a typical passive-margin sedimentary sequence. This sequence commences with quartz sandstone, evaporite, and dolomite (Triassic), and is followed by a thick, muddy, shallow-water limestone sequence (Jurassic to Eocene). An active margin environment preceding the continental collision was established only after late Eocene time, when turbiditic sandstone and then andesitic volcanism appeared. On the other hand, circum-Pacific ophiolites are underlain by accretionary complexes consisting of "oceanic plate stratigraphy" commencing with chert or a seamount-limestone sequence, followed by shale, sandstone, olistostrome, and considerable arc-volcanic material. In this respect, most circum-Pacific ophiolites were emplaced along active margin environments. Continental collision may have little importance in the emplacement of the circum-Pacific ophiolites.

The idea that the ophiolites in Papua New Guinea, New Caledonia, and Celebes were emplaced on passive margins is questionable. The sedimentary sequence underlying the Papuan ophiolite consists mainly of Cretaceous metapelite; older continental crust is not exposed. Permo-Triassic conglomerate, Pre-Permian(?) metatuff, and Eocene flysch underlying the New Caledonian ophiolite and may represent accretionary complexes rather than passive continental margins. Most ophiolites in the circum–Pacific region were likely emplaced on the active continental margins.

3. Geotectonic significance of the Cenozoic ophiolite belts in Southwest Pacific

Late Cretaceous and Tertiary ophiolites form a continuous belt extending from Kamchatka through Japan, the Philippines, Indonesia, Papua New Guinea, New Caledonia, and the North Island of New Zealand to Macquarie Island. Tertiary ophiolites are almost restricted to this belt, and are virtually absent elsewhere. This belt is also characterized by vast marginal sea floors underlain by Tertiary oceanic lithosphere as large as that on both sides of the mid–ocean ridges (Fig. 9). Moreover, this belt is the site where some of the most active volcanism and tectonism (in island arcs and back–arc rift zones) occur at present. Miyashiro [60] argues that a "hot region" originated in the shallow mantle, and moved northward from eastern Australia (Late Cretaceous) to the Japan Sea (Miocene), producing marginal sea lithosphere. The marginal basins of island arc systems may not be secondary by–products of plate tectonics, but may have essential, active roles in geotectonics.

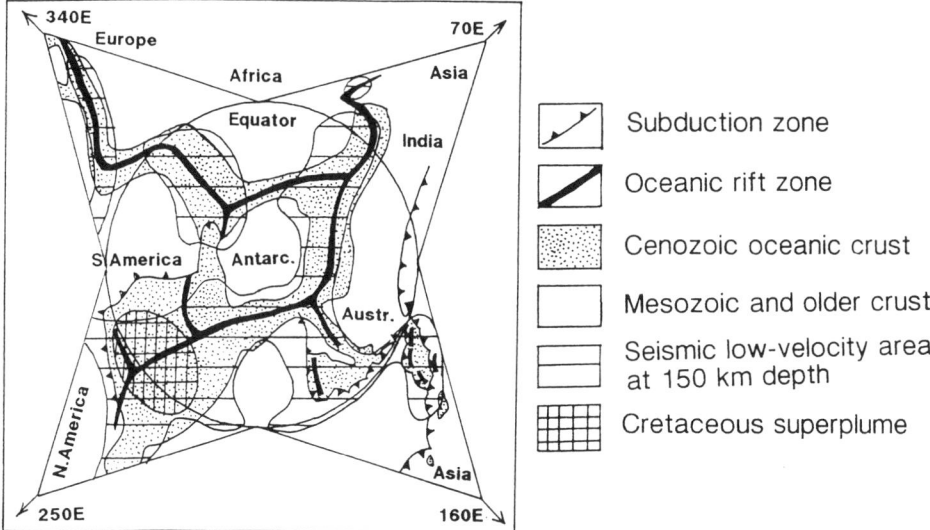

Figure 9. Distribution of Cenozoic oceanic crust (stippled area) on the map centered at the South Pole after Milanovsky [57]. The four major rift systems (thick lines) radiates in four directions with 90° intervals from the circum–Antarctic rift system, and are commonly displaced 30° to the left along the equator [57]. The major rift system and nearby Cenozoic oceanic crust is underlain by mantle low–velocity zones at 150 km depth (lined area) [83]. Approximate position of Larson's Cretaceous superplume [47] is also marked (crosshatched area).

Milanovsky [57] views the Western Pacific area as an oceanic rift zone comparable to the other mid–ocean ridges both in size and magnitude of magmatic activity. The mid–ocean ridges encircle Antarctica, then branch to the north in three directions with 90 degree intervals (250, 340, and 70 degrees east of the Greenwich meridian), and are commonly displaced 30 degrees to the left (to the west) along the equator before further extending toward the north pole. Milanovsky considers that West Pacific Cenozoic magmatism represents that of the fourth oceanic rift zone, branching along the 160 degree east meridian (90 degrees away from nearby ridges) with a 30 degrees westward swing along the equator in common with the other oceanic ridges (Fig. 9).

The synchronous formation of ophiolites and marginal basins in the Western Pacific during late Cretaceous and early Tertiary time suggests the presence of a vast region of oceanic rifting. The subsequent emplacement of the ophiolite may represent collapse of the oceanic rift zone into the small oceanic basins separated from the adjacent oceanic basins by subduction zones.

4. Ophiolite pulses, superplumes, and surge tectonics
A histogram of the generation ages of ophiolites shows distinct peaks in Jurassic–Cretaceous (about 150 Ma), Ordovician (about 450 Ma), and Late Proterozoic (about 750 Ma) (Fig. 2). Ophiolites may have formed as a result of the earth's major igneous pulses with a 300 m.y. interval. Ophiolites in an intra–continental collisional orogenic belt may have formed during a single pulse. For example, Appalachia–Caledonia–Ural orogenic belt is dominated by Ordovician ophiolites [20, 70], though ophiolite ages may range from the latest Proterozoic to Devonian in some areas [84]. The Alpine–Himalayan orogenic belt includes Jurassic and Cretaceous ophiolites only [1]. However, circum–Pacific orogenic belts include ophiolites formed by the two major (Ordovician and Jurassic–Cretaceous) and one minor (Permian) pulses [32, 33].

The intermittent nature of the earth's magmatism was well demonstrated by Larson [47, 48], who showed that ocean crust production rate sharply increased in Cretaceous time between 120 and 80 Ma. Most oceanic plateaus in the Pacific and Indian oceans also formed during this interval. He proposed that this increase in global magmatism resulted from the burst of a "superplume", which ascended from the core/mantle boundary and was emplaced in the upper mantle beneath the South Pacific Ocean. The origin of the superplume at the core/mantle boundary is inferred from the "Long Cretaceous Normal", the long period without geomagnetic reversals lasted from 120 to 80 Ma, exactly the same period as the magmatic pulse. Although the mechanism of geomagnetic reversals is not yet known, their long absence suggests some dynamic changes in the earth's core, where most of the earth's magnetic field is produced.

The 40 m.y. interval in the Cretaceous is also characterized by global high sea level. Higher production rate of the oceanic crust must have resulted in shallower oceans and hence higher sea level. Geomagnetic stratigraphy reveals that long–term "quiet periods" were present also in Carboniferous–Permian time and possibly in Ordovician–Cambrian time, corresponding to the Mississippian and Ordovician transgressions (high sea levels), respectively [47, 48]. These stratigraphic events may correspond to the Permian (minor) and Ordovician (major) ophiolite pulses, which possibly represent previous superplumes.

The hot deep–mantle material transported by the superplume may have surged through the upper mantle. As ocean crust production increased over the global oceanic rift system (circum–Antarctic and four other radiating rifts as discussed above), it is likely that the rift system may have worked as the paths of this surge. That is, the superplume material may have been distributed through big "channels" beneath the rift zones. This idea is analogous to the "surge tectonics" proposed by Myerhoff et al. [55, 56], who insist that the flow of the mantle material beneath a mid–oceanic ridge is parallel to the ridge axis. Their primary evidence is the topographic resemblance between the surface of a viscous flow (glacier, lava flow, etc.) and an oceanic rift zone, both with stripes parallel to the flow direction. Flow lines perpendicular to a ridge axis, characteristic of the plate–tectonic divergent zones on the surface of a lava lake, are not observed along any mid–ocean ridges [55, 56].

The surge tectonic model also proposes that mid–ocean ridges and orogenic belts have a common origin in surge channels. A rift zone, either continental or oceanic, may immediately change into an orogenic belt, when the surge channels collapse into fragments by shrinkage of the channel itself due to shortage of the superplume supply, or by horizontal pushing by the neighboring, more powerful channels. The Cenozoic magmatic-tectonic history of the Western Pacific area may provide a convincing evidence for this theory. Evidence of the earth's thermal pulses may be recorded as ophiolites if some surge channels run and collapse along active continental margins as in the case of Cenozoic Western Pacific, or run beneath continents and collapse into collision zones as in the case of Mesozoic Tethys.

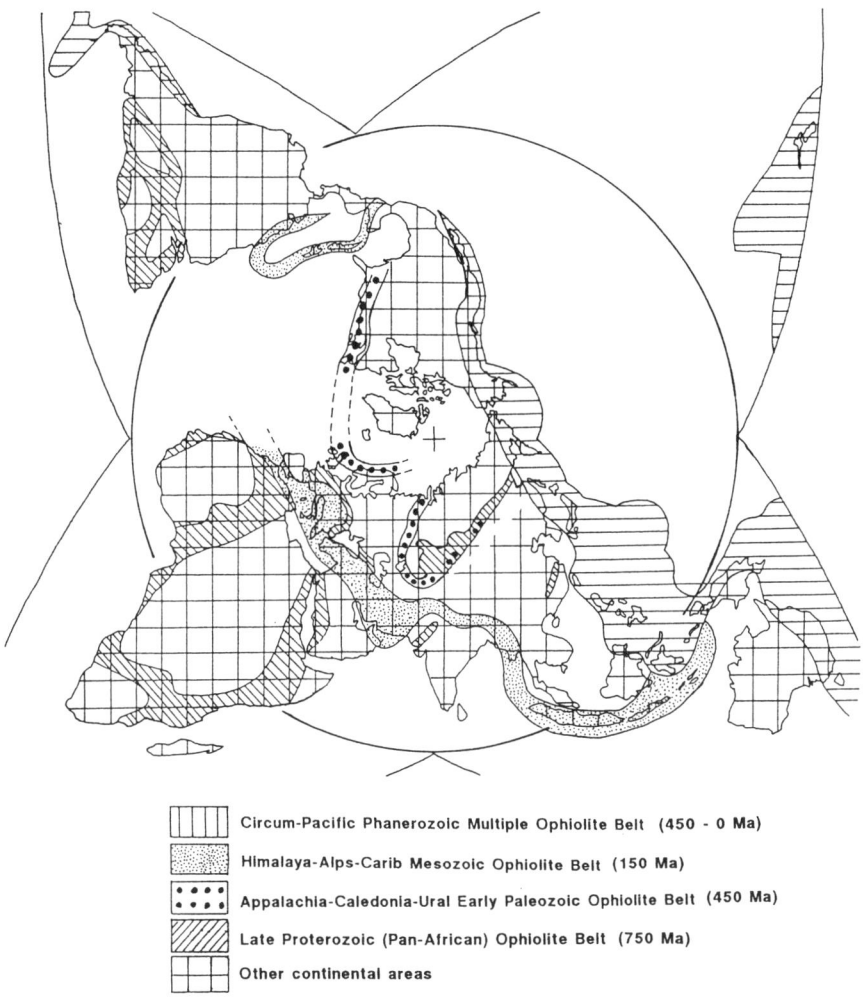

Figure 10. Earth's major ophiolite belts.

CONCLUSION

Ophiolites in the circum-Pacific area formed simultaneously with the world-wide pulses (Ordovician, Permian (minor), and Jurassic-Cretaceous times) of ophiolite generation. The ophiolites occur as nappes and melanges among accretionary complexes in which younger material has been underplated beneath older crust through episodic but continuous subduction-accretion processes. The Cretaceous-Tertiary ophiolite belts in the Western Pacific area are the latest demonstration of a smaller ophiolite pulse, possibly the last phase of the major Jurassic-Cretaceous pulse.

Periods of ophiolite generation may be related to the explosion and subsequent surge of a superplume, which may have ascended from the core-mantle boundary as enriched mantle material of more than thousand kilometers diameter. The formation and emplacement of the ophiolites may be related to the collapse of major rift systems developed along surge channels in the asthenosphere propagating from the explosion site of the superplume. A rift system without rigid lithosphere is vulnerable to horizontal stress, and will easily collapse under horizontal compression. The surge channels (rift zone) passing near the ocean/continent boundaries may be easier to collapse than those between continents or in the middle of the ocean, and the oceanic lithosphere formed and fragmented there may soon be swept and accumulated along the continental margins as parts of accretionary complexes. Less powerful surge channels developed within continents may also collapse resulting in continental collision zones.

The tectonic model involving superplumes and surge tectonics can thus explain ophiolite pulses. The Tertiary history of the western Pacific area vividly shows us how a global-scale oceanic rift zone collapsed and was fragmented into small island arc-marginal basin systems fringed with ophiolites. However, steady-state plate tectonics is also necessary to explain episodic but continuous accretion of oceanic material into subduction zones. The circum-Pacific multiple ophiolite belts are thus interpreted to be the results of the combination of the superplume-induced, episodic ophiolite pulses and steady-state plate tectonics.

Acknowledgements. This is an extended version of my paper read in the 29th IGC Ophiolite Symposium in Kyoto. The paper has been improved through the discussion in the symposia in Hokuriku Geology Institute (HGI), Niigata University and Tokyo Institute of Technology. I thank Professors Y. Kaseno, S. Miyashita, S. Maruyama and Y. Isozaki of these institutes. A preliminary, Japanese version of this paper was published in HGI Report, No. 3, pp. 1-31 (1993) under the title "Ophiolites of East Asia". Inoue Foundation for Science is thanked for supporting my participation in the Geodynamics Field Seminar in the Koryak Mountains organized by Drs. A.P. Stavsky, S.D. Sokolov, and others in July 1990. Prof. S.A. Shcheka and Dr. S. Vysotskiy of FEGI-RAS in Vladivostok kindly guided us to the Sikhote-Alin ophiolites in April 1993. I also thank Professor J.G. Malpas of Memorial University of Newfoundland (Canada), Dr. K.R. Wirth of Macalester College (U.S.A.), and Dr. C. Xenophontos of Geological Survey Deptartment (Cyprus) for critical reading and grammatical corrections of the manuscript.

References

1. E. Abbate, V. Bortolotti, P. Passerini and G. Principi. The rhythm of Phanerozoic ophiolites. *Ofioliti*, **10**, 109-138. (1985)
2. J.C. Aitchison, T.R. Ireland, M.C. Blake, Jr. and P.G. Flood. 530 Ma zircon age for ophiolite from the New England orogen: Oldest rocks known from eastern Australia. *Geology*, **20**, 125-128. (1992).
3. S. Arai. The circum-Izu massif peridotite, central Japan, as back-arc mantle fragments of the Izu-Bonin arc system. In: *Ophiolite Genesis and Evolution of Oceanic Lithosphere*. T. Peters et al. (Eds). pp. 807-822. Kluwer Academic Publ., Dordrecht. (1991).
4. P. Ballantyne. Petrology and geochemistry of the plutonic rocks of the Halmahera ophiolite, eastern Indonesia, and analogue of modern oceanic forearcs. In: *Ophiolites and their Modern Oceanic Analogues*. L.M. Parson, B.J. Murton, and Browning, P. (Eds), pp. 179-202, Geol. Soc. (London) Spec. Publ., **60**. (1992).
5. V.G. Batanova and O.V. Astrakhantsev. Island-arc mafic-ultramafic plutonic complexes of North Kamchatka. In: *Circum-Pacific Ophiolites: Proceedings of the 29th IGC Ophiolite Symposium*, A. Ishiwatari *et al*. (Eds). pp. 129-144. VSP Publisher, Netherlands (1994)
6. J.B. Beard. Characteristic mineralogy of arc-related cumulate gabbros: Implications for the tectonic setting of gabbroic plutons and for andesite genesis. *Geology*, **14**, 848-851. (1986).

7. B. Bessho. Geological structure of the Halmahera Island (preliminary report). *J. Geogr.* (Tokyo), **56**, 195–203. (1944). (in Japanese).
8. M. Beurrier, M. Ohnenstetter, B. Cabanis, J.-L. Lescuyer, M. Tegyey and J. Le Metour. Géochimie des filons doléritiques et des roches volcaniques ophiolitiques de la nappe de Semail: contraintes sur leur origine géotectonique au Crétace supérieur. *Bull. Soc. Géol. France*, **1989**, 205–219. (1989)
9. S. Bloomer and R.L. Fisher. Petrology and geochemistry of igneous rocks from the Tonga Trench: A non-accreting plate boundary. *J. Geol.*, **95**, 469–495. (1987).
10. S. Bloomer and J. Hawkins. Gabbroic and ultramafic rocks from the Mariana Trench: An island arc ophiolite. *AGU Geophysical Monograph*, **27**, 294–317.
11. W.E. Cameron, E.G. Nisbet, V.J. Dietrich. Boninites, komatiites, and ophiolitic basalts. *Nature*, **280**, 550–553. (1979).
12. N.L. Christensen and M.H. Salisbury. Structure and constitution of the lower oceanic crust. *Rev. Geophys. Space Physics*, **13**, 57–86. (1975)
13. R.G. Coleman, Plate tectonic emplacement of upper mantle peridotites along continental edges. *J. Geophys. Res.*, **76**, 1212–1222. (1971)
14. R.G. Coleman, *Ophiolites: Ancient Oceanic Lithosphere?*, Springer-Verlag, Berlin. (1977).
15. R.G. Coleman, Ophiolites and accretion of the North American Cordillera. *Bull. Soc. Géol. France*, **1986**, 961–968. (1986).
16. H.L. Davies and I.E. Smith. Geology of eastern Papua. *Geol. Soc. Amer. Bull.*, **82**, 3299–3312.(1971)
17. S.M. DeBari and R.G. Coleman. Examination of the deep levels of an island arc: Evidences from the Tonsina ultramafic–mafic assemblage, Tonsina, Alaska. *J. Geophys. Res.*, 94, 4373–4391. (1989).
18. W.P. De Roever. Sind die alpinotypen Peridotitmassen vielleicht tektonisch verfrachtete Bruchstücke der Peridotitschale? *Geol. Rdsch.*, **46**, 137–146. (1957)
19. H.J.B. Dick and T. Bullen. Chromian spinel as a petrogenetic indicator in abyssal and alpine-type peridotites and spatially associated lavas. *Contrib. Mineral. Petrol.*, **86**, 54–76. (1984).
20. G.R. Dunning and R.B. Pedersen. U–Pb ages of ophiolites and arc-related plutons of the Norwegian Caledonides: implications for the development of Iapetus. *Contrib. Mineral. Petrol.*, **98**, 13–23. (1988).
21. N. Filatova and V. Vishnevskaya. Tectonic position of the Mesozoic ophiolitic and island arc formations in the Koryak region (northeastern Russia). In: *Circum–Pacific Ophiolites: Proceedings of the 29th IGC Ophiolite Symposium*, A. Ishiwatari et al. (Eds). pp. 109–128. VSP Publisher, Netherlands (1994).
22. J.W. Hawkins and C.A. Evans. Geology of the Zambales Range, Luzon, Philippine Islands: Ophiolite derived from an island arc–back arc basin pair. In: *The Tectonic and Geologic Evolution of Southeast Asian Seas and Islands*, Part 2 (Geophys. Monogr. 27), Hayes, D.E. (Ed.). pp. 95–123. Amer. Geophys. Union. (1983).
23. F. Hervé, E. Godoy, M.A. Parada, V. Ramos, C. Rapela, C. Mpodozis and J. Davidson. A general view on the Chilean–Argentine Andes, with emphasis on their early history. In: *Circum–Pacific orogenic belts and evolution of the Pacific Ocean basin*. (AGU Geodynamics Series, Vol. 18). J.W.H. Monger and J. Francheteau (Eds). pp. 97–113. (1987).
24. E. Honza and H. Kagami. A possible accretion accompanied by ophiolite in the Mariana Trench. *J. Geography* (Tokyo), **86**, 80–91. (1977)
25. C.S. Hutchison. Ophiolite in Southeast Asia. *Geol. Soc. Amer. Bull.*, **86**, 797–806. (1975).
26. C.S. Hutchison. Ophiolite metamorphism in northeast Borneo. *Lithos*, **11**, 195–208. (1978).
27. W.P. Irwin. Ophiolite terranes of California, Oregon, and Nevada. In: *North American Ophiolites*. R.G. Coleman and W.P. Irwin (Eds). pp. 75–92. State of Oregon, Dept. Geol. Min. Indst. Bull. No. 95. (1977).
28. W.P. Irwin. Ophiolitic terranes of part of the western United States. In: *International Atlas of Ophiolites*. pp. 2–4. Geol. Soc. Am., Map and Chart Ser. MC-33. (1979).
29. W.P. Irwin. Tectonic accretion of the Klamath Mountains. In: *The Geotectonic Development of California*. W.G. Ernst et al. (Eds). Rubey Vol. 1, pp. 29–49. Princeton Hall, Jersey. (1981).
30. A. Ishiwatari. Granulite-facies metacumulates of the Yakuno ophiolite, Japan: Evidence for unusually thick oceanic crust. *J. Petrology*, **26**, 1–30. (1985a).
31. A. Ishiwatari. Igneous petrogenesis of the Yakuno ophiolite in the context of the diversity of ophiolites. *Contrib. Mineral. Petrol.*, **89**, 155–167. (1985b).
32. A. Ishiwatari. Ophiolites in the Japanese islands: Typical segment of the circum–Pacific multiple ophiolite belts. *Episodes*, **14**, 274–279. (1991a).
33. A. Ishiwatari. Time–space distribution and petrologic diversity of Japanese ophiolites. In: *Ophiolite Genesis and Evolution of Oceanic Lithosphere*. Tj. Peters et al. (Eds.). pp. 723–743, Ministry of Petroleum and Minerals, Sultanate of Oman. (Kluwer Acad. Publ.). (1991b)

34. A. Ishiwatari and Y. Hayasaka. Ophiolite nappes and blueschists of the Inner Zone of Southwest Japan. In: *29th IGC Field Trip Guide Book*, Vol. 5, pp. 285-325. Geological Survey of Japan. (1992).
35. A. Ishiwatari, Y. Ikeda and Y. Koide. The Yakuno ophiolite, Japan: Fragments of Permian island arc and marginal basin crust with a hot spot. In: *Ophiolites: Oceanic Crust Analogues* (Proceedings of the Troodos '87 Symposium). J. Malpas *et al.* (Eds). pp. 497-506. Geol. Surv. Dept., Cyprus. (1990).
36. H. Ishizuka. Igneous and metamorphic petrology of the Horokanai ophiolite in the Kamuikotan zone, Hokkaido, Japan: A synthetic thesis. *Mem. Fac. Sci., Kochi Univ., Ser. E, Geology*, **8**, 1-70. (1987).
37. Y. Isozaki, S. Maruyama and F. Furuoka. Accreted oceanic materials in Japan. *Tectonophysics*, **181**, 179-205. (1990).
38. B.-M. Jahn. Mid-ocean ridge or marginal basin origin of the East Taiwan ophiolite: chemical and isotopic evidence. *Contrib. Mineral. Petrol.*, **92**, 194-206. (1986)
39. A.L. Jaques. Petrology and petrogenesis of cumulate peridotites and gabbros from the Marum ophiolite complex, northern Papua New Guinea. *J. Petrology*, **22**, 1-40. (1981)
40. A.L. Jaques and B.W. Chappell. Petrology and trace element geochemistry of the Papuan Ultramafic Belt. *Contrib. Mineral. Petrol.*, **75**, 55-70. (1980)
41. G. Jones, A.H.F. Robertson and J.R. Cann. Genesis and emplacement of the supra-subduction zone Pindos ophiolite, northwestern Greece. In: *Ophiolite Genesis and Evolution of the Oceanic Lithosphere*, (Proceedings of Oman '90 Symposium), Tj. Peters *et al.* (Eds). pp. 771-799. Ministry of Petroleum and Minerals, Sultanate of Oman (Kluwer Acad. Publ.). (1991).
42. D.E. Karig. Origin and development of marginal basins in the Western Pacific. *J. Geophys. Res.*, **76**, 2542-2561. (1971)
43. D.E. Karig. Remnant arcs. *Geol. Soc. Am. Bull.* **83**, 1057-1068. (1972)
44. A.I. Khanchuk and I.V. Panchenko. Garnet gabbro in the southern Sikhote Alin ophiolites. *Dokl. Akad. Nauk SSSR*, **321**, 800-803. (1991). (in Russian).
45. S. Kojima. Mesozoic terrane accretion in northeast China, Sikhote Alin and Japan regions. *Palaeogeogr. Palaeoclimat. Palaeontol.*, **69**, 213-232. (1989)
46. E. Kündig. Geology and ophiolite problems of East-Celebes. *Verhandelingen van het Koninklijk Nederlandsch Geologisch-Mijnbouw-kundig Genootschap*, Geol. Ser., **16**, 210-235. (1956)
47. R.L. Larson. Latest pulse of Earth: Evidence for a mid-Cretaceous superplume. *Geology*, **19**, 547-550 (1991a)
48. R.L. Larson. Geological consequences of superplumes. *Geology*, **19**, 963-966. (1991b)
49. L.D. Lavrova. Genesis of ultramafic rocks of the Maynits tectonic zone, Koryak Mountains. *Doklady Akad. Nauk SSSR* (English translation), **253**, 173-176. (1982)
50. E.C. Leitch and E. Scheibner. Stratotectonic terranes of the eastern Australian Tasmanides. In: *Terrane Accretion and Orogenic Belts*. E.C. Leitch and E. Scheibner (Eds.). pp. 1-19. Amer. Geophys. Union, Geodynamics Series **19**. (1987).
51. N. Lindsley-Griffin. The Cambrian Trinity ophiolite and related rocks of the lower Paleozoic Trinity complex, northern California, U.S.A. In: *Circum-Pacific Ophiolites: Proceedings of the 29th IGC Ophiolite Symposium*. A. Ishiwatari *et al.* (Eds.). pp. 47-68. VSP Publisher, Netherlands. (1994)
52. H. Maekawa., M. Shozui, T. Ishii, K.L. Saboda. and Y. Ogawa. Metamorphic rocks from the serpentinite seamounts in the Mariana and Izu-Ogasawara forearcs. In: *Proc. ODP, Sci. Res.* Vol. 125, P. Fryer, J.A. Pearce, L.B. Stokking et al. pp. 415-430. (1992).
53. J.G. Malpas, I.E.M. Smith and D. Williams. Comparative genesis and tectonic setting of ophiolitic rocks of the South and North Islands of New Zealand. In: *Circum-Pacific Ophiolites: Proceedings of the 29th IGC Ophiolite Symposium*. A. Ishiwatari *et al.* (Eds.). pp. 29-46. VSP Publisher, Netherlands. (1994).
54. A.O. Mazarovich. Ophiolite allochthons of the Maritime region. *Dokl. Akad. Nauk SSSR* (English translation), **249**, 49-51. (1982)
55. A.A. Meyerhoff, W.B. Agocs, I. Taner, A.E.L. Morris and B.D. Martin. Origin of midocean ridges. In: *New Concepts in Global Tectonics*, S. Chatterjee, and N. Hutton III (Eds.), pp. 151-178, Texas Tech Univ. Press. (1992a).
56. A.A. Meyerhoff, I. Taner, A.E.L. Morris, B.D. Martin, W.B. Agocs and H.A. Meyerhoff. Surge tectonics: a new hypothesis of Earth dynamics. In: *New Concepts in Global Tectonics*, S. Chatterjee, and N. Hutton III (Eds.), pp. 309-409, Texas Tech Univ. Press. (1992b).
57. E.E. Milanovsky. Rifting and its role in tectonic structure and Meso-Cenozoic geodynamics of the earth. *Hokuriku Geol. Inst. Report* (Kanazawa), No. 1, 37-55. (1991)
58. J.S. Milsom. Papuan ultramafic belt: gravity anomalies and the emplacement of ophiolites. *Geol. Soc. Am. Bull.* **84**, 2243-2258. (1973)

59. A. Miyashiro. The Troodos ophiolite complex was probably formed in an island arc. *Earth Planet. Sci. Lett.*, **19**, 218-224. (1973).
60. A. Miyashiro. Hot regions and the origin of marginal basins in the western Pacific. *Tectonophysics*, **122**, 195-216. (1986).
61. S. Miyashita and A. Yoshida. Pre-Cretaceous and Cretaceous ophiolites in Hokkaido, Japan. *Bull. Géol. Soc. France,* 8th series, **4**, 251-260. (1988).
62. S. Miyashita and A. Yoshida. Geology and petrology of the Shimokawa ophiolite (Hokkaido, Japan): Ophiolite possibly generated at R-T-T triple junction. In: *Circum-Pacific Ophiolites: Proceedings of the 29th IGC Ophiolite Symposium.* A. Ishiwatari *et al.* (Eds.). pp. 163-182. VSP Publisher, Netherlands. (1994).
63. T. Nakajima, A. Ishiwatari, S. Sano, K. Kunugiza, M. Okamura, T. Kano, T. Sohma and Y. Hayasaka. Geotraverse across the Southwest Japan arc: An overview of tectonic setting of Southwest Japan. In: *29th IGC Field Trip Guide Book,* Vol. 5, pp. 171-253. Geological Survey of Japan. (1992).
64. A. Nicolas. *Structures of Ophiolites and Dynamics of Oceanic Lithosphere,* Kluwer Academic Publisher, Dordrecht. (1989)
65. K. Ozawa. Ultramafic tectonite of the Miyamori ophiolitic complex in the Kitakami Mountains, Northeast Japan: Hydrous upper mantle in an island arc. *Contrib. Mineral. Petrol.*, **99**, 159-175. (1988).
66. S.A. Palandzjan. Ophiolite belts in the Koryak upland, Northeast Asia. *Tectonophysics*, **127**, 341-360. (1986)
67. J.P. Paris. *Géologie de la Nouvelle-Calédonie; un essai de synthese.* (with geologic map of scale 1:200,000). Mémoire du B.R.G.M. No. 113, Orléans, France. (1981).
68. J.F. Parrot and F. Dugas. The disrupted ophiolitic belt of the Southwest Pacific: Evidence of an Eocene subduction zone. *Tectonophysics*, **66**, 349-372. (1980).
69. W.W. Patton, Jr., I.L. Tailleur, W.P. Brosgé and M.A. Lanphere. Preliminary report on the ophiolites of northern and western Alaska. In: *North American Ophiolites.* R.G. Coleman and W.P. Irwin (Eds). pp. 51-57. State of Oregon, Dept. Geol. Min. Indst. Bull. No. 95. (1977).
70. A. Perfiliev. Ophiolitic belt of the Urals. In: *International Atlas of Ophiolites.* pp. 9-12. Geol. Soc. Am., Map and Chart Ser. MC-33. (1979).
71. Yu. M. Pushcharovskiy, S.V. Ruzhentsev and S.D. Sokolov. Tectonic thrust sheets and geologic mapping. *Geotectonics*, **22**, 1-8. (1988). (Original Russian paper published in *Geotektonika*, **1969**, No. 4, 5-23).
72. H. Raschka, E. Nacario, D. Rammlmair, G. Samonte and L. Steiner. Geology of the ophiolite of Central Palawan Island, Philippines. *Ofioliti*, **10**, 375-390. (1985)
73. J.W. Shervais. Ti-V plots and the petrogenesis of modern and ophiolitic lavas. *Earth Planet. Sci. Lett.*, **59**, 101-118. (1982).
74. N.A. Shilo, F.V. Kaminskiy, S.A. Palandzhyan, S.M. Til'man, L.A. Tkachenko, L.D. Lavrova and K.A. Shepeleva. First diamond finds in alpine-type ultramafic rocks of the northeastern USSR. *Doklady Akad. Nauk SSSR* (English translation), **241**, 179-182. (1981)
75. E.A. Silver and R. McCaffrey. Ophiolite emplacement by collision between the Sula platform and the Sulawesi island arc, Indonesia. *J. Geophys. Res*, **88**, 9419-9435. (1983).
76. A.P. Stavsky, V.D. Chekhovitch, M.V. Kononov and L.P. Zonenshain. Plate tectonics and palinspastic reconstructions of the Anadyr-Koryak region, Northeast USSR. *Tectonics*, **9**, 81-101. (1990).
77. K. Tamaki and E. Honza. Global tectonics and formation of marginal basins: Role of the western Pacific. *Episodes*, **14**, 224-230. (1991).
78. R. Varne and M.J. Rubenach. Geology of Macquarie Island and its relationship to oceanic crust. In: *Antarctic Oceanography*, II. (Antarctic Research Series 19). pp. 251-266. Am. Geophys. Union, Washington. (1972).
79. S.V. Vysotskiy. High and low pressure cumulates of Paleozoic ophiolites in Primorye, eastern Russia. In: *Circum-Pacific Ophiolites: Proceedings of the 29th IGC Ophiolite Symposium,* A. Ishiwatari *et al.* (Eds). pp. 145-162. VSP Publisher, Netherlands (1994)
80. X. Wang and Z. Hao. Time-space distribution and tectonic types of ophiolites in China. In: *Circum-Pacific Ophiolites: Proceedings of the 29th IGC Ophiolite Symposium.* A. Ishiwatari *et al.* (Eds). pp. 183-204. VSP Publisher, Netherlands. (1994)
81. Q. Wang, A. Ishiwatari, Z. Zhao, T. Hirajima, N. Hiramatsu, M. Enami, M. Zhai, J. Li and B. Cong. Coesite-bearing granulite retrograded from eclogite in Weihai, eastern China. *Eur. J. Min.*, **5**, 141-152. (1993).
82. K.R. Wirth, J.M. Bird and J.N. Wessels. The diversity of accreted oceanic lithosphere in the Brooks Range, Alaska. In: *Circum-Pacific ophiolites: Proceedings of the 29th IGC Ophiolite Symposium.* A.

Ishiwatari et al. (Eds). pp. 89–108. VSP Publisher, Netherlands. (1994).
83. G.H. Woodhouse and A.M. Dziewonski. Mapping the upper mantle: Three dimensional modelling of Earth structure by inversion of seismic wave forms. *J. Geophys. Res.*, **89**, 5953–5986. (1984)
84. A.S. Yakubchuk. Polychronous ophiolite belts of central Kazakhstan and their evolution. In: *Circum-Pacific Ophiolites: Proceedings of the 29th IGC Ophiolite Symposium*. A. Ishiwatari *et al.* (Eds). pp. 235–254. VSP Publisher, Netherlands. (1994)
85. A.S. Yakubchuk, A.M. Nikishin and A. Ishiwatari. Late Proterozoic ophiolite pulse. In: *Circum-Pacific Ophiolites: Proceedings of the 29th IGC Ophiolite Symposium*. A. Ishiwatari *et al.* (Eds). pp. 273–286. VSP Publisher, Netherlands. (1994)
86. Yu.D. Zakharov, I.V. Panchenko, and A.I. Khanchuk. *A field guide to the late Paleozoic and early Mesozoic circum-Pacific bio- and geologic events*. (IGCP 272 and 321). Russian Acad. Sci., Far-East Geol. Inst., Vladivostok, 88 pp. (1992)
87. L.P. Zonenshain, M.I. Kuzmin and L.M. Natapov. *Geology of the USSR: A plate-tectonic synthesis*. Geodynamics Series, Vol. 21, (B.M. Page (Ed.)). Amer. Geophys. Union. (1990)

Comparative genesis and tectonic setting of ophiolitic rocks of the South and North Islands of New Zealand

J. MALPAS[1], I. E. M. SMITH[2] and D. WILLIAMS[1]
[1] Department of Earth Sciences, Memorial University, St. John's, Newfoundland, Canada, A1B3X5
[2] Department of Geology, Auckland University, Auckland, New Zealand

Abstract: This paper describes the origin, mode of emplacement and tectonic setting of rocks from two major ophiolite belts in New Zealand. In the East Nelson area of South Island, three separate ophiolite complexes exist. These are the Dun Mountain Ophiolite, the Patuki melange and the Croisilles melange. The Dun Mountain Ophiolite represents a semi-complete ophiolite suite that was produced in the early Permian by magmatic activity in a suprasubduction zone environment. Its disrupted nature is considered to be the result of tectonism associated with obduction and later orogenic events. Rocks of the Patuki and Croisilles melanges lie in fault contact beneath the Dun Mountain Ophiolite and, although highly dismembered, are considered vestiges of true ophiolitic assemblages. Basaltic rocks occurring as blocks in these melanges can be divided into two petrographically and geochemically defined suites; a mid-ocean ridge suite and an alkaline within-plate suite. These rocks likely represent fragments of oceanic crust otherwise subducted beneath the Dun Mountain Ophiolite. Indeed a fore-arc environment of formation is favoured for the Dun Mountain Ophiolite, and the Patuki and Croisilles rocks are considered as fragments of ocean crust and accompanying seamounts sheared off the down-going slab during subduction. The Northland Ophiolite of North Island, was probably emplaced as a single sheet during the late Oligocene. Chemically, the bulk of the igneous rocks of the ophiolite are mid-ocean ridge basalts, but an associated younger suite of hornblende-modal alkalic rocks is believed to represent seamounts built upon this MORB ocean crust. The contrasting relationships between ocean crust and seamount volcanism displayed by these two major ophiolite occurrences in New Zealand can be used to model their tectonic environments of formation.

INTRODUCTION

Ophiolitic rocks occur on both the North and South Islands of New Zealand. Although intact complete ophiolite stratigraphies are not preserved, the variety of rock types in these associations allows for identification of the tectonic environment of their formation. We here summarise the regional geology and tectonic setting of two major ophiolite occurrences, the Dun Mountain ophiolite belt of South Island and the Northland Ophiolite of North Island, review their geochemical characteristics, and contrast models for their genesis and emplacement.

The early Permian Dun Mountain ophiolite belt (Coombs et al., 1976) marks an important crustal suture in the South Island of New Zealand. It outcrops from D'Urville Island in Nelson province for 150 km south to the Alpine Fault (Figure 1). It reappears 480 km to the southwest on the opposite side of the Alpine Fault a little north of Red Mountain and continues 185 km southward to near Lumsden. Some felsic igneous rocks extend for another 150 km southeastward to the coast, and although there are no true ophiolite sequences, essential continuity is inferred for this segment with the remainder of the belt.

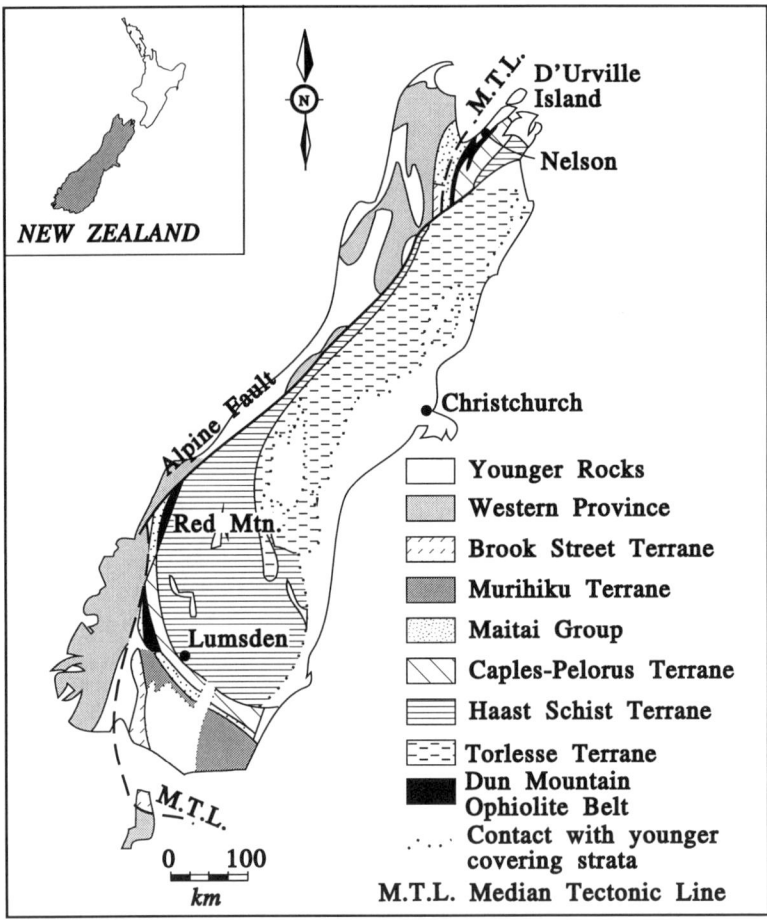

Figure 1. General geology of South Island, New Zealand.

The whole belt is interpreted as marking a major crustal suture in excess of 1000 km in length. In this paper, the Nelson segment of the belt is discussed in more detail.

The Northland Ophiolite (Malpas et al., 1992a) is one of a number of ophiolites with a lower Tertiary emplacement age which occur on the southwest Pacific rim; the others include the Papuan Ultramafic Belt (Davies, 1968) and the New Caledonia ophiolite (Prinzhofer et al., 1980). The ophiolites in Papua New Guinea and New Caledonia are major belts of ultramafic and plutonic rocks representing the deeper levels of oceanic lithosphere sections. In contrast the Northland Ophiolite comprises isolated massifs consisting mainly of rocks representing the uppermost levels of the igneous oceanic crust. Northland ophiolite massifs occur in the northern half of the Northland peninsula and rocks of comparable age and composition are found at the northern tip of the East Cape peninsula (Figure 2).

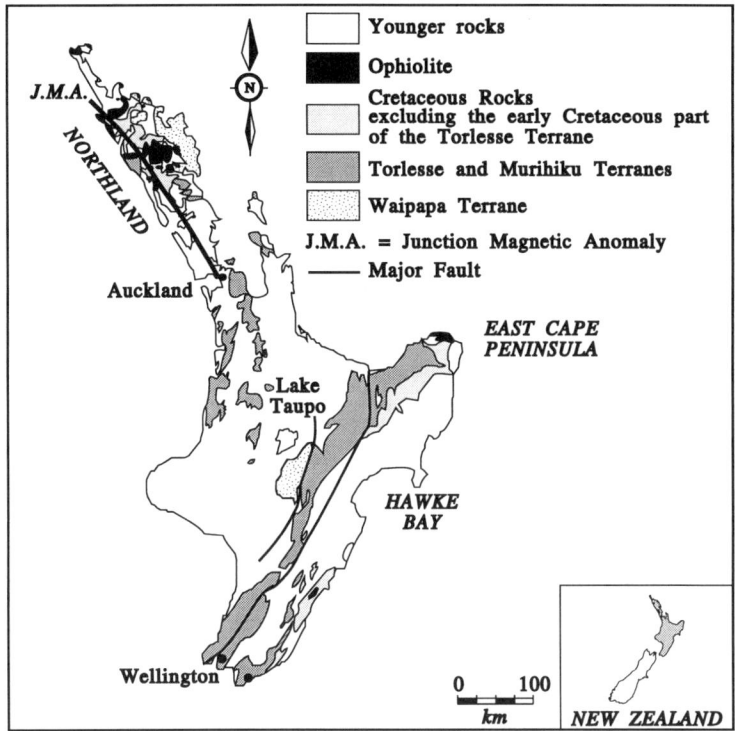

Figure 2. General Geology of North Island, New Zealand.

STRUCTURAL SETTING

a) Dun Mountain ophiolite belt

Two distinct provinces define the geological structure of South Island, the Eastern and Western Provinces (Landis and Coombs, 1967). The older rocks of the Western Province are late Precambrian to Ordovician in age, while rocks of the Eastern Province are Upper Carboniferous to Middle Cretaceous in age (Korsch and Wellman, 1988). The provinces are separated by tectonically complex zones marked by faulting and magmatism referred to as the Median Tectonic Line (MTL) and the younger, still active, Alpine Fault (Figure 1). The MTL is considered to mark the change from continental (west) to oceanic (east) basement (Landis and Coombs, 1967). The Dun Mountain ophiolite belt lies within the Eastern Province, which also contains the mainly quartzo-feldspathic sediments of the Rangitata orogen. The stratigraphy of the Eastern Province is best documented as a series of lithologic units which can be referred to as "terranes", "ophiolite belts" and "melanges". The scheme adopted here is modified after Coombs et al. (1976) and Bishop et al. (1985) and includes from the Pacific side and going to the west (Figure 3):

        i) Torlesse terrane
        ii) Caples-Pelorus terrane
        iii) Greenstone (south) and Croisilles (north) ophiolitic melange
        iv) Patuki ophiolitic melange
        v) Dun Mountain-Maitai terrane
        vi) Murihiku terrane
        vii) Brook Street terrane

Coombs (1976) regarded the Dun Mountain-Maitai, Murihiku, and Brook Street terranes as belonging to an upper crustal plate, the upturned eastern edge of which is marked by the Dun Mountain Ophiolite. In terms of this model, sediments of the Maitai Group and Murihiku Supergroup might be regarded as fore-arc or back-arc basin deposits. The Caples-Pelorus, Haast Schist, and Torlesse terranes were rafted in from a greater or lesser distance, their leading edge underthrusting the ophiolite belt as part of a subduction process. A model involving oblique or strike-slip convergence has been proposed by MacKinnon (1983). By late Mesozoic times the various terranes had been accreted onto the margin of the Gondwana landmass. Each terrane is fault bounded and regional in its extent and is defined by differences in lithology, structure and metamorphism.

The Dun Mountain ophiolite belt therefore represents a major crustal suture, perhaps 1000 km or more long; the date of ophiolite formation as established by U-Pb dating of plagiogranite zircons is 280 ± 5 Ma (Kimbrough et al., 1992). The belt is severely disrupted, though relatively complete ophiolite sequences can be recognized at localities on its western or upper side. The ophiolitic rocks in the Nelson portion of the Dun Mountain-Maitai terrane are generally steeply dipping or slightly overturned with tops to the west and are overlain, in some places in clear sedimentary contact, in other places in tectonic contact, by the Maitai Group, 6 km thick, of Permian to early Middle Triassic age. A turbiditic limestone (the Wooded Peak Limestone), locally over 1000 m thick, containing atomodesmatinid bivalves, occurs near the base of the Maitai Group. If the mid-Permian (Tae Weian = 260 Ma) fossil age (Waterhouse, 1964) is correct, then, together with the 280 ± 5 Ma age from the ophiolite, this suggests a time gap of ~ 20 Ma between ophiolite formation and burial by Maitai Group sediments. Submarine volcanic activity both basaltic and felsic, continued briefly after sedimentation had commenced and is followed upsequence by a quartz-bearing feldspathic sandstone, the Tramway Sandstone. Otherwise the Maitai Group consists essentially of the epiclastic marine derivatives of a largely inactive basaltic to andesitic volcanic arc.

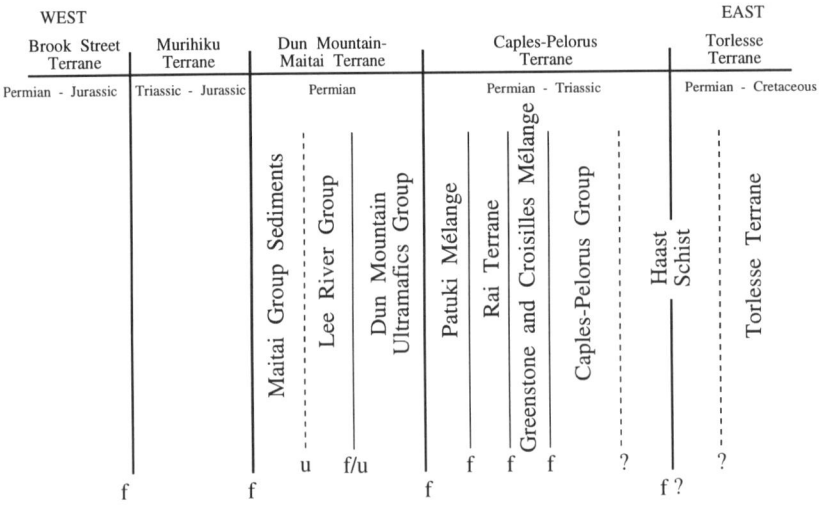

Figure 3. Terranes of the Eastern Province, South Island, New Zealand.

Maitai rocks are succeeded to the west, commonly in demonstrable fault contact, by the Murihiku terrane, containing a sequence, 10 km thick, of epiclastic mostly marine sediments of the Murihiku Supergroup. This includes innumerable thin ash beds ranging in composition from andesitic to rhyolitic. The source was clearly an active volcanic arc of Triassic to Jurassic age.

The west flank of the Murihiku terrane is in tectonic contact with the Brook Street terrane. The latter consists of a Permian assemblage up to 15 km thick of epiclastic and pyroclastic volcanogenic sediments, together with interbedded volcanic breccias, basaltic pillow lavas, ankaramites, porphyritic andesites, and associated intrusives. The Brook Street rocks are interpreted as representing the upturned submarine flank of a Permian volcanic arc in which igneous activity may well have continued into Mesozoic times (Devereux et al., 1968).

In the Nelson area, the Dun Mountain ophiolite belt is dismembered by numerous faults subparallel to its general strike (035 degrees). It is bounded on its east side by the Livingstone suture, and includes tectonic slices of low grade metasediments locally several kilometres thick in addition to the ophiolitic rocks themselves. To the east of the Livingstone suture is the Caples-Pelorus terrane of probably Permian to Triassic age. It includes the Pelorus Group north of the Alpine Fault and the Caples and Tuaneka Groups in the south. The Caples and Pelorus Groups contain much andesitic and rhyolitic arc-derived psammitic and pelitic sediment as well as mafic volcanic and non-volcanogenic quartzofeldspathic components. The terrane is interpreted by Turnbull (1979) as a submarine fan complex that formed in a trench-slope and possibly trench-floor environment.

The Caples-Pelorus terrane passes eastward with increasing metamorphic grade into the Haast Schists, best developed in Otago. The Haast Schists then pass into Carboniferous to early Cretaceous Torlesse terrane. This terrane is less metamorphosed than the Haast Schists and consists of sparsely fossiliferous greywackes and argillites, with occasional pillow lava, chert and limestone horizons.

b) The Northland Ophiolite
In North Island two metagreywacke terranes (the Torlesse and Murihiku) form the exposed basement beneath late Mesozoic and Cenozoic basin sequences and late Cenozoic arc-type and intraplate-type volcanic sequences (Figure 2). The Junction Magnetic Anomaly (JMA, Hatherton, 1969) separates the Torlesse and Murihiku terranes and is interpreted as the northward continuation of the South Island (Dun Mountain) ophiolite belt. In the northern part of North Island a third metagreywacke basement terrane, the Waipapa terrane, occurs between the Torlesse terrane and the JMA.

The Northland Ophiolite outcrops in the northern part of the Northland peninsula as part of the Northland Allochthon (Ballance and Sporli, 1979; Malpas et al., 1992a). The Northland Allochthon overlies an autochthonous sequence of Lower Tertiary sediments which rest unconformably on basement metagreywacke of the Waipapa terrane. Ophiolitic rocks are dominated by volcanics with subordinate plutonic and sedimentary lithologies; ultramafic rocks occur only at North Cape (Bennett, 1976). The volcanics have been assigned a variety of stratigraphic names but are most commonly referred to as the Tangihua Volcanics. The ophiolite was emplaced at the end of the Oligocene as part of a series of thrust sheets of volcanic and sedimentary rocks. The subsequent late Cenozoic geological development of the Northland peninsula has been dominated by arc-type volcanism followed by the eruption of intraplate basaltic volcanic rocks (Smith et al., 1989; Briggs et al., 1989).

The Northland Ophiolite occurs as twelve large and thirteen smaller massifs overlying the deep water Lower Tertiary sediments. Geophysical studies (Sharp et al., 1989) have shown that the massifs have no continuity at depth and they are modelled as rootless blocks within the sedimentary thrust sheets of the Northland Allochthon (Sporli and Kear, 1989). Sedimentary rocks directly associated with the ophiolitic volcanics have yielded Cretaceous and Palaeocene ages; K-Ar dates of the igneous rocks of the ophiolite range from Mesozoic to Pliocene (Brothers and Delaloye, 1982). The Northland Ophiolite is therefore interpreted as a dismembered slice of late Cretaceous and Palaeocene oceanic crust thrust onto the quasi-continental crust of northern New Zealand at the end of the Oligocene.

OPHIOLITE STRATIGRAPHY

a) East Nelson ophiolites
The Nelson segment of the Dun Mountain ophiolite belt includes three separate ophiolite occurrences, the Dun Mountain Ophiolite, the Patuki Melange and the Croisilles Melange (Figure 4). Because of extensive faulting and melange development, a complete ophiolite succession is not preserved intact. In the Dun Mountain Ophiolite however, a structural stratigraphy can be reconstructed from lower protoclastic harzburgite through harzburgite and dunite tectonites to layered peridotites and pyroxenites. These are followed westward by the gabbros of the Tinline Formation and diabases, basalts and basaltic breccias of the Glennie Formation and the Lee River Group. However, in many places, the Lee River Group rests in fault contact against the Dun Mountain Ultramafics. Immediately above the Lee River Group, conglomerates of the Upukerora Formation contain basaltic clasts apparently derived from the immediately underlying volcanics.

Along their eastern side, the Dun Mountain Ultramafics are bounded by the Patuki Melange. Although much of the melange material may be derived from tectonised ophiolite, other materials such as abundant blocks of dense metasomatised argillite are from an apparently distal sedimentary suite, unrecognised *in situ*. In the Croisilles Harbour region 40 km north of Dun Mountain, the Lee River Group volcanics and Dun Mountain Ultramafics are reduced to thin screens or are absent entirely, Maitai strata then lie in tectonic contact with the Patuki Melange. The Patuki Melange is succeeded to the east by a 5 km wide belt of the volcanogenic Rai Formation sandstone, regarded by Landis and Blake (1987) as a separate terrane fragment (Figure 5).

Farther east lies another ophiolite melange, the Croisilles Melange, here a 2 km-thick zone. This melange contains blocks of metasomatized argillites, arenites and conglomerates, gabbros, pyroxenites, basalts, amphibolites, rodingites and plagiogranites in a sheared serpentinite matrix. A late Permian fossil age for a sedimentary block establishes the earliest date at which the melange itself could have been formed (Dickins et al. 1986).

East of the Croisilles Melange a belt of Caples-Pelorus rock some tens of kilometres wide passes gradationally from greywacke and argillite to schist. The entire assemblage can be regarded as a sequence of imbricated slabs forming part of an accretionary prism the formation of which involved undetermined strike-slip as well as convergent components of movement.

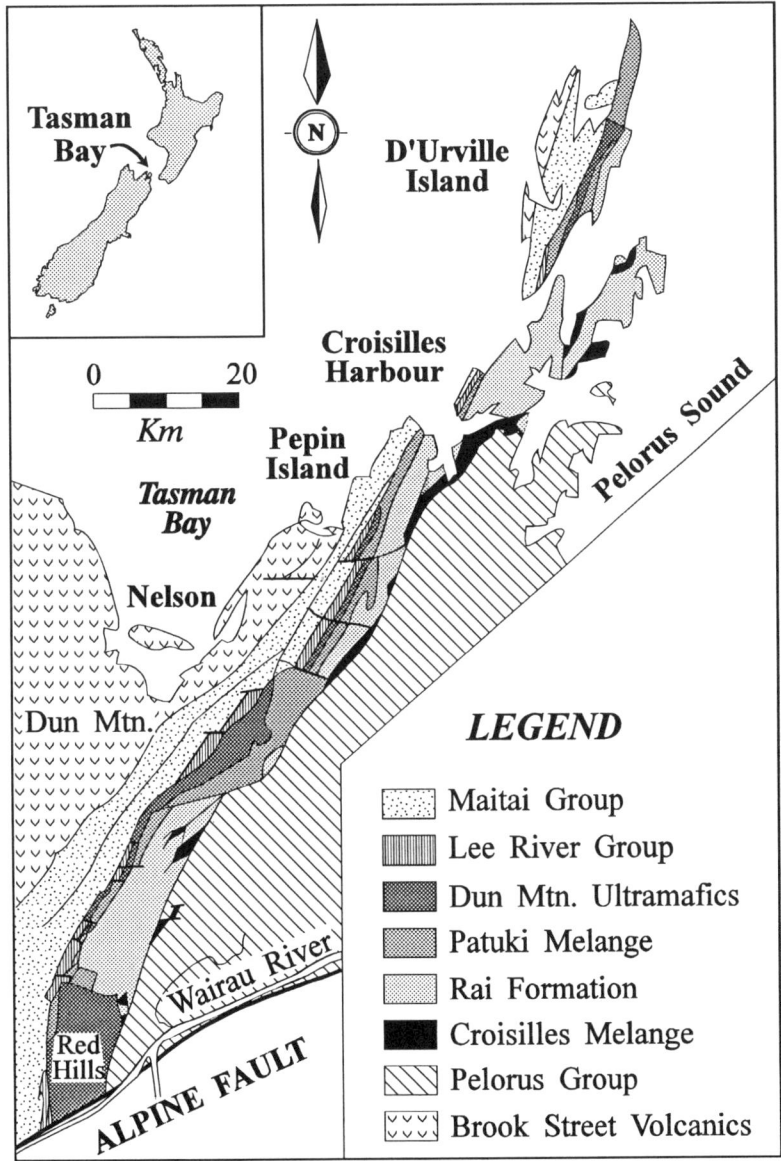

Figure 4. Geology of the Nelson segment of the Dun Mountain ophiolite belt, north of the Alpine Fault, New Zealand.

b) Northland Ophiolite

Individual massifs of the Northland Ophiolite contain structurally coherent sequences of mainly volcanic rocks up to several kilometres thick. Within these sequences lithologic units are generally shallow dipping and structurally upright. Such coherent sequences are typically bounded by steeply dipping faults or shear zones. In some massifs plutonic intrusive rocks are juxtaposed with pillow lavas and sedimentary units. All of these features indicate that the Northland Ophiolite is not a single sheet of oceanic crust dismembered into individual massifs; rather each massif is in fact a structural melange in which different components of an oceanic lithosphere section have been assembled (Figure 5). Nevertheless, the relatively shallow dips of coherent sequences indicate that the processes of disruption and emplacement have not been chaotic.

A group of volcanic rocks that are closely comparable to those forming the Northland Ophiolite occur in the East Cape region of North Island. Although these rocks are not a contiguous part of the Northland sequence, available information indicates that they represent oceanic crust of similar age which was emplaced at the same time.

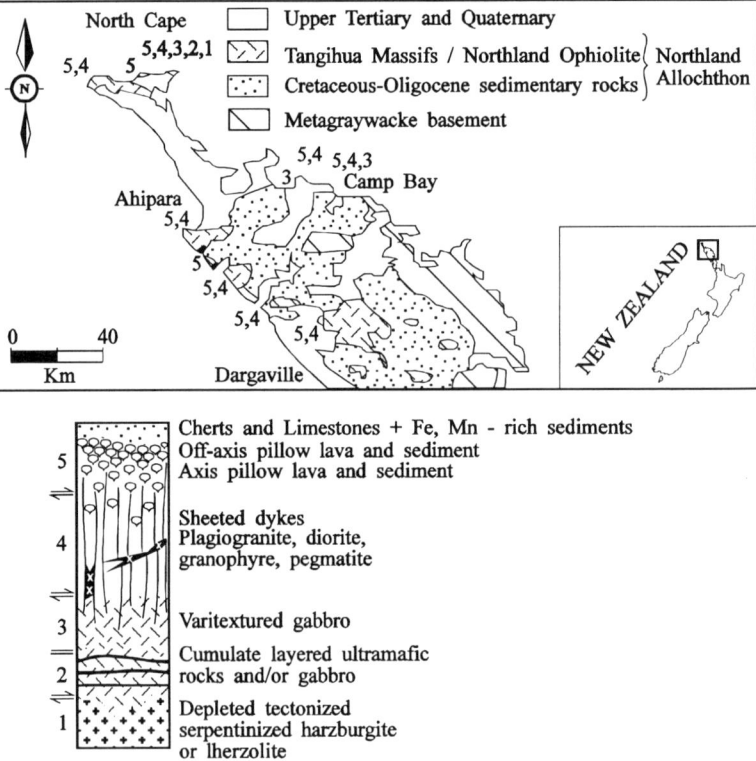

Figure 5. Distribution of ophiolite massifs in Northland, North Island, and identification (with numbers) of rock types present in each massif in terms of a decoupled column of oceanic crust (after Malpas et al., 1992a)

PETROLOGY AND GEOCHEMISTRY OF MAFIC VOLCANIC ROCKS

Mafic volcanic and subvolcanic rocks occur in each of the ophiolite occurrences studied, indeed in the North Island they are the dominant rock type.

a) East Nelson ophiolites
In the East Nelson ophiolites the fine-grained mafic rocks can be subdivided into a number of distinct petrographic suites. In the Dun Mountain Ophiolite the Lee River Group basalts (and clasts in the Upukerora Formation) are dominantly glassy, fine-grained, clinopyroxene-phyric lavas which have undergone moderate degrees of lower greenschist and sub-greenschist facies metamorphism. Associated diabase dykes and microgabbros are subvolcanic equivalents of these lavas but in addition include an aphyric suite and a younger plagioclase-phyric suite which are both metamorphosed to amphibolite grade assemblages.

Volcanic rocks of the Patuki Melange are unlike those of the Dun Mountain Ophiolite and Upukerora Formation and can be divided into 'olivine-poor' and an 'olivine-rich' suites, the former containing less than 1% olivine phenocrysts and the latter approximately 5%. These basalts are generally glassy and typically have a quenched, variolitic to intersertal texture. Volcanic rocks of the Croisilles Melange are similar to the 'olivine-poor' basalts of the Patuki Melange and like those rocks have undergone lower greenschist facies metamorphism.

b) Northland Ophiolite
The predominant rock type of the Northland Ophiolite is tholeiitic basalt occurring as pillow lava and less commonly as thin sheet flows and hyaloclastite. Intrusive equivalents (dolerites) occur as thin sills within the lava sequences. The essential mineralogy of these rocks is augite plus plagioclase; olivine or less common quartz are accessories and iron-titanium oxides are ubiquitous minor phases. Coarse-grained plutonic equivalents occur in a few of the massifs and while their intrusive relationship is clear in some places, their juxtaposition with volcanic rocks is structural in others. The intrusive rocks show considerable petrologic diversity but they are obviously fractionation and cumulate products of magmas represented by the associated basaltic volcanics (Thompson et al., 1993). The ultramafic rocks at North Cape comprise a faulted sequence of serpentinites and cumulate peridotites which occur together with layered gabbro and sheeted dolerite sills and dykes (Bennett, 1976). Minor but not insignificant components of the Northland Ophiolite are volcanic and intrusive rocks of clear alkaline character. These alkaline volcanic rocks are amphibole-bearing basalts, described as lamprophyres by Brothers (1983), which form pillow lavas and minor sheet flows, generally overlying the tholeiitic sequences. Their intrusive equivalents range from mafic to felsic compositions and form small stocks in a few of the massifs (Thompson et al., 1993).

Rocks of the Northland Ophiolite are pervasively altered, the grade being generally low, olivine replaced by iddingsite/bowlingtonite and a general development of chlorite, sericite and epidote. Adjacent to shear zones more intense development of secondary minerals to lower greenschist facies conditions is apparent.

Immobile trace elements can be used to determine tectonic environments of formation of mafic rocks from each of the ophiolite occurences and it is interesting to compare the geochemical characteristics of those in South and North Islands. To facilitate this comparison, we present the geochemistry of the East Nelson ophiolites and the Northland Ophiolite on combined diagrams.

Two commonly used discrimination diagrams are Zr-Ti-Y (Pearce and Cann, 1973) and Zr/Y-log Zr (Pearce and Norry, 1979) which are depicted in Figures 6 and 7. For the East Nelson ophiolites there is a clear distinction between the 'olivine-rich' rocks which plot in the within plate basalt (WPB) field, and the other suites which plot either in the mid-ocean ridge basalt (OFB) field or island-arc basalt (CAB) field. These distinctions also hold for the Northland Ophiolite where alkaline rocks plot as within plate basalts and the remainder plot mainly in the OFB field. The different suites of the East Nelson ophiolites are clearly distinguished on a plot of Th/Yb versus Ta/Yb (Figure 8) in which basalts of the Patuki and Croisilles Melanges do not appear to contain a 'subduction component'. Rocks of the Lee River Group however, plot within the island arc compositional field and amongst these rocks, the plagioclase-phyric suite appears to have been derived from a more depleted mantle source and contains lower Th/Yb and Ta/Yb ratios. Only limited data for these elements are available from Northland rocks and show no evidence for a subduction component.

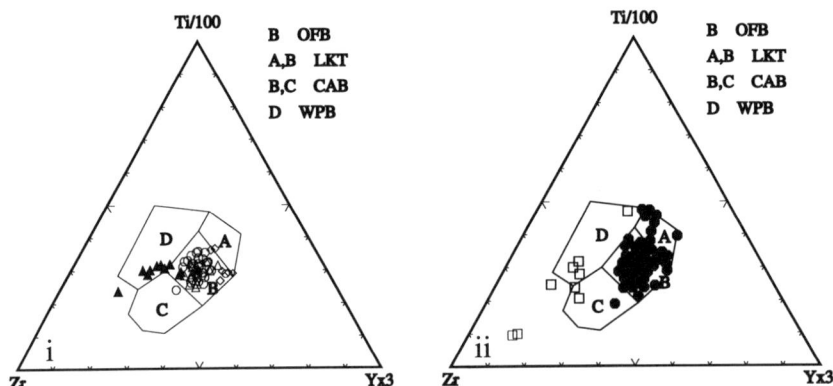

Figure 6. Zr-Ti-Y diagram showing fields of ocean floor basalts (OFB), low potassium tholeiites (LKT), calc-alkaline basalts (CAB), and within plate basalts (WPB). i) East Nelson ophiolites (see figure 11i for symbols) ii) Northland Ophiolite, □ = MORB suite, ● = alkaline suite.

On the Cr-Y variation diagram (Figure 9), despite some overlap into the compositional fields defined for MORB and WPB, basalts of the Lee River Group are typically more depleted than Patuki or Croisilles basalts with similar Cr contents. The younger plagioclase-phyric units again appear to be more primitive. On this diagram, the rocks of each suite tend to plot in distinct fields; however, a number of samples contain relatively high Y presumably as a result of high degrees of fractional crystallisation. In Figure 9, the

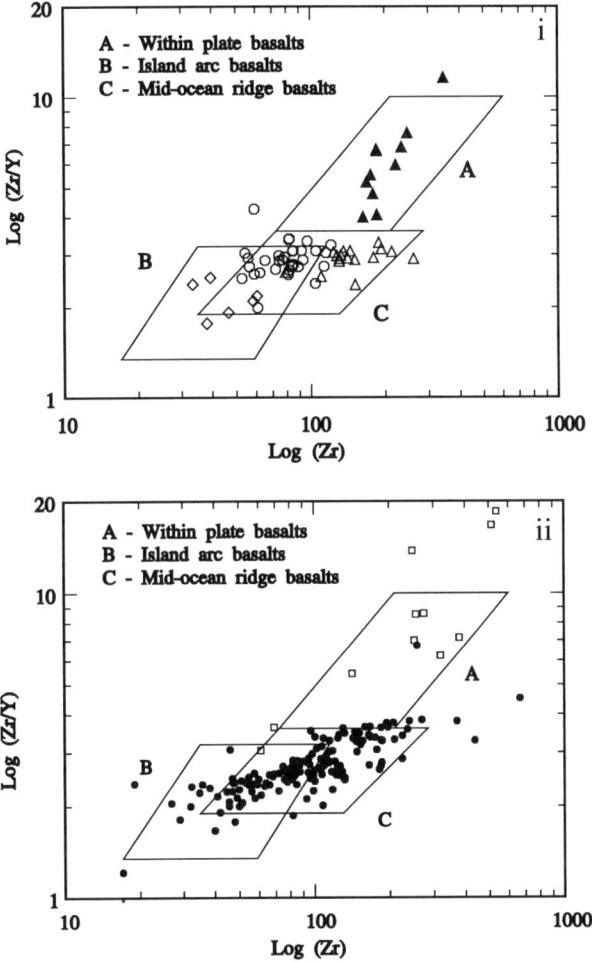

Figure 7. Zr/Y-Zr diagrams for i) East Nelson ophiolites and ii) Northland Ophiolite. Symbols as for figures 11i and 6.

two suites of the Northland Ophiolite cannot be distinguished within a broad field which overlaps that of the East Nelson rocks. The latter samples plot within three distinct, slightly overlapping groups on a plot of Ti versus V (Figure 10). Lee River basalts have ratios which range between 14 and 26, whereas basalts of the Patuki and Croisilles 'olivine-poor' and 'olivine-rich' suites show values between 23 and 37, and 32 and 88 respectively. These ratios suggest that rocks of the Lee River Group are transitional in composition between IAB and MORB, but that the 'olivine-poor' and 'olivine-rich' basalts have Ti/V ratios similar to N-MORB and WPB respectively. The alkaline rocks of the Northland Ophiolite are comparatively low in vanadium and plot separately from most of the samples which overlap the boundary between IAB and MORB as defined by Shervais (1982).

MORB normalised average geochemical patterns for basaltic rocks of the East Nelson ophiolites are shown in Figure 11. Samples are normalised to average MORB as estimated by Pearce (1981), and only the relatively immobile elements are plotted. The Lee River aphyric and clinopyroxene-phyric basalts are indistinguishable on this plot and exhibit flat patterns with consistent Nb depletions and normalised Th/Nb ratios greater than 1. The plagioclase-phyric basalts are not dissimilar but contain lower elemental abundances and marked light rare earth element depletions. Patterns for the 'olivine-poor' basalts whether from the Patuki or Croisilles Melanges compare closely to N-MORB and do not display Nb depletions. Rocks of the 'olivine-rich' suite are enriched in Th, Nb and LREE and are slightly HREE depleted. The normalised plot for the Northland rocks shows a contrast between the incompatible element enriched alkaline rocks and the relatively depleted predominant MORB. Neither suite shows the Nb depletions of the Lee River basalts from Nelson.

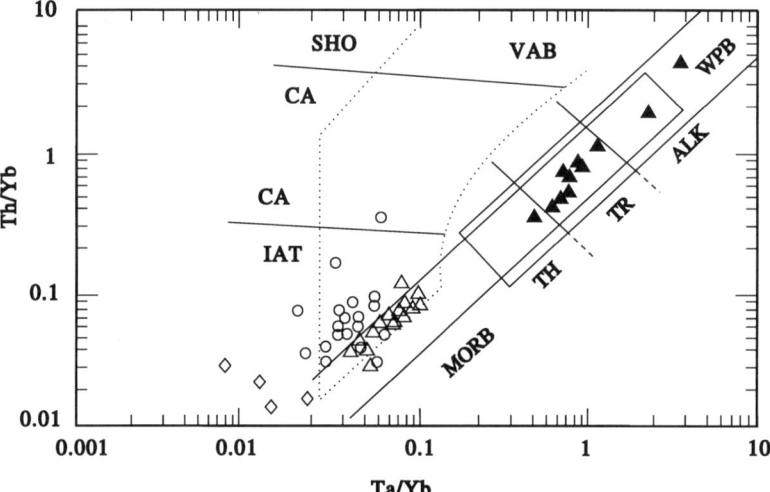

Figure 8. Th/Yb vs Ta/Yb for the East Nelson ophiolites. Relative enrichment of Th is a result of suprasubduction zone magmatism. Symbols as for figure 11i.

The geochemical data presented in Figures 6 to 11 show features of the East Nelson and Northland ophiolites which constrain both the primary nature and the structural relationships of their constituent components. In general terms, the rocks from the East Nelson ophiolites exhibit a greater variety and there is some geochemical evidence for a suprasubduction zone (SSZ) component. The Northland rocks are mainly N-MORB with a minor association of alkaline rocks which are here attributed to seamount volcanism, although the discriminant function diagrams do not confine them to this environment alone.

Geochemistry suggests that the petrographic subdivision of the East Nelson ophiolites can be redefined, i.e. the Lee River aphyric and clinopyroxene-phyric basalts are likely a single suite which appears somewhat transitional between depleted MORB and IAT. The Nb-Ta depletion is characteristic of suprasubduction zone magmatism. The 'olivine-poor' basalts of both Patuki and Croisilles melanges are compositionally the same and probably represent a single suite of rocks which has been tectonically divided and incorporated into two discrete melange units. They are similar to N-MORB, showing no selective enrichment in Th, Ta, Nb, Ce or P_2O_5 typical of E-MORB. The geochemical pattern for the 'olivine-rich' basalts is analogous to that of ocean island alkali basalts. Most of the basaltic rocks of the Northland Ophiolite show N-MORB chemical characteristics with some overlap into the arc field on some discriminant diagrams. However, there is no indication in the available data for the relative HFSE depletion characteristic of arc-associated rocks. The alkaline rocks are a minor part of the association and show wide variations in chemical composition. Both the MORB and alkaline suites include differentiated rocks interpreted as the products of shallow fractionation processes.

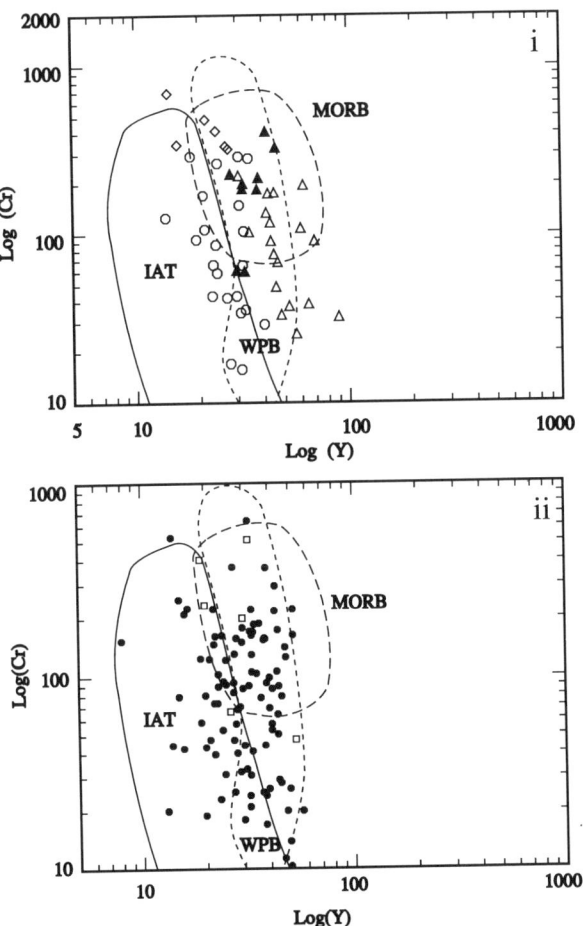

Figure 9. Cr-Y variation for East Nelson ophiolites (i) and Northland Ophiolite (ii). IAT = Island Arc tholeiite, MORB = Midocean ridge basalt, WPB = Within plate basalt. Other symbols as figures 11i and 6.

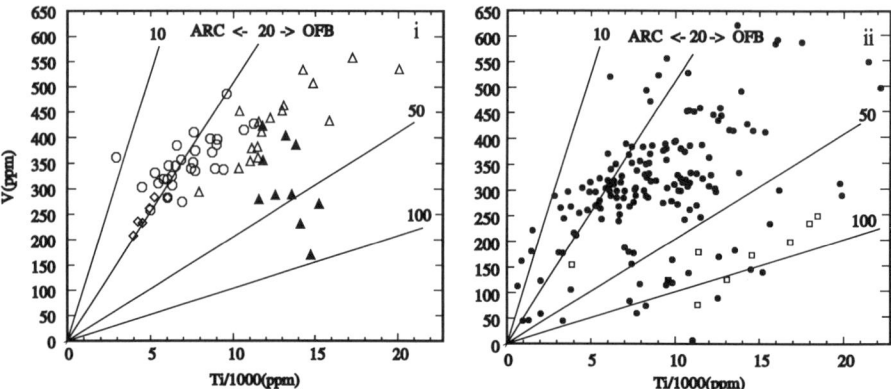

Figure 10. Ti vs V for East Nelson ophiolites (i) and Northland ophiolites (ii). A ratio of 20 delineates arc basalts from ocean floor basalts. Symbols as for figures 11i and 6.

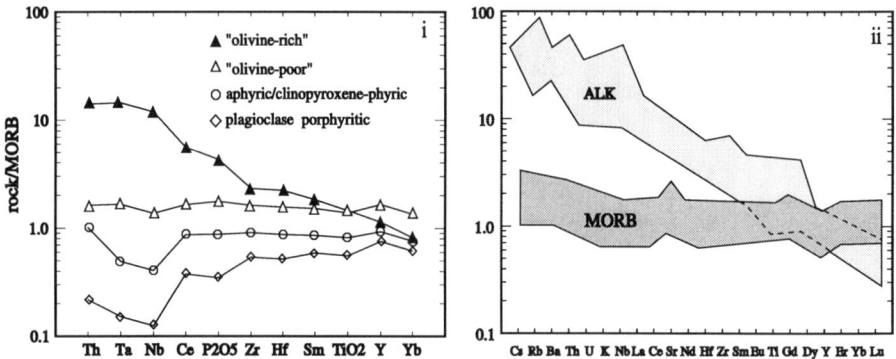

Figure 11. MORB normalized geochemistry for East Nelson ophiolites (i) and Northland Ophiolite (ii).

MODELS FOR THE ORIGIN AND EMPLACEMENT OF THE OPHIOLITES

Although detailed models of ophiolite emplacement are as numerous as ophiolite complexes themselves, these models can be broadly divided into two categories. The first invokes the genesis and emplacement of ophiolite complexes as parts of oceanic plates lying immediately above subduction zones. These ophiolites are generally characterised by a suprasubduction zone (SSZ) component in their geochemistry and appear to be the most common of the two categories (Searle and Stevens, 1984; Malpas and Stevens, 1977). Other ophiolites appear to have been obducted as thin slivers shaved off a subducting plate in a process which has been called "flake tectonics" (Oxburgh, 1972). These two categories, although fundamentally different, are not easy to distinguish once the ophiolite complexes have been emplaced and, in certain cases, transported some considerable distance by late gravity sliding (Cawood, 1990). However, we believe that both types can be identified in New Zealand, exemplified by the ophiolites we have described above. One key factor in the identification of the relevant mechanisms of generation and emplacement lies in the relationship between ophiolitic and associated alkaline, within-plate basaltic rocks. The common association of alkaline volcanic rocks, thought to represent ocean island volcanism, and ophiolitic rocks has been noted in Oman, Cyprus and Newfoundland

(Searle and Malpas, 1992; Malpas et al., 1992b; Jenner et al., 1992). In each of these cases the alkaline volcanics are found in thrust sheets structurally below and separated from a major overriding ophiolite complex. Such is also the case for the East Nelson ophiolites, a model for the origin and emplacement of which is outlined in Figure 12.

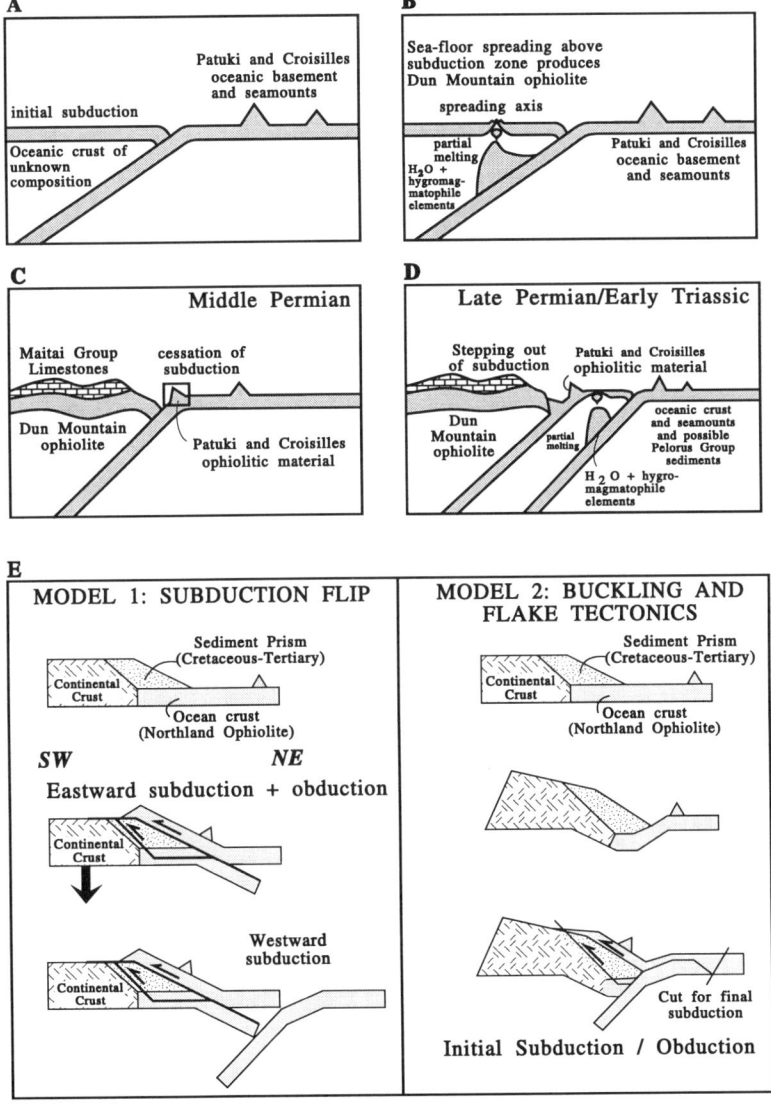

Figure 12. Models for the emplacement of the East Nelson ophiolites (A-D) and the Northland Ophiolite (E). For explanation, see text.

a) East Nelson ophiolites

It has been proposed by previous workers (Coombs et al., 1976; Davis et al., 1980) that ophiolites of the East Nelson area were produced during the evolution of a westward dipping subduction zone, now represented by the faulted contact at the base of the Dun Mountain Ophiolite. Clearly, the evidence presented above suggests that these ophiolites were formed in at least two different ocean basins during the early Permian, and whereas the Dun Mountain Ophiolite represents the ocean crust formed above a subduction zone, rocks of the Patuki and Croisilles melanges are considered to structurally underlie it and likely represent part of the subducted plate. The oceanic crust in both cases is of similar age, and older parts of the subducted plate, produced at a mid-ocean spreading centre east of the trench, must have been destroyed prior to the Middle Permian. Indeed, available radiometric age dates suggest that ophiolitic rocks of the Patuki and Croisilles melanges are slightly older than those of the Dun Mountain Ophiolite. Subduction of this older crust induced partial melting of depleted mantle material, enriched in hygromagmatophile elements such as Th, Rb and K, producing volcanism similar to that of early island arc development (Natland and Tarney, 1981). At some time prior to late Middle Permian time (Tae Weian), rocks of the Dun Mountain Ophiolite underwent a period of extensive uplift and erosion which produced the conglomeratic Upukerora Formation upon which limestones of the Wooded Peak Formation were later deposited. This period of emergence may have resulted in response to jamming of the subduction zone as seamounts on the Patuki ocean crust entered the trench. Concomitant stepping out of the subduction zone likely induced arc-related igneous activity to the east of the Dun Mountain Ophiolite within the small stranded basins of Patuki and Croisilles crust. Such activity is now represented by island arc tholeiitic rocks within the Croisilles melange as reported by Sivell (1988).

b) Northland Ophiolite

The observed MORB geochemical signature of the Northland Ophiolite is consistent with its generation either at a major spreading centre or in a mature extensive back-arc basin. The alkalic basalts are considered to represent seamounts built on this crust. Assuming no major rotation of the massifs during emplacement, the generating spreading ridge, if preserved, would lie to the northeast and trend north to northwest relative to the present orientation of the ophiolites. Two models have been proposed to explain the origin and emplacement of the ophiolites from this ridge (Malpas et al., 1992 and Figure 12). The principal difference in these models is the polarity of pre-early Miocene subduction: model 1 involves collision of an east-facing continental margin with an east-dipping subduction zone, whereas model 2 invokes delamination of oceanic crust and mantle at a west-dipping subduction zone.

Model 1 corresponds to the obduction and initiation of subduction mechanisms frequently proposed in other areas (Muller and Phillips, 1991). However, there are no arc rocks that can be tied to the early eastward subduction. This may be because the early subduction event was too brief to generate volcanic activity. If, however, the initial subduction was to the west (model 2), it is not necessary to identify such arc rocks but it is clear that the ophiolite must have been emplaced by a flake tectonic mechanism such as proposed by Oxburgh (*op.cit.*).

Acknowledgements

The authors would like to thank their graduate students who have been involved in the investigations of New Zealand ophiolites, particularly P. Moore and G. Thompson. Also, the late Professor R.N. Brothers whose enthusiasm stimulated this work. Thanks to Nancy Fagan for manuscript preparation.

References

P.F. Ballance and K.B. Sporli. Northland Allochthon, *Roy. Soc. N.Z. Jour.* **9**, 259-275 (1979).

M.C. Bennett. The Ultramafic-Mafic Complex at Northcape, Northernmost New Zealand, *Geol. Mag.* **113**(1), 61-76 (1976).

D.G. Bishop, J.D. Bradshaw and C.A. Landis. Provisional Terrane Map of South Island, New Zealand. In: *Tectonostratigraphic Terranes*. D.G. Howell,(ed.). Circumpacific Council for Energy and Minerals Resources. Earth Sciences Series, No. 1. Houston, Texas. (1985).

R.M. Briggs, T. Itaya, D.J. Lowe and A.J. Kean. Ages of the Pliocene-Pleistocene Alexandra and Ngatutura Volcanics, Western Northland, New Zealand, and some Geological Implications. *N.Z. Jour. Geol. Geophys.*, **32**(4), 417-427 (1989).

R.N. Brothers and M. Delaloye. Obducted Ophiolites of North Island, New Zealand: Origin, Age, Emplacement and Tectonic Implications for Tertiary Volcanicity. *N.Z. Jour. Geol. Geophys.* **25**, 257-274 (1982).

R.N. Brothers. Tertiary Accretion of Ophiolite Seamounts, North Island, New Zealand. In: *Accretion Tectonics in the CircumPacific Regions*. M. Hashimoto and S. Uyeda (eds). Tokyo, Terrapub. 307-318 (1983).

P. Cawood. Late-Stage Gravity Sliding of Ophiolite Thrust Sheets in Oman and Western Newfoundland. In: *Ophiolites: Oceanic Crustal Analogues*. J. Malpas, E. Moores, A. Panayiotou and C. Xenophontos (eds). Geological Survey Department, Cyprus (1990).

D.S. Coombs, C.A. Landis, R.J. Norris, J.M. Sinton, D.J. Borns and D. Craw. The Dun Mountain Ophiolite Belt, New Zealand, its Tectonic Setting, Constitution and Origin, With Special Reference to the Southern Portion. *Am. Jour. Sci.*, **276**, 561-603 (1976).

H.L. Davies, Papuan Ultramafic Belt. *23rd International Geological Congress*, Section 1, 209-220 (1968).

T.E. Davis, M.R. Johnston, P.C. Rankin, R.J. Stull. The Dun Mountain Ophiolite Belt in East Nelson, New Zealand. In: *Ophiolites. Proceedings of the International Ophiolite Symposium, Cyprus, 1979*. A. Panayioutou (ed.) Cyprus Ministry of Agriculture and Natural Resources, Geological Survey Department, 480-498 (1980).

I. Devereux, I. MacDougall, W.A. Watters. Potassium - Argon Mineral Dates on Intrusive Rocks from the Foveaux Strait Area. *N.Z. Jour. Geol. Geophys.*, **11**, 1230-1235 (1968).

J.M. Dickins, M.R. Johnston, D.L. Kimbrough and C.A. Landis. The Stratigraphic and Structural Position and Age of the Croisilles Melange, East Nelson, New Zealand. *N.Z. Jour. Geol. Geophys.*, **29**, 291-301 (1986).

T. Hatherton. Geophysical Anomalies over the Eu- and Miogeosynclinal Systems of California and New Zealand. *Geol. Soc. Am. Bull.*, **80**, 213-230 (1969).

G.A. Jenner, G.R. Dunning, J. Malpas, M. Brown and T. Brace. Bay of Islands and Little Port Complexes, Revisited: Age, Geochemical and Isotopic Evidence Confirm Suprasubduction-Zone Origin. *Can. Jour. Ear. Sci.*, **28**, 1635-1652 (1991).

D.L. Kimbrough, J.M. Mattinson, D.S. Coombs, C.A. Landis and M.R. Johnston. Uranium-Lead Ages from the Dun Mountain Ophiolite Belt and Brook Street Terrane, South Island, New Zealand. *Geol. Soc. Am. Bull.*, **104**(4), 429-443 (1992).

R.J. Korsh and H.W. Wellman. The Geological Evolution of New Zealand and the New Zealand Region. In: *The Ocean Basins and Margins. Vol. 7B: The Pacific Ocean*. A.M. Nairn, F.G. Stehli and S. Uyeda (eds) 411-482 (1988).

C.A. Landis and M.C. Blake Jr.. Tectonostratigraphic Terranes of the Croisilles Harbour Region, South Island, New Zealand. In: *Terrane Accretion and Orogenic Belts Geodynamic Series*. E.C. Leitch and E. Scheiber (eds) **19**, AGU, Washington, D.C., 179-198 (1987).

C.A. Landis and D.S. Coombs. Metamorphic Belts and Orogenesis in Southern New Zealand. Tectonophysics, **4**, 501-518 (1967).

T.C. MacKinnon. Origin of the Torlesse Terrane and Coeval Rocks, South Island, New Zealand. *Geol. Soc. Am. Bull.*, **94**, 969-985 (1983).

J. Malpas, K.B. Sporli, P.M. Black and I.E.M. Smith. Northland Ophiolite, New Zealand, and Implications for Plate Tectonic Evolution of the S.W. Pacific. *Geology*, **20**, 149-152 (1992a).

J. Malpas and R.K. Stevens. The Origin and Emplacement of the Ophiolite Suite with Examples from Western Newfoundland. *Geotectonics*, **11**(6), 453-466 (1977).

J. Malpas, C. Xenophontos and D. Williams. The Ayia Varvara Formation of SW Cyprus: A Product of Complex Collisional Tectonics, *Tectonophysics*, **212**, 193-211 (1992b).

S. Mueller and R.J. Phillips. On the Initiation of Subduction, *Jour. Geophys. Res.*, **96**, 651-665 (1991).

J.H. Natland and J. Tarney. Petrologic Evolution of the Mariana Arc and Back-Arc Basin System: A. Synthesis of Drilling Results in the South Philippine Sea. In: *Initial Reports of the Ocean Drilling Project*, D.M. Hussong, S. Uyeda, et al. (eds), **60**, 877-908 (1981).

E.R. Oxburgh. Flake Tectonics and Continental Collision, *Nature*, **239**, 202-204 (1972).

J.A. Pearce. Geochemical Evidence for the Genesis and Eruptive Setting of Lavas from Tethyan Ophiolites. In: *Ophiolites: Proceedings of the International Ophiolite Symposium, Cyprus, 1979, Panayiotou, A. (ed.)*, 261-272 (1981).

J.A. Pearce and J.R. Cann. Tectonic Setting of Basic Volcanic Rocks Determined Using Trace Element Analyses. *Ear. Plan. Sci. Lett.* **19**, 290-300 (1973).

J. Pearce and M.J. Norry. Petrogenetic Implications of Ti, Zr, Y, and Nb Variations in Volcanic Rocks. *Contrib. Min. Pet.* **69**, 33-47 (1979).

A. Prinzhofer, A. Nicolas, D. Cassard, J. Moutte, M. Leblanc, P. Paris and M. Rabinovitch. Structures in the New Caledonia Peridotites-Gabbros: Implications for Oceanic Mantle and Crust, *Tectonophysics*, **69**, 85-112 (1980).

M.P. Searle and J. Malpas. Petrochemistry and Origin of Sub-Ophiolitic Metamorphic and Related Rocks in the Oman Mountains. *Jour. Geol. Soc. Lond.* **139**, 235-248 (1982).

M.P. Searle and R.K. Stevens. Obduction Processes in Ancient, Modern and Future Ophiolites. *Geol. Soc. Lond., Spec. Pub.* **13**, 303-320 (1984).

B.M. Sharp, C.A. Locke and J.A. Cassidy. Gravity Investigations of the Maungataniwha and Ahipara Ophiolite Massifs, Northland, New Zealand. *Roy. Soc. N.Z. Bull.* **26**, 175-181 (1989).

J.W. Shervais. Ti-V Plots and the Petrogenesis of Modern and Ophiolitic Lavas. *Ear. Plan. Sci. Lett.* **59**, 101-118 (1982).

W.J. Sivell. Geochemical Constraints on the Origin of Croisilles and Patuki Ophiolites: Implications for Late Paleozoic - Mesozoic Tectonics in New Zealand. *Tectonics*, **7**, 1015-1032 (1988).

I.E.M. Smith, R.S. Ruddock and R.A. Day.. Miocene Arc-Type Volcanic/Plutonic Complexes of the Northland Peninsula, New Zealand. In: *Geology of Northland - Accretion, Allocthons and Arcs at the Edge of the New Zealand Micro-Continent*, K.B. Sporli and D. Kear (eds), *Roy. Soc. N.Z. Bull.*, **26**, 205-213 (1989).

B. Sporli and D. Kear (eds). Geology of Northland: Accretion, Allochthons and Arcs at the Edge of the New Zealand Microcontinent, *Roy. Soc. N.Z. Bull.*, **26** (1989).

G.M. Thompson, J. Malpas and I.E.M. Smith. The Tectonic Implications of Contrasting Suites of Plutonic Rocks within the Northland Ophiolites. *Tectonophysics*, (submitted) (1993).

I.M. Turnbull. Petrography of the Caples Terrane of the Thompson Mountains, Northern Southland, New Zealand. *N.Z. Jour. Geol. Geophys.*, **22**, 709-727 (1979).

J.B. Waterhouse. Permian Stratigraphy and Faunas of New Zealand. *N.Z. Geol. Surv. Bull.*, **72**, 101p (1964).

The Cambrian Trinity Ophiolite and Related Rocks of the Lower Paleozoic Trinity Complex, Northern California, U.S.A.

NANCY LINDSLEY-GRIFFIN
Department of Geology, 214 Bessey Hall, University of Nebraska, Lincoln, NE 68588, U.S.A.

Abstract

The Trinity Complex of western North America consists of the Cambrian Trinity ophiolite, pre-Late Ordovician peridotite, and Ordovician and Silurian intrusions. It is a polygenetic composite terrane of ophiolitic rocks within a 1900 km² subhorizontal sheet that formed over a period of 165 million years. The highly deformed Trinity ophiolite consists of harzburgite, gneissic gabbro, plagiogranite, and pillow basalt; ophiolite stratigraphy is partially preserved within fault blocks. Cambrian ages are based on zircons from the plagiogranite and gabbro. The pre-Late Ordovician Trinity peridotite is faulted against the Cambrian Trinity ophiolite along a broad, nearly vertical zone of mylonitic schist. No Ordovician or younger ophiolite succession is preserved; Ordovician gabbro and plagiogranite are present only as fault slivers or as small intrusions and no Ordovician basalt is present. Both Cambrian and Ordovician rocks and the intervening fault zone are crosscut by voluminous Lower Silurian pegmatitic gabbro and pyroxenite intrusions. These undeformed intrusions post-date amalgamation of the Cambrian and Ordovician parts of the complex. No volcanic rocks of Silurian age are known from the Trinity Complex, but it is overlain by Devonian? pillow basalts. Some mafic dikes may be feeders for the Devonian? lavas, but sheeted dike packets that were previously interpreted as Devonian are actually Jurassic in age.

Keywords: ophiolite, Trinity Complex, Klamath Mountains, Cambrian, Ordovician, Silurian, Devonian, Jurassic

INTRODUCTION

The Trinity ophiolite and related rocks of the Trinity Complex lie within the eastern Klamath Mountains of western North America (Fig. 1). Previously thought to be Ordovician [49], the Trinity ophiolite is now known to be of earliest Cambrian age [51, 70-72]. The Trinity ophiolite is faulted against a large body of pre-Late Ordovician peridotite known as the Trinity peridotite; both have been crosscut by voluminous younger magmatic suites of Silurian age (Figs. 1, 2) to form the lower Paleozoic Trinity Complex. The lower Paleozoic rocks of the Trinity Complex are overlain by Devonian? lavas and dikes of the adjacent Redding terrane; both are crosscut by Jurassic and Cretaceous intrusions, some of which are petrologically similar to the older rocks. Until the last few years, the multiple crosscutting relationships of similar rock types, lack of radiometric ages, and insufficient geologic mapping allowed many contradictory tectonic models to be developed [52, 5]. However, all previous models are suspect until they can be re-evaluated in the context of new isotopic data. This paper summarizes our present knowledge of the Trinity ophiolite and related rocks of the Trinity Complex in light of the new isotopic ages and recent field work.

The Cambrian Trinity ophiolite is not an "ideal" ophiolite [1, 9]. It does not represent a simple section of lithosphere formed at a typical oceanic ridge. Instead, it consists of mafic

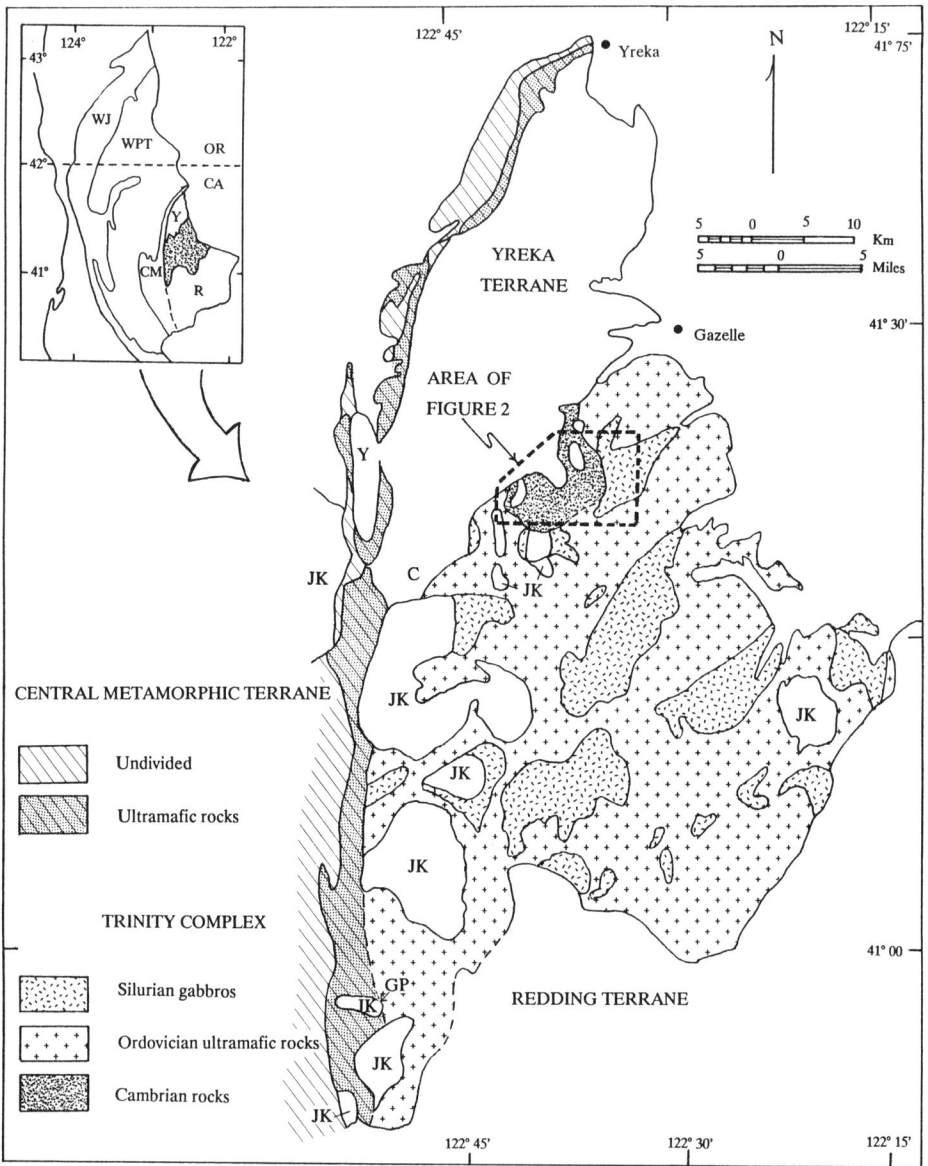

Figure 1. Geologic map of Trinity Complex. Inset shows relationship of Trinity terrane (stippled) to other terranes of the Klamath Mountains in western North America. Symbols: CA, state of California, U.S.A.; OR, state of Oregon, U.S.A.; R, Redding terrane; Y, Yreka terrane; CM, central metamorphic terrane; WPT, western Paleozoic and Triassic terrane; WJ, western Jurassic terrane; JK, Jurassic and Cretaceous plutons; GP, Gibson Peak. Geology modified after Lindsley-Griffin and Griffin [52], Strand [67], Wagner and Saucedo [69].

and ultramafic rocks that formed in three magmatic pulses from Ordovician through Devonian (Fig. 3), later overprinted by Jurassic-Cretaceous intrusions. Such a prolonged evolution for the Trinity Complex is evidence of its polygenetic, composite character. Although the atypical character of this ophiolitic complex has been recognized before [47-49,

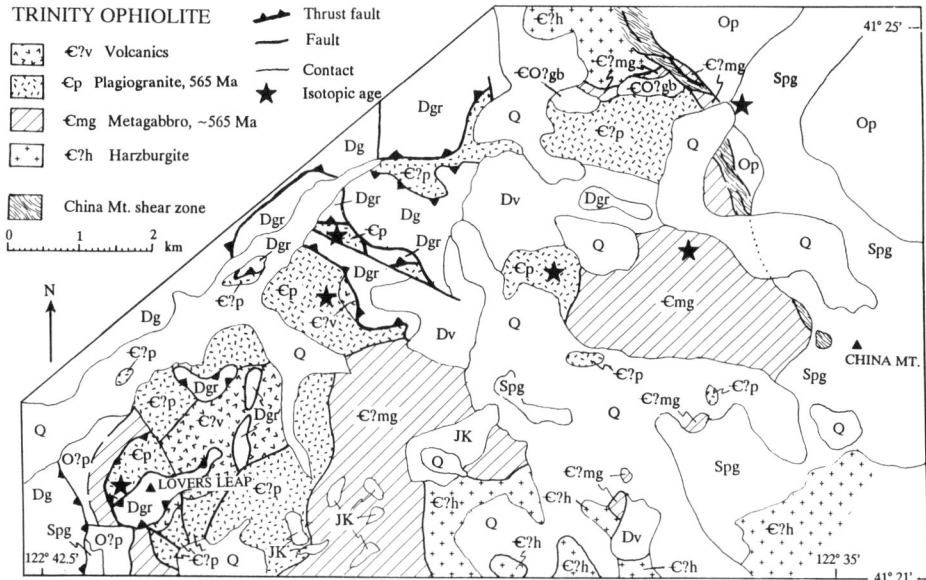

Figure 2. Geologic map of northwestern edge of Trinity Complex, showing Cambrian rocks of the Trinity ophiolite. Stars indicate locations of Cambrian isotopic ages and of the Silurian isotopic age for the China Mountain pluton. Other units of the Trinity Complex, CO?gb, gabbros of Cambrian or Ordovician age; Op, Ordovician peridotite; Spg, Silurian pegmatitic gabbros. Yreka terrane units: Dg, Gazelle Formation; Dgr, Gregg Ranch Complex. Dv, Devonian? volcanic rocks; JK, Jurassic-Cretaceous intrusions; Q, Quaternary sediments.

64], the precise origin of its components is not well understood.

Regional Geologic Setting

A *terrane* is "a fault-bounded body of rock of regional extent, characterized by a geologic history different from that of contiguous terranes" [2: p. 679]. In this paper, the term is used for allochthonous bodies accreted to the continent at an active margin [32, 39]. The Klamath Mountains comprise an eastwardly-concave set of arcuate terranes within the western North American Cordillera (Fig. 1). The *Trinity terrane* consists of igneous and meta-igneous rocks of Cambrian, Ordovician, and Silurian age that appear to have a unique history not shared by rocks in adjoining terranes until after the end of the Silurian. The lower Paleozoic Trinity terrane (Fig. 1) is a fault-bounded, subhorizontal sheetlike terrane of mafic and ultramafic rocks, over 1900 km² in extent, within the eastern Klamath terrane or eastern Klamath plate of Irwin [35, 36]. The *Trinity Complex* consists of three major components [51]: 1) the *Trinity ophiolite*, a sequence of Cambrian metagabbro and plagiogranite, and possibly Cambrian metabasalt, faulted against Cambrian? harzburgite (Fig. 2); 2) ophiolitic pre-Late Ordovician peridotite (*Trinity peridotite*) and minor Ordovician plagiogranites and gabbros; and 3) Silurian pegmatitic gabbros and related rocks (Fig. 2). As shown in Figure 1, the Trinity Complex is bounded by the Yreka and central metamorphic terranes to the northwest and the Redding terrane to the southeast.

The *Yreka terrane* (Fig. 1) consists of fault-bounded sheets of lower Paleozoic marine sedimentary and metasedimentary rocks and melange that have been thrust over the Trinity Complex [27, 28, 52]. Geophysical data suggest that the Yreka terrane overlies a basement

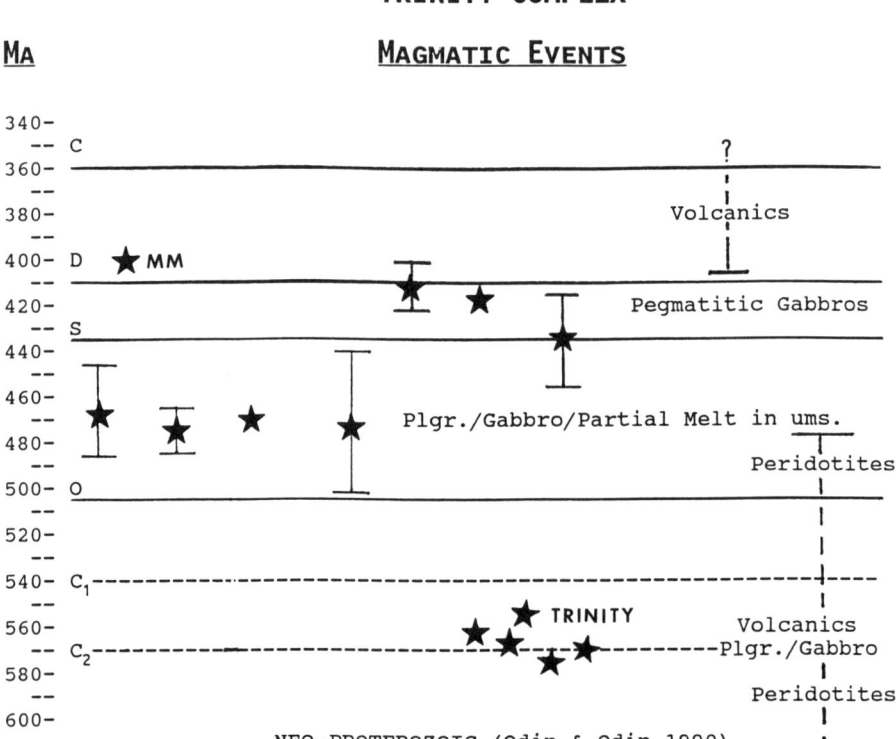

Figure 3. Magmatic events of the Trinity Complex. Three distinct clusters of early Paleozoic isotopic ages in the Trinity Complex highlight its composite nature. The Neo-Proterozoic/Cambrian ages (depending on which time scale is used) are from the Trinity ophiolite. Ordovician ages are from intrusions within the Trinity ophiolite and the pre-late Ordovician peridotite of the Trinity Complex. Silurian ages represent pegmatitic gabbros of the Trinity Complex; Devonian ages include the Mule Mountain Stock (MM) and Devonian? volcanic rocks of the Trinity-Yreka composite terrane.

of dense rocks of probable ultramafic composition, generally interpreted to be a subsurface continuation of the Trinity terrane [19, 37, 41, 16]. The *Trinity-Yreka composite terrane* includes the Trinity Complex, Yreka terrane, and rocks of Devonian, Jurassic, and Cretaceous age that intrude or overlie either terrane.

West of the Yreka terrane, the *central metamorphic terrane* (Fig. 1) consists of an elongate, steeply dipping ultramafic slab, with minor amphibolite and metasedimentary melange, of approximately 375 km^2 extent. The north end of the central metamorphic terrane bounds the Yreka terrane and its south end abuts the Trinity terrane. To the south, the central metamorphic terrane widens and consists of Devonian schists [35]. Ultramafic rocks of the central metamorphic terrane are not part of the Trinity terrane because they have experienced a different tectonic history [51].

The *Redding terrane* (Fig. 1) comprises a sequence of marine metavolcanics and metasediments that ranges in age from Early Devonian to Middle Jurassic; it is roughly homoclinal with a regional dip to the east [35]. Although rocks of the Redding terrane are generally regarded as having been deposited on a basement consisting of Trinity Complex [35], the contact between Redding terrane and Trinity terrane is now a fault.

Previous Work
Early use of the name "Trinity" for these rocks is summarized by Lindsley-Griffin [51]. The name was originally applied [23] to ultramafic rocks in a region that straddles the boundary between the Trinity terrane and the central metamorphic terrane. Later workers extended the name to the entire ultramafic body of the Trinity Complex [11-14, 18, 24-26, 34, 35, 42, 46, 54, 57].

Brewer [4] first recognized the close relationship between serpentinized peridotite and hornblende diorite, hornblende gabbro, and dikes in the China Mountain area (Fig. 2). Recognition of the Trinity Complex as an ophiolitic assemblage came in the 1970's [33, 46, 57, 58] and resulted in a number of detailed field studies of the mafic and ultramafic rocks [18; 49, 50, 63-66, 68].

Regional relationships between the Trinity Complex and surrounding areas have been extensively discussed in the literature. The contributions of W.P. Irwin [29-37] are of special note because he recognized that the Trinity terrane is the oldest part of the Klamath Mountains, and that terranes to the west become progressively younger due to sequential accretion against the Trinity terrane.

Isotopic Ages
Isotopic ages have been reported for the Trinity Complex by a number of authors, but the significance of many was previously uncertain because the samples were collected without the benefit of detailed structural maps. Recently published isotopic data [70, 72], combined with my detailed geologic and structural surveys of the sample sites (1989-1992) permit new insights into the origin and history of the Trinity Complex. Wallin *et al.* [70] discuss isotopic data for the Trinity Complex and other igneous rocks in the eastern Klamath region. Valid isotopic ages for the Trinity terrane are summarized in Table 1.

The difficulty of interpreting isotopic ages from the eastern Klamaths is illustrated by geological relationships at Lovers Leap (Fig. 2). Here, plagiogranites have been found to be both Ordovician [58, 72] and Cambrian [72]. The Cambrian sample site lies structurally above at least one of the Ordovician sites, which suggests a fault, but poor exposure prevents a clear understanding of the relationship. Mattinson and Hopson [58] describe one of their Ordovician sites as within a fault slice. It is probable that Lovers Leap consists of a tectonic melange of the Cambrian and Ordovician basement rocks overlain by polygenetic melange of the Lower Devonian Gregg Ranch Complex (Fig. 2) [53]. Both melanges are penetrated by several different ages of dikes, yielding at least one Late Jurassic age [7]. Other dikes at Lovers Leap are lithologically similar to rocks of Silurian, Devonian?, and Cretaceous age. The only sheeted dikes within the Trinity terrane that have been isotopically dated are Jurassic; because of lithologic similarities and cross-cutting relationships, it is likely that all sheeted dikes within the Trinity terrane are Jurassic.

The Trinity Complex is of early Paleozoic age, but is intruded or overlain by igneous rocks of middle Paleozoic to late Mesozoic age. Thus, within the Trinity terrane (Figs. 1, 2) may

Table 1. Isotopic ages for Trinity terrane and adjacent regions

Rock Unit	Isotopic Age	Method
Cambrian metagabbro of Trinity ophiolite (CM)	556-579 [70]	U-Pb zircon
Cambrian plagiogranites of Trinity ophiolite (4 sites: GR, LL, HM)	565-571 [72]	U-Pb zircon
Ordovician plagiogranite or gabbro of Trinity Complex (LL)	475 +/- 10 [72]	U-Pb zircon
Ordovician late-stage partial melt segregate in plagioclase lherzolite of Trinity Complex	472 +/- 32 [38]	Sm-Nd isochron
Ordovician hornblende gabbro dike of Trinity Complex (LL)	470-480 [58,72]	U-Pb zircon
Ordovician plagiogranite or gabbro of Trinity Complex (LL)	469 +/- 21 [72]	U-Pb zircon
Ordovician tonalite block in Lower Devonian melange of Gregg Ranch Complex (LL)	440, 455 [58,72]	U-Pb zircon
Ordovician or Silurian microgabbro of Trinity Complex	435 +/- 21 [38]	Sm-Nd isochron
Silurian pegmatitic gabbro of Trinity Complex (CM)	415 +/- 3 [70]	U-Pb zircon
Silurian pegmatitic trondhjemite dikes of Trinity Complex (KL)	412 +/- 10 [72]	U-Pb zircon
Devonian Mule Mountain Stock	400 +/- 3 [40]	U-Pb zircon
Devonian Schneider Hill stock	400 [*]	U-Pb zircon
Jurassic sheeted dikes in Trinity Complex (CM)	161 +/- 4 [*]	$^{40}Ar/^{39}Ar$
Cretaceous diabase dike cutting Devonian? pillow basalts (CM)	140 [*]	U-Pb zircon
Cretaceous tonalite dike (GM)	125 [*]	U-Pb zircon

Numbers in brackets indicate reference, [*] Wallin, Martin, Lindsley-Griffin, unpub. data. Localities indicated by: CM, China Mountain; GM, Gazelle Mountain; GR, old Gregg Ranch; HM, Houston Mine; KL, Kangaroo Lake; LL, Lovers Leap.

be found gabbroic rocks of Cambrian, Ordovician, Silurian, Devonian?, and Jurassic-Cretaceous age (Table 1). Plagiogranites, granitoids and other intermediate rock types have been dated as Cambrian, Ordovician, and Cretaceous; geologic relationships suggest that some intermediate composition rocks in the terrane are also Silurian, Jurassic, and possibly Devonian. Thus, results based on Sm-Nd isochron determined from a suite of samples, like the age published by Brouxel and Lapierre [5], cannot be considered valid because they might include a mixture of Ordovician, Silurian, and Devonian samples.

The only age available for ultramafic rocks of the complex is based on an Ordovician late-stage gabbroic segregation [38]. However, at least five distinct map units of ultramafic rocks can be recognized within the Trinity Complex; geologic relationships with gabbros whose ages are known suggest that both Cambrian and Silurian ultramafic rocks may be present in addition to the pre-Late Ordovician unit. Basaltic lavas of two ages are present within the boundaries of the Trinity terrane. Neither volcanic unit has been isotopically dated, but geological restrictions on their age are discussed below.

Because each major rock type within the Trinity terrane (ultramafic rocks, gabbros, plagiogranites, basalts) includes several different ages of similar rock compositions, age and geochemical data not linked to a detailed geologic map are of limited utility. Interpretations based on geochemical data from melange zones, such as the melanges and dike complexes at Lovers Leap and Gregg Ranch [5, 8], are misleading because they do not correctly distinguish between suites of different ages and origins.

CAMBRIAN TRINITY OPHIOLITE

The *Trinity ophiolite* consists of a partially dismembered sequence of Cambrian metagabbro and Cambrian plagiogranite, overlain by Cambrian? volcanic rocks and faulted against deformed harzburgites of pre-Late Ordovician age. The ophiolite sequence is preserved only along the northwestern edge of the Trinity Complex (Figs. 1, 2). The ophiolite was previously considered to be Ordovician [49, 50] but has yielded isotopic ages of 565 Ma for plagiogranites of the dike complex unit and for the "amphibolitic gabbro" unit, a deformed metagabbro [70-72]. Numerous isotopic ages from these two units confirm that much of the ophiolite sequence is actually Early Cambrian in age. The highly deformed ophiolite is faulted against less deformed gabbro and peridotite fragments of Ordovician age, and intruded by dikes and small stocks of Ordovician (?), Silurian, Devonian?, Jurassic, and Cretaceous age (Fig. 3, Table 1).

Cambrian Ophiolite Stratigraphy
Stratigraphic relationships within the Cambrian ophiolite sequence are poorly preserved, and most contacts within this assemblage appear to be faults. East of Lovers Leap (Fig. 2), the best stratigraphic sequence from metagabbro to plagiogranite to basaltic lavas appears to be undisrupted by faulting. Although no isotopic ages have been obtained for the Lovers Leap metagabbro, it is correlated with the Cambrian ophiolite because its composition and structural style are similar to those of the isotopically dated Cambrian metagabbro block.

The stratigraphic relationship between Cambrian plagiogranites and the overlying Cambrian? basaltic dikes and lavas is well established by field relationships and map pattern (Fig. 2). The lavas overlie the plagiogranites; basaltic feeder dikes intrude the plagiogranites and locally can be traced upward into the lavas [50]. Isolated screens of plagiogranite occur within the basaltic lavas. Because no dikes of the Cambrian? basalts cut harzburgite or Cambrian metagabbro, the basalts must have been erupted through the Cambrian plagiogranites before the ophiolite sequence was dismembered. The lavas, dikes, and plagiogranites exhibit the same deformation style, suggesting that they were deformed together.

The Trinity ophiolite does not have a sheeted dike complex. Instead, the hypabyssal intrusions consist of irregular individual intrusions to small packets of subparallel sills or dikes [50]. Small sheeted dike complexes do intrude the Trinity Complex. Previously thought to be related to eruption of the Devonian? basalts [51], new isotopic data confirm that at least one such sheeted dike complex is Jurassic in age (E.T. Wallin, written comm., 10-22-92). Although these sheeted dike complexes and volcanics have been included in the "Trinity ophiolite" by some authors [5, 6, 8], they postdate the juxtaposition of both the Trinity and the Yreka terranes and therefore cannot be part of either the Trinity ophiolite or the Trinity Complex.

Harzburgite

No isotopic ages have been obtained for the fault-bounded block of harzburgite that I correlate with the Trinity ophiolite. It is tentatively assigned to the Cambrian ophiolite because: 1) it occurs along the northwestern edge of the Trinity terrane adjacent to the blocks of known Cambrian age (Fig. 2); 2) its textures and structures suggest that it is more intensely deformed than other pre-Late Ordovician peridotites and thus is likely to be pre-Ordovician; 3) it exhibits a structural style similar to that of rocks known to be Cambrian; and 4) unlike the Ordovician peridotite, it lacks feldspar clots formed by partial fusion, suggesting it may be a different block of mantle which has undergone a different history.

The Trinity harzburgite unit contains an average of 30%-40% orthopyroxene (enstatite), suggesting its fertile nature, although orthopyroxene content ranges from 10% to 75%. No clinopyroxene has been observed. Layers and lenses of dunite are present where orthopyroxene content is low, and local concentrations of chrome spinels form podiform chromite bodies and thin layers. The harzburgite is 50%-95% serpentinized; the olivines are completely altered to serpentine and the orthopyroxenes extensively altered with either rims or cores of serpentine.

The original texture of the harzburgite is recognizable because secondary magnetite grains outline the olivine grains. The grains of olivine "ghosts", magnetite, and chrome spinel are elongated into trains to produce a planar fabric faintly visible in outcrop and readily visible in thin section. The large flattened orthopyroxene oikocrysts are aligned in trains parallel to this fabric; typically grains within the same train have nearly the same optical orientation, suggesting they once may have been part of a single large grain which has been dismembered with little or no rotation. Strain lamellae within orthopyroxene grains also indicate the extensive deformation that characterizes this unit.

In addition to the mineral foliation, compositional banding or layering is locally present, although much of the harzburgite unit is massive. The layers, typically 20-50 cm thick, are caused by changes in the proportions of pyroxene to olivine. Compositional layering is subparallel to the mineral foliation where both can be observed in the same outcrop.

Metagabbro

The Cambrian age of 556-579 Ma [70-72] was determined for the large fault-bounded block of metagabbro that comprises the western slope of China Mountain (Fig. 2). This block is bounded on the east by a broad, steep fault zone; both the fault zone and the metagabbro are intruded by undeformed Silurian pegmatitic gabbro of the China Mountain pluton. Additional areas underlain by metagabbro are correlated with the ophiolite because of similar composition and structural style.

The Cambrian metagabbro (Fig. 4) consists of hornblende gneiss with hornblendite layers, locally interlayered with leucocratic gneiss. Less deformed parts of the unit are uralitized layered gabbro. Hornblendite layers consist of up to 95% hornblende, whereas leucocratic layers may contain as little as 20% hornblende. Leucocratic layers contain albite and locally common quartz. The metagabbro unit is hydrothermally altered; epidote coats joints and fracture surfaces and occurs as veins up to 5 cm thick. Slivers and blocks of clinopyroxenite, consisting of about 97% augitic diopside and 2-3% magnetite are associated with the China Mountain block of metagabbro around its faulted boundaries.

The metagabbro is strongly deformed. Although it contains some zones of nearly pristine layered gabbro with graded bedding and relict cumulate textures, much of the metagabbro is characterized by ductilely necked and thinned bands and schlieren probably derived from original cumulate layering. Near the boundaries of the Cambrian metagabbro block, the rock is mylonitized, with the more leucocratic layers pulled apart into boudins.

In the clinopyroxenite fragments associated with the metagabbro, textures reveal a history of intense deformation. The recrystallized matrix consists of small, clean clinopyroxene grains with a uniformly even grain size and equiangular grain boundaries, which surround sparse relict clinopyroxene grains. The large relict grains exhibit internal strain lamellae and fractures, and are aligned in trains with nearly the same optical orientation, suggesting they are dismembered parts of the same original grain.

Along fault contacts between the Cambrian metagabbro and the Cambrian? harzburgite (Fig. 2), blocks of ductilely deformed clinopyroxenite interlayered with hornblendite occur in serpentinite and serpentine schist matrix. The larger blocks of clinopyroxenite are locally intercalated with boudins and stringers of hornblendite and amphibolitic metagabbro, suggesting the clinopyroxenite may have originated as the basal ultramafic part of a cumulate

Figure 4. Hornblende gneiss and hornblendite of the Cambrian metagabbro, Trinity ophiolite. Scale is 15 cm.

gabbro.

Plagiogranite
Cambrian isotopic ages of 565 Ma [70-72] have been obtained for most of the fault-bounded blocks of plagiogranite that form a substrate for basaltic dikes and lavas of the Trinity ophiolite sequence (Fig. 2).

The Cambrian plagiogranites consist of plagioclase feldspar (40%-60%), quartz (15%-35%), and altered mafic minerals (trace to 25%). Proportions of these minerals are highly variable, hence the rocks range from quartz-rich to quartz-poor, and from leucocratic to locally mafic. The feldspar is extensively saussuritized and typically consists of a felted mass of zoisite, actinolite, chlorite, prehnite, and pumpellyite. Unaltered plagioclase ranges in composition from An_{25} to An_{40}, well within the reported range for oceanic plagiogranites [9]. No alkali feldspar has been observed. The mafic minerals were probably hornblende originally; some of the plagiogranites still contain up to 20% recognizable hornblende. However, in most samples the mafic minerals have been altered to chlorite and actinolite.

Like the metagabbro unit, the plagiogranites are intensely deformed (Fig. 5). In these rocks, evidence of extensive ductile to brittle deformation is seen in stretched and dismembered clots of feldspar and quartz grains, and the streaked foliated appearance of the rock in outcrop. In thin section, clots of quartz and plagioclase grains exhibit ductile stretching and boudinage, overprinted by brittle tension fractures oriented perpendicular to the stretching direction (Fig. 5). Large grains are broken into subgrains that have been rotated slightly relative to each other, with small recrystallized grains along subgrain boundaries. The deformed quartzo-feldspathic clots are surrounded by a fragmental matrix composed of quartz, plagioclase, and altered hornblende. Some of the larger matrix fragments would fit back together if the intervening matrix were removed (Fig. 5). The finest grain sizes within the matrix exhibit the clean, fresh appearance and equiangular grain junctions typical of

Figure 5. Cambrian plagiogranite of the Trinity ophiolite. Photomicrograph shows quartz and plagioclase boudins in mylonitic matrix of quartz, plagioclase, and chlorite after hornblende. View is 3 mm across, cross-polarized light.

recrystallization textures. This evidence suggests the intense ductile deformation was followed by brittle deformation and partial recrystallization.

Metabasalt Dikes and Pillow Lavas
In several localities, altered basaltic dikes penetrate the Cambrian plagiogranites and can be traced stratigraphically upward into Cambrian? metabasalt lavas interbedded with sparse metadacite lavas and cut by rare metadacite dikes. Some dikes are sheared into phacoids with plagiogranite wrapping around them, and the volcanic rocks are also fractured and sheared in appearance, although pillow lavas are locally well preserved at Lovers Leap. The structural style and intimate spatial relationship to the Cambrian plagiogranites support the assignment of the basaltic rocks to the Cambrian ophiolite sequence. Dikes of microgabbro and plagiogranite are associated with the metabasaltic dikes where they intrude the plagiogranite. Well developed sheeted structure is lacking in dikes of the Trinity ophiolite.

As noted by Lindsley-Griffin [49, 50], plagiogranites are spatially associated with these Cambrian? metabasalts at Lovers Leap, Gregg Ranch, and Crater Creek (Figs. 1, 2). At all three localities the ophiolitic rocks are overlain by a thrust sheet of melange that contains Middle Ordovician through Middle Devonian blocks, the Gregg Ranch Complex of Lindsley-Griffin *et al.* [53]. At Gregg Ranch and Crater Creek the plagiogranite-basalt-melange sequence also is overlain by a second, post-Lower Devonian? basalt sequence, a circumstance which has caused much confusion in the literature. The Devonian? basalt (discussed below) is not part of the Trinity ophiolite, nor is it part of the Trinity Complex; instead, it overlaps the Trinity-Yreka composite terrane.

The Cambrian? volcanic unit consists of metabasalts and minor metadacites. The most typical rock type contains 1-5 mm euhedral to subhedral phenocrysts of feldspar (partially saussuritized albite) and sparse hornblende in an aphanitic matrix. Phenocrysts comprise 5%-50% of the rock. Rocks of this unit are extensively altered to an assemblage of albite, saussurite, red and green silica, calcite, chlorite, and epidote. The rocks also contain ubiquitous pyrite, chalcopyrite, bornite, and a variety of iron and copper oxides. Calcite veins are common. Locally, so much calcite has replaced the original matrix that the rock fizzes on contact with HCl: a useful technique for distinguishing it from the younger pillow lavas of Devonian? age.

REE and trace element data from the Cambrian? metabasalt unit are sparse and of poor analytical quality [50]. REE patterns are nearly flat and slightly depleted in LREE. Ti-Zr values fall within the overlap field of Pearce and Cann [61, 62], near the average value for ophiolitic extrusives [9]. These rocks are currently being studied in more detail.

Brouxel [6, 8] analyzed samples from dikes and volcanics at Lovers Leap and found two geochemical suites, one LREE-depleted and the other LREE-enriched. The LREE-depleted suite undoubtedly represents the Cambrian? volcanic unit, as the geochemistry is very similar to that reported by Lindsley-Griffin [50] for this unit. Brouxel's LREE-enriched suite from Lovers Leap probably represents dikes of Devonian? or Jurassic-Cretaceous age. Brouxel, Lecuyer and Lapierre [8: p. 260] interpreted the LREE-depleted suite as low-K tholeiites typical of "immature island-arc volcanics". The data for the Cambrian? volcanic unit also could be interpreted as typical of other tectonic settings, such as a backarc marginal basin or a triple junction. Because of extensive alteration in this unit, geochemical data should be interpreted with caution, but the evidence does support an oceanic origin for these rocks.

The Cambrian? metabasalts range from a feldspar-rich porphyritic pillow lava to massive aphyric lava flows to breccia. Pillows are difficult to recognize because of the extensive fracturing and shearing characteristic of the unit, but are best preserved at Lovers Leap (Fig. 2). The pillow structures exhibit both concentric and radial fractures, aphanitic chilled margins, and a coarser grained porphyritic interior. Locally, clots of chert and siliceous mudstone occur in the interstices between pillows. Metabasaltic dikes crosscut the pillows. The pillow structures are clear evidence of a submarine origin for metabasaltic lavas of the Trinity ophiolite.

ORDOVICIAN OPHIOLITIC ROCKS

A complete ophiolite sequence of Ordovician age is not present within the Trinity Complex. In the area shown on Figure 2, where the best ophiolite sequence is displayed [49, 50], most of the plagiogranites and all of the volcanic rocks are now known to be Cambrian. Deformed gabbros of unknown age are probably Cambrian as well, although further isotopic dating is necessary to confirm this hypothesis. Thus, ophiolitic rocks of Ordovician age appear to consist only of fault slivers and small intrusions (Table 1).

Pre-Late Ordovician Ultramafic Rocks
East of the China Mountain fault zone (Figs. 1, 2) lies a unit of relatively unserpentinized peridotites and dunites that were mapped by Quick [65] and by Lindsley-Griffin [50]; most of the northeastern and north-central part of the Trinity Complex consists of this unit, but its extent to the south is not known. Jacobsen *et al.* [38] obtained a Sm-Nd mineral isochron age of 472 ± 32 Ma for a feldspathic pocket within this peridotite (Fig. 6). They interpreted this age as the time when small pockets of basaltic magma formed by adiabatic partial melting. Thus, this is a minimum age for the peridotite, rather than the time of crystallization. The description of composition, textures, and structures given below is drawn from Quick [63-65]

Figure 6. Feldspathic lherzolite of the Trinity Complex. Clots of plagioclase-orthopyroxene-clinopyroxene partial melt segregations that crosscut mineral foliation and compositional layering have been dated isotopically as Late Ordovician (Table 1). Scale is 15 cm.

and Lindsley-Griffin [49, 50] and is supplemented by Lindsley-Griffin's field and lab work in 1988-1990.

The pre-Late Ordovician ultramafic rocks of the Trinity Complex are unusual because of their high plagioclase content (Fig. 6). Peridotite comprises 60-70% of the unit [64] with dunite and pyroxenite comprising the remainder. Lithologic types include lherzolite (ol + opx + cpx) and plagioclase lherzolite, harzburgite (ol + opx) and plagioclase harzburgite, dunite, and spinel pyroxenite (cpx + opx + sp). Chrome spinel is a ubiquitous accessory, which locally forms small layers or pods. The orthopyroxene is enstatite; the clinopyroxene is diopside or chrome diopside [50]. Serpentinization is extensive near the Silurian gabbroic plutons and other intrusions and along fault zones, but elsewhere serpentine represents only about 20-50% of the rock [64]. Contacts between different rock types are commonly gradational, caused by changes in proportions of pyroxene to olivine, or abundance of feldspar.

Pyroxene grains within the pre-Late Ordovician lherzolite, harzburgite, and pyroxenite have been stretched and partially dismembered into a planar fabric by ductile deformation. Mineral foliation and compositional layering are characteristic of most of the Ordovician ultramafic rocks (Fig. 6); such structures are typical of mantle tectonites and may have formed by cumulate processes or by pressure solution during mantle creep [9, 15]. Some dunite bodies appear to crosscut peridotite structures; these bodies appear less deformed than the surrounding peridotite. Quick [63, 64] interpreted these as having formed after most of the ductile deformation by a complex process related to the generation and expulsion of basaltic magma from suboceanic mantle.

Crosscutting the tectonic fabric are stringers and pockets of feldspar, now altered to albite or saussurite, about which are clustered tiny recrystallized grains of clinopyroxene and orthopyroxene (Fig. 6). Several authors have concluded that these distinctive textures represent a partial melting event in which small amounts of basaltic melt formed during pressure-release melting during or after ductile deformation [48-50, 59, 64, 65].

Ordovician Gabbro and Plagiogranite
A number of Ordovician ages have been obtained for gabbro and plagiogranite of the Trinity Complex (Table 1). Some of the older data have been recalculated using new constants [72]. At Lovers Leap (Fig. 2), one 440 Ma tonalite boulder lies within a conglomeratic melange block; a 470 Ma fault slice may also represent melange, as it is in fault contact with Cambrian plagiogranites. Wallin *et al.* [72] reported two new isotopic ages of 469 and 475 Ma for Ordovician "hornblende tonalite of the Trinity ophiolite". Both these samples are also from Lovers Leap, but their structural setting is uncertain.

Ordovician plagiogranites within the Trinity Complex are highly variable in composition, accounting for the variety of rock names applied to them in the literature: "tonalite", "hornblende gabbro", "trondhjemite". Because of this range of compositional types, the best term for the rocks is "plagiogranite" [9, 10]. These quartz-bearing rocks consist of plagioclase (albite), with hornblende ranging from less than 10% (trondhjemitic) to as much as 30% (tonalitic if quartz-rich; gabbroic if quartz-poor). Quartz abundances vary from 15% to 35%, and alkali feldspar is absent or present only in trace amounts.

Ordovician plagiogranites and gabbros lack textures and fabric caused by deformation. In contrast to the highly deformed Cambrian plagiogranites and metagabbros, the Ordovician rocks exhibit a nonfoliated appearance, with a well preserved hypidiomorphic-granular texture

(Fig. 7).

Ordovician Stratigraphic Relationships

No Ordovician ophiolite succession is preserved in the Trinity Complex. In some cases, hornblende gabbro and plagiogranite with Ordovician ages (Table 1) lie within fault blocks juxtaposed against serpentinized lherzolite and plagioclase lherzolite at Lovers Leap, on a scale much smaller than can be shown on Figure 2. In other cases, small intrusions of undeformed, nearly massive, fresh-looking hornblende gabbro and plagiogranite (Fig. 7) occur within the Cambrian? metagabbro unit that is located east of Lovers Leap. These small dikes resemble Ordovician gabbro and plagiogranite in composition and structural style, and may also be Ordovician. In addition, two small stocks of hornblende gabbro intrude Cambrian plagiogranite west of the China Mountain shear zone (Fig. 2). Although no isotopic data are available for these two bodies, they are probably Ordovician or Late Cambrian, because they postdate deformation within the Cambrian rocks and are cut off by the pre-Late Silurian China Mountain shear zone. Thus, the Ordovician gabbro and plagiogranite are either fault slices or intrusive into the Trinity Complex.

SILURIAN PEGMATITIC GABBRO SUITE

Isotopic data establish the age of the China Mountain pluton (Fig. 2) as 415 Ma, Late Silurian [70-72]. This large stock, which intrudes ophiolitic Cambrian and Ordovician rocks on the northwestern edge of the Trinity Complex, is part of a distinctive suite of mafic and ultramafic dikes and stocks that is widespread throughout the Trinity Complex [71]. The intrusive nature of these rocks has long been recognized [18, 36, 42, 47, 49, 50, 54, 66, 68].

Although included in the Trinity ophiolite by some authors [5, 6, 8], the Silurian gabbroic suite is not part of an ophiolite sequence because it does not comprise a continuous sheet above the peridotite or below the dike complex-basalt. Instead, it consists of individual stocks

Figure 7. Nonfoliated Ordovician? leucogabbro dike of the Trinity Complex (top) cutting foliated Cambrian hornblende metagabbro of Trinity ophiolite.

and dikes, as well as dike swarms, that crosscut the older deformed ophiolitic assemblages [49, 50]. The Silurian plutons are, however, included in the Trinity Complex and the Trinity terrane because they intrude only rocks of the Trinity terrane; they do not intrude rocks of the Yreka terrane or the central metamorphic terrane.

The dominant rock types in the Silurian suite are pyroxene gabbro and hornblende pyroxene gabbro (Figs. 8, 9). The pyroxene is green clinopyroxene (diopside) that weathers bronze to reddish brown and comprises 30-40% of the rock. Plagioclase comprises 40-60% of the rock

Figure 8. Silurian pegmatitic gabbro suite of the Trinity Complex: medium to coarse grained and pegmatitic hornblende-pyroxene gabbro exhibiting a characteristic layered but non-foliated appearance.

Figure 9. Silurian pegmatitic gabbro suite of the Trinity Complex, showing characteristic coarse-grained, non-foliated appearance of the pyroxene gabbro.

and is typically saussuritized; fresh plagioclase is calcic, An_{60} to An_{96} [66]. Black hornblende is present as interstitial material or as rims on pyroxene grains (Fig. 8); it comprises 5-15% of the hornblende pyroxene gabbros. Quartz is locally present as an accessory.

Subordinate rock types in the Silurian suite include dikes of pyroxenite, hornblende diorite, quartz diorite, and aplite. The pyroxenites include clinopyroxenite (green diopside), websterite (cpx + opx + ol), and wehrlite (cpx + ol). Cumulate pyroxenites occur as individual stocks and as parts of gabbroic stocks; pyroxenite dikes intrude rocks of the Cambro-Ordovician basement complex. Diorites range from hornblende diorite to quartz diorite and quartz-feldspar aplite, and typically occur as dikes and dike swarms that intrude the basement complex, the pyroxenites, and the gabbros.

The Silurian gabbroic suite is termed "pegmatitic gabbro" because the abundance of pegmatitic grain sizes is the major distinguishing feature of the unit; however, regions of fine to medium grained rocks are common. Grain size in the gabbros varies abruptly; these variations commonly define a crude layering (Fig. 8) that Schwindinger and Anderson [66] attributed to episodic water saturation of the magma. The larger gabbro and clinopyroxenite plutons exhibit cumulate textures and structures: poikilitic pyroxene grains and mineral layering, as well as grain-size layering. Most of the pyroxenite, gabbro, and hornblende diorite dikes are pegmatitic, with an average grain size of 2-5 cm; locally, individual crystals attain a size of >5 cm by 20 cm.

The lack of a penetrative deformational fabric in rocks of the Silurian suite suggests that these rocks formed after deformation ceased in the Cambro-Ordovician rocks. Mutual crosscutting relationships indicate the following sequence of intrusion: pyroxenite dikes and stocks, gabbro dikes and stocks, hornblende diorite dikes, diorite and quartz diorite dikes, aplite dikes.

POST-EARLY DEVONIAN ROCKS AND THE TRINITY COMPLEX

Devonian? Lavas and Jurassic Sheeted Dike Swarms
Altered basaltic pillow lavas and pillow breccias, andesites, pyroclastics, hyaloclastites, and massive lava-flow rocks overlie both the Trinity and Yreka terranes (Fig. 2) in a number of localities [51, 55, 56]. The extrusive rocks and associated basaltic dikes and dike swarms contain xenoliths of the Late Silurian pegmatitic gabbro suite, hence, they must be post-Late Silurian in age. These assemblages are not part of the Trinity ophiolite, but may be correlative with Devonian Copley and Kennett Formations of the Redding terrane [55, 56].

At least 4 different pulses of dike intrusion that postdate the Silurian gabbros can be recognized [50; P.H. Masson, oral comm., 7-15-90]. Along the northwestern edge of the Trinity Complex, the oldest suite of dikes, containing large phenocrysts of green diopside and pink hydrogrossular, is post-Early Devonian and could be Middle (to Late?) Devonian in age, because it cuts rocks containing Early Devonian (Emsian) fossils [51, 53]. These green-diopside basalt dikes represent the oldest intrusion in the dike assemblage [P.H. Masson, oral comm., 7-15-90]. The second and third suites of dikes are basaltic to andesitic in composition. Their age is unknown but falls between the post-Early Devonian and the Jurassic intrusive events.

Jurassic Sheeted Dikes

The youngest suite of dikes consists of sheeted packets of hornblende microgabbro and fine grained basalt. This suite has yielded one $^{40}Ar/^{39}Ar$ isotopic age of 161 +/- 4 Ma, on a sample which had minor excess radiogenic Ar and a well-defined plateau over the last 70% of the ^{39}Ar released [E.T. Wallin, written comm., 10-22-92]. Small to large sheeted packets of dikes, which are of similar composition and which exhibit the same crosscutting relationships to older rocks, are abundant throughout the Trinity Complex [55]. In the absence of evidence to the contrary, it appears likely that many, if not all, of the sheeted dikes within the Trinity Complex are Jurassic.

Although included in the Trinity ophiolite by some authors [5, 6, 8, 44], the Devonian? dikes and lavas and the Jurassic sheeted dikes are not part of the Cambrian Trinity ophiolite or the lower Paleozoic Trinity Complex. Because they intrude both Trinity and Yreka terranes, crosscutting faults which juxtapose the two terranes, the Devonian? and Jurassic rocks are an overlap sequence which is part of the Trinity-Yreka composite terrane.

Jurassic and Cretaceous Plutons
Plutonic rocks of Jurassic and Cretaceous age (Table 1) intrude the Trinity Complex (Figs. 1, 2) and stitch together the Trinity-Yreka composite terrane, the central metamorphic terrane, and other terranes of the Klamath Mountains [36]. These plutons are intermediate in composition, ranging from granodiorite to quartz diorite and diorite to trondhjemite.

ORIGIN OF THE TRINITY COMPLEX

The Cambrian ophiolite assemblage is of oceanic origin. The harzburgite is typical of suboceanic mantle tectonites, metagabbro is typical of deformed oceanic gabbros, and plagiogranites meet the petrographic criteria for oceanic plagiogranites [9, 10]. For samples that I collected within the Cambrian? basaltic rocks, the Zr-Ti-Y values [50] plot within the overlap field of Pearce and Cann [61, 62] and thus are permissive of either island-arc or ocean-ridge origin. The extensive development of quartz-rich plagiogranite and the textural evidence for pre-Ordovician ductile deformation within the basalt-plagiogranite-gabbro sequence both suggest that the Trinity ophiolite is not a typical oceanic ridge complex, but may have formed in an uncommon tectonic setting such as a migrating triple junction or a "leaky" transform fault.

The Ordovician ophiolitic fragments are clearly of oceanic origin. The initial Sr^{87}/Sr^{86} ratios that Mattinson and Hopson [58] report for Ordovician gabbroic rocks are consistent with an origin at an oceanic ridge. Undeformed gabbros of Ordovician? (or Cambrian?) age may represent intrusion of gabbro into older deformed ocean crust during an abortive rifting episode. Quick [64] interpreted the Ordovician ultramafic rocks as a mantle diapir; such an origin is consistent with formation in a marginal basin beneath a backarc spreading center as suggested by Lindsley-Griffin [49]. Quick [64] hypothesized that the mantle diapir, because of its feldspar-rich composition, may have been located near a continental margin. He postulated a tectonic setting within either a volcanic arc or back-arc basin for the Trinity peridotite. Thus, the Ordovician peridotites, dunites, and minor gabbros of the Trinity Complex formed in an oceanic setting which included mantle diapirism with partial fusion and generation of basaltic magmas.

Map relationships demonstrate that the Cambrian plagiogranites and the Cambrian metagabbro are faulted against each other and against adjoining ultramafic rock units of both Ordovician and unknown ages along the China Mountain fault zone (Fig. 2). This broad, sinuous fault zone exhibits textural and structural evidence of a repeated ductile to brittle deformation, recrystallization, intrusion of mafic dikelets (< 1 cm width), and continued deformation [50, 51]. These features suggest a major fault in oceanic crust along which deformation was associated with minor magmatism. A likely tectonic setting for such a sequence of events would be along an oceanic fracture zone or transform fault. Alternatively, the China Mountain fault zone may represent a suture between two oceanic terranes, one Cambrian and the other Ordovician, that amalgamated in Late Ordovician to Silurian time.

The China Mountain fault zone is intruded by undeformed Silurian gabbro of the China Mountain pluton (Fig. 2). Thus, the Cambrian and Ordovician assemblages were juxtaposed before intrusion in the Late Silurian. The present attitude of this fault is nearly vertical, but its original orientation is uncertain. The association of pyroxene gabbro, hornblende pyroxene gabbro, and pyroxenite with only minor intermediate composition rocks suggests these Silurian magmas were produced in an oceanic setting. Jacobsen *et al.* [38] studied a post-deformation intrusive pyroxenite dike that is probably part of the Silurian suite. They concluded that the pyroxenite was not derived from partial melting of the Ordovician peridotites, but from a different source area in the mantle. Schwindinger and Anderson [66] concluded that the pegmatitic gabbros at Castle Lake crystallized at a relatively low pressure. They hypothesized that the gabbros may have formed "in a shallow, oceanic body of serpentinizing peridotite and were metamorphosed by a sodium-poor hydrothermal fluid...derived either from deserpentinization of peridotite or by natural boiling (distillation) of sea water at a pressure <800 bars" (p. 372).

Previously published hypotheses [6, 8, 43, 44] that advocated an island-arc origin for the Trinity ophiolite/Trinity Complex apparently were based on samples from the Silurian gabbros, Jurassic sheeted dikes, and Devonian? basalts rather than samples from the Cambrian ophiolite sequence. (However, the low-K tholeiites reported from Lovers Leap [8] may be Cambrian? basalts of the Trinity ophiolite.)

Features that are generally considered to be characteristic of magmatic-arc assemblages [3, 17, 20] include the following: overlain by thick sequence of volcanogenic rocks, with volcaniclastics much greater than lavas; volcanic rocks interbedded with graywackes, mudstones; lavas that develop from primitive to evolved: calc-alkalic basalt, andesite, dacite; LREE-enriched basalts. For ophiolites believed to have formed in arcs, such as the Eocene Acoje ophiolite, Luzon [21] or the Jurassic Smartville ophiolite [73] the following characteristics can be listed: overlain by thick volcaniclastics; 1-2 km of dikes, sills, pillow basalt; chemistry ranges from LKT to MORB; may have abundant plagiogranite and a thick cumulate gabbro sequence; basement of depleted harzburgite; much thicker crustal section than ocean ridges. The Trinity Complex does not exhibit any of the features characteristic of magmatic arcs, and although the Trinity ophiolite does appear to have abundant plagiogranite, it lacks the other characteristics of arc ophiolites.

CONCLUSIONS

The Trinity Complex formed in an oceanic setting over a time span of 165 Ma, from earliest Cambrian (570 Ma) to Late Silurian. Such a long-lived oceanic terrane is unlikely to represent a simple slice of oceanic crust and upper mantle. Instead, the polygenetic Trinity Complex probably occupied a series of different tectonic settings in succession.

A reasonable model is that the Cambrian Trinity ophiolite formed as a sequence of harzburgite, layered gabbro, plagiogranite, and basalt in a slowly rifting environment. The data could be interpreted as permissive of such tectonic settings as: a marginal basin undergoing backarc rifting, a rifting continent undergoing transition into an oceanic spreading center, localized rifting along a curvilinear transform fault, or extension at a triple junction located within oceanic crust. The Taitao Ridge would be a modern analog for a triple junction complex [45].

The Ordovician components of the Trinity Complex formed at an ocean ridge of unknown character. In the Silurian, the Cambrian Trinity ophiolite was dismembered, deformed, and juxtaposed against fragments of Ordovician ophiolitic rocks, possibly within an oceanic transform fault complex. The fault had become inactive by Late Silurian, when voluminous gabbroic magmatism occurred.

From Late Silurian to early Middle Devonian, the Trinity Complex probably formed the leading edge of a plate that was overriding the east-dipping subducting slab of the central metamorphic terrane. This produced the accretionary complex of the Yreka terrane. During the amalgamation of these terranes, isolated near-trench volcanic centers erupted minor pillow lavas over the composite Trinity-Yreka terrane.

This paleogeographic setting was not two-dimensional, and may have included a number of microplates and triple junctions. For example, both the Devonian volcanism and the cessation of Devonian subduction could have been caused by migration of a triple junction along the active subduction zone. A triple junction containing at least one divergent or convergent plate margin component would have provided a heat source for the near-trench extension and volcanism, and passage of the triple junction would have changed plate motions so that the subduction zone became inactive.

By Middle Devonian, the Trinity terrane had amalgamated with the Yreka and central metamorphic terranes to form an oceanic plateau. This oceanic plateau was only marginally affected by development of island arc volcanism and volcaniclastic sedimentation in the Redding terrane to the east, but it served as a high-standing nucleus against which younger terranes to the west were accreted in the Jurassic. Calc-alkaline stocks and dikes, including sheeted dike swarms, invaded the Trinity-Yreka composite terrane in the Jurassic and Cretaceous.

Acknowledgements

I thank J.R. Griffin and E.T. Wallin for their helpful reviews of an early version of the manuscript. Suggestions by Dr. A. Ishiwatari and an anonymous reviewer substantially improved the final manuscript. Thanks also are due to the many field geologists who have shared constructive criticism and their best outcrops with me, especially P.H. Masson, Joanne Danielson, A.W. Potter, J.R. Griffin, and E.T. Wallin. Field and laboratory work during 1987-1992 were supported by the University of Nebraska-Lincoln Research Council and the UNL Department of Geology. I am grateful to E.T. Wallin for permission to include his unpublished isotopic data on the Jurassic sheeted dikes.

REFERENCES

1. Anonymous. Ophiolites--Penrose Conference report, *Geotimes* **17:12**, 24-25 (1972).
2. R.L. Bates and J.A. Jackson (Eds). *Glossary of geology, third edition*. American Geological Institute, Alexandria, Virginia (1987).
3. S.H. Bloomer and J.W. Hawkins. Petrology and geochemistry of boninite series volcanic rocks from the Mariana trench, *Contr. Mineral. Petrol.* **97**, 361-377 (1983).
4. W.A. Brewer III. *The geology of a portion of the China Mountain quadrangle, California* [unpublished M.A. thesis]. University of California, Berkeley, California (1955).
5. M. Brouxel and H. Lapierre. Geochemical study of an early Paleozoic island-arc--back-arc basin system. Part 1: The Trinity ophiolite (northern California), *Geol. Soc. Amer. Bull.* **100**, 1111-1119 (1988).
6. M. Brouxel, H. Lapierre, A. Michard and F. Albarede. Geochemical study of an early Paleozoic island-arc--back-arc basin system. Part 2: Eastern Klamath, early to middle Paleozoic island-arc volcanic rocks (northern California), *Geol. Soc. Amer. Bull.* **100**, 1120-1130 (1988).
7. M. Brouxel, H. Lapierre and J. Zimmerman. Upper Jurassic mafic magmatic rocks of the eastern Klamath Mountains, northern California: Remnant of a volcanic arc built on young continental crust, *Geology* **17**, 273-276 (1989).
8. M. Brouxel, C. Lecuyer and H. Lapierre. Diversity of magma types in a lower Paleozoic island arc--marginal basin system (eastern Klamath Mountains, California, U.S.A.), *Chem. Geol.* **77**, 251-264 (1989).
9. R.G. Coleman. *Ophiolites: Ancient oceanic lithosphere?* Springer-Verlag, New York (1977).
10. R.G. Coleman and Z.E. Peterman. Oceanic plagiogranites. *Jour. Geophys. Res.* **80**, 1099-1108 (1975).
11. G.A. Davis. Metamorphic and granitic history of the Klamath Mountains, California, *Calif. Div. Mines Geol. Bull.* **190**, 39-50 (1966).
12. G.A. Davis. Westward thrust faulting in the south-central Klamath Mountains, California, *Geol. Soc. Amer. Bull.* **79**, 911-934 (1968).
13. G.A. Davis. Tectonic correlations, Klamath Mountains and western Sierra Nevada, California, *Geol. Soc. Amer. Bull.* **80**, 1095-1108 (1969).
14. G.A. Davis, M.J. Holdaway, P.W. Lipman and W.D. Romey. Structure, metamorphism, and plutonism in the south-central Klamath Mountains, California, *Geol. Soc. Amer. Bull.* **76**, 933-966 (1965).
15. H.J.B. Dick and J.M. Sinton. Compositional layering in alpine peridotites: Evidence for pressure solution creep in the mantle, *Jour. Geol.* **87**, 403-416 (1979).
16. G.S. Fuis and J.J. Zucca. A geologic cross section of northeastern California from seismic refraction results. In: *Geology of the Upper Cretaceous Hornbrook Formation, Oregon and California*. T.H. Nilsen (Ed.). pp. 203-209. SEPM Pacific Sect. Pub. **42**, Bakersfield (1984).
17. M.O. Garcia. Criteria for the identification of ancient volcanic arcs, *Ear. Sci. Rev.* **14**, 147-165 (1978).
18. L. Goullaud. Structure and petrology in the Trinity mafic-ultramafic complex, Klamath Mountains, northern California. In: *Geology of the Klamath Mountains, northern California*. N. Lindsley-Griffin and J.C. Kramer (Eds). pp. 112-133. Geol. Soc. Amer. Cordill. Sect. Guidebook (1977).
19. A. Griscom. Aeromagnetic and gravity interpretation of the Trinity ophiolite complex, northern California (abs.), *Geol. Soc. Amer. Abs. with Progs.* **9**, 426-427 (1977).
20. W.B. Hamilton. Plate tectonics and island arcs, *Geol. Soc. Amer. Bull.* **100**, 1503-1527 (1988).
21. J.W. Hawkins and C.A. Evans. Geology of the Zambales Range, Luzon, Philippine Islands--Ophiolite derived from an island arc-back arc basin pair, *Amer. Geophys. Un. Geophys. Mon.* **27**, 95-123 (1983).
22. W.B. Harland, R.L. Armstrong, A.V. Cox, L.E. Craig, A.G. Smith, and D.G. Smith. *A geologic time scale*. Cambridge University Press, Cambridge (1989).
23. O.H. Hershey. Metamorphic formations of northwestern California, *The Amer. Geol.* **27**, 225-245 (1901).
24. N.E.A. Hinds. Paleozoic eruptive rocks of southern Klamath Mountains, *Univ. Calif. Publ. Geol. Sci.* **23**, 375-410 (1932).
25. C.A. Hopson and J.M. Mattinson. Ordovician and Late Jurassic ophiolitic assemblages in the Pacific northwest (abs.), *Geol. Soc. Amer. Abs. with Progs.* **5**, 57 (1973).
26. P.E. Hotz. Geology of lode gold districts in the Klamath Mountains, California and Oregon, *U.S.G.S. Bull.* **1290**, 1-91 (1971).
27. P.E. Hotz. Geology of the Yreka Quadrangle, Siskiyou County, California, *U.S.G.S. Bull.* **1436**, 1-72 (1977).
28. P.E. Hotz. Geologic map of the Yreka Quadrangle and parts of the Fort Jones, Etna, and China Mountain Quadrangles, California, *U.S.G.S. Open File Rept.* **78-12**, 1:62,500 (1978).
29. W.P. Irwin. Geologic reconnaissance of the northern Coast Ranges and Klamath Mountains, California, with a summary of the mineral resources, *Calif. Div. Mines Bull.* **179**, 1-80 (1960).

30. W.P. Irwin. Late Mesozoic orogenies in the ultramafic belts of northwestern California and southwestern Oregon, *U.S.G.S. Prof. Pap.* **501-C**, 1-9 (1964).
31. W.P. Irwin. Geology of the Klamath Mountains province, *Calif. Div. Mines Geol. Bull.* **190**, 19-38 (1966).
32. W.P. Irwin. Terranes of the western Paleozoic and Triassic belt in the southern Klamath Mountains, California, *U.S.G.S. Prof. Pap.* **800-C**, 103-111 (1972).
33. W.P. Irwin. Sequential minimum ages of oceanic crust in accreted tectonic plates of northern California and southern Oregon (abs.), *Geol. Soc. Amer. Abs. with Progs.* **5**, 62-63 (1973).
34. W.P. Irwin. Review of Paleozoic rocks of the Klamath Mountains. In: *Paleozoic paleogeography of the western United States. Pacific Coast Paleogeography Symposium 1.* J.H. Stewart, C.H. Stevens and A.E. Fritsche (Eds). vol.1, pp. 441-454., SEPM Pacific Sect. Pub. 7, Bakersfield (1977).
35. W.P. Irwin. Tectonic accretion of the Klamath Mountains. In: *The geotectonic development of California, Rubey Volume 1.* W.G. Ernst (Ed.). pp. 29-49. Prentice-Hall, Englewood Cliffs (1981).
36. W.P. Irwin. Age and tectonics of plutonic belts in accreted terranes of the Klamath Mountains, California and Oregon. In: *Tectonostratigraphic terranes of the circum-Pacific region.* D.G. Howell (Ed.). no.1, pp. 187-199. Circum-Pacific Council for Energy and Mineral Resources, Earth Science Series (1985).
37. W.P. Irwin and G.D. Bath. Magnetic anomalies and ultramafic rock in northern California, *U.S.G.S. Prof. Pap.* **450-B**, 65-67 (1962).
38. S.B. Jacobsen, J.E. Quick and G.J. Wasserburg. A Nd and Sr isotopic study of the Trinity peridotite: implications for mantle evolution, *Ear. Planet. Sci. Lett.* **68**, 361-378 (1984).
39. D.L. Jones, D.G. Howell, P.J. Coney and J.W.H. Monger. Recognition, character, and analysis of tectonostratigraphic terranes in western North America. In: *Accretion tectonics in the Circum-Pacific regions.* M. Hashimoto and S. Uyeda (Eds). pp. 21-35. Terra Scientific Publishing Co., Tokyo (1983).
40. R.W. Kistler, E.H. Mckee, K. Futa, Z.E. Peterman and R.E. Zartman. A reconnaissance Rb-Sr, Sm-Nd, U-Pb, and K-Ar study of some host rocks and ore minerals in the West Shasta Cu-Zn district, California, *Econ. Geol.* **80**, 2128-2135 (1985).
41. T.R. LaFehr. Gravity in the eastern Klamath Mountains, California, *Geol. Soc. Amer. Bull.* **77**, 1177-1190 (1966).
42. M.A. Lanphere, W.P. Irwin and P.E. Hotz. Isotopic age of the Nevadan orogeny and older plutonic and metamorphic events in the Klamath Mountains, California, *Geol. Soc. Amer. Bull.* **79**, 1027-1052 (1968).
43. H. Lapierre, F. Albarede, J. Albers, B. Cabanis and C. Coulon. Early Devonian volcanism in the eastern Klamath Mountains, California: evidence for an immature island arc, *Can. Jour. Ear. Sci.* **22**, 214-226 (1985).
44. H. Lapierre, M. Brouxel, F. Albarede, C. Coulon, C. Lecuyer, P. Martin, G. Mascle and O. Rouer. Paleozoic and lower Mesozoic magmas from the eastern Klamath Mountains (North California) and the geodynamic evolution of northwestern America, *Tectonophys.* **140**, 155-177 (1987).
45. S.D. Lewis, J.H. Behrmann, R. Musgrave and ODP Leg 141 Scientific Staff. Geology and tectonics of the Chile Triple Junction, *EOS, Amer. Geophys. Un.* **73**, n. 38, 404-405, 410 (1992).
46. N. Lindsley-Griffin. Lower Paleozoic ophiolite of the Scott Mountains, eastern Klamath Mountains, California (abs.), *Geol. Soc. Amer. Abs. with Progs.* **5**, 71-72 (1973).
47. N. Lindsley-Griffin. Geology of the northwestern edge of the Trinity ophiolite, eastern Klamath Mountains, California (abs.), *EOS, Amer. Geophys. Un.* **56**, 1079 (1975).
48. N. Lindsley-Griffin. Feldspathic lherzolites of the Trinity ophiolite complex, eastern Klamath Mountains, California (abs.), *EOS, Amer. Geophys. Un.* **57**, 1025 (1976).
49. N. Lindsley-Griffin. The Trinity ophiolite, Klamath Mountains, California. In: *North American ophiolites.* R.G. Coleman and W.P. Irwin (Eds). pp. 107-120. Oregon Dept. Geol. Min. Ind. Bull. **95** (1977).
50. N. Lindsley-Griffin. *Structure, stratigraphy, petrology and regional relationships of the Trinity ophiolite, eastern Klamath Mountains, California* [unpublished Ph.D. dissertation]. University of California, Davis (1982).
51. N. Lindsley-Griffin. The Trinity complex, A polygenetic ophiolitic assemblage. In: *Paleozoic paleogeography of the western U.S..* J.D. Cooper and C.H. Stevens (Eds). vol.2, pp. 589-607. SEPM Pacific Sect. Pub. 67, Bakersfield (1991).
52. N. Lindsley-Griffin and J.R. Griffin. The Trinity terrane: An early Paleozoic microplate assemblage. In: *Pre-Jurassic rocks in western North American suspect terranes.* C.H. Stevens (Ed.). pp. 63-75. SEPM Pacific Sect. Pub. 32, Bakersfield (1983).
53. N. Lindsley-Griffin, J.R. Griffin and E.T. Wallin. Redefinition of the Gazelle Formation of the Yreka terrane, Klamath Mountains, California: Paleogeographic implications. In: *Paleozoic paleogeography of the western U.S.* J.D. Cooper and C.H. Stevens (Eds). vol.2, pp. 609-624. SEPM Pacific Sect. Pub. **67**, Bakersfield (1991).
54. P.W. Lipman. Structure and origin of an ultramafic pluton in the Klamath Mountains, California, *Amer. Jour. Sci.* **262**, 199-222 (1964).
55. P.H. Masson. Sheeted dikes in the Trinity ophiolite, northern California (abs.), *Geol. Soc. Amer. Abs. with Progs.* **22:3**, 64 (1990).

56. P.H. Masson. Grey Rocks, A remnant of a submarine volcano in the Trinity Complex, northern California (abs.), *Geol. Soc. Amer. Abs. with Progs.* **23:2**, 76 (1991).
57. J.M. Mattinson and C.A. Hopson. Paleozoic ages of rocks from ophiolitic complexes in Washington and northern California (abs.), *EOS, Amer. Geophys. Un.* **53**, 543 (1972a).
58. J.M. Mattinson and C.A. Hopson. Paleozoic ophiolitic complexes in Washington and northern California, *Carnegie Inst. Yearbook* **71**, 578-583 (1972b).
59. M. Menzies and C. Allen. Plagioclase lherzolite--residual mantle relationships within two eastern Mediterranean ophiolites, *Contr. Mineral. Petrol.* **45**, 197-213 (1974).
60. G.S. Odin and C. Odin. Echelle numerique des temps geologiques, *Geochron.* **35**, 12-21 (1990).
61. J.A. Pearce and J.R. Cann. Ophiolite origin investigated by discriminant analysis using Ti, Zr, and Y, *Ear. Planet. Sci. Lett.* **19**, 290-300 (1971).
62. J.A. Pearce and J.R. Cann. Tectonic setting of basic volcanic rocks determined using trace element analyses, *Ear. Planet. Sci. Lett.* **24**, 419-426 (1973).
63. J.E. Quick. The origin and significance of large, tabular dunite bodies in the Trinity peridotite, northern California, *Contr. Mineral. Petrol.* **78**, 413-422 (1981a).
64. J.E. Quick. Petrology and petrogenesis of the Trinity peridotite, an upper mantle diapir in the eastern Klamath Mountains, northern California, *Jour. Geophys. Res.* **86**, 11837-11863 (1981b).
65. J.E. Quick. *Petrology and petrogenesis of the Trinity peridotite, northern California, part I* [unpublished Ph.D. dissertation]. Calif. Inst. of Tech., Pasadena (1981c).
66. K.R. Schwindinger and A.T. Anderson Jr. Probable low-pressure intrusion of gabbro into serpentinized peridotite, northern California, *Geol. Soc. Amer. Bull.* **98**, 364-372 (1987).
67. R.G. Strand. Geologic atlas of California, Redding sheet, *Calif. Div. Mines Geol.* 1:250,000 (1962).
68. M.L. Throckmorton. *Petrology of the Castle Lake peridotite-gabbro mass, eastern Klamath Mountains, California* [unpublished M.S. thesis]. Univ. Calif., Santa Barbara (1978).
69. D.L. Wagner and G.J. Saucedo (compilers). Geologic map of the Weed quadrangle, California, *Calif. Div. Mines Geol.* Regional Geol. Map Series, Map 4A, 1:250,000 (1987).
70. E.T. Wallin, N. Lindsley-Griffin and J.R. Griffin. Overview of early Paleozoic magmatism in the eastern Klamath Mountains, California: An isotopic perspective. In: *Paleozoic paleogeography of the western U.S..* J.D. Cooper and C.H. Stevens (Eds). vol.2, pp. 581-588. SEPM Pacific Sect. Pub. **67**, (1991).
71. E.T. Wallin, N. Lindsley-Griffin and A.W. Potter. Trinity ultramafic complex, Klamath Mountains, California: Cambro-Ordovician ophiolite or composite terrane? (abs.), *Geol. Soc. Amer. Abs. with Progs.* **22**, 91-92 (1990).
72. E.T. Wallin, J.M. Mattinson and A.W. Potter. Early Paleozoic magmatic events in the eastern Klamath Mountains, northern California, *Geology* **16**, 144-148 (1988).
73. C. Xenophontos and G.C. Bond. Petrology, sedimentation and paleogeography of the Smartville terrane (Jurassic)--Bearing on the genesis of the Smartville ophiolite. In: *Mesozoic Paleogeography of the Western United States*, D.G. Howell and K.A. McDougal (Eds). vol.1, pp. 291-302. SEPM Pacific Sect. Pub. **8** (1978).

The Shulaps Ophiolite Complex of British Columbia, Canada: a Palaeozoic/Mesozoic arc-related microterrane.

J. MALPAS[1], T.J. CALON[1] and R.W.J. MACDONALD[2]

[1]Department of Earth Sciences, Memorial University, St. John's, NF, Canada, A1B 3X5
[2]Mineral Development Research Unit, University of British Columbia, Vancouver, BC, Canada, V6T 1Z4

Abstract: The Shulaps Ophiolite Complex is situated at the western boundary of the Intermontane superterrane along the Yalakom and Marshall Creek fault systems and is part of a collage of late Palaeozoic to Mesozoic suspect mini-terranes. It consists of a highly dismembered ophiolite succession divided into three lithotectonic units which define a southwesterly verging linked thrust system. The Shulaps peridotite suite constitutes the upper part of the thrust system and comprises layered harzburgite tectonites with abundant intrusive dunite, interpreted as a section of residual oceanic upper mantle. This unit overlies a large duplex of ophiolitic melange. The serpentinite matrix of this melange is subdivided into an upper unit derived from harzburgite protoliths of the overlying upper mantle sheet and a lower unit derived from ultramafic cumulates representing the basal part of layer 3 of the Shulaps oceanic crust. The melange contains blocks of variable size including up to 1 km-thick ultramafic to gabbroic plutonic complexes interpreted as coherent sections of oceanic layer 3. It also contains exotic blocks of sedimentary and volcanic rocks. A large block along the structural base of the melange comprises a complete section from high level gabbro to pillowed basaltic lavas cut by abundant mafic dikes. The transition between gabbros and volcanic rocks in this block occurs along a zone of dikes and screens and represents the transition between oceanic layers 2 and 3 without the development of a well-formed sheeted dike complex.

The ophiolitic units show a tectonic stacking order in which the original ophiolite stratigraphy is reversed. They were emplaced over a footwall of arc-related sediments of the Upper Triassic Cadwallader Group and oceanic sedimentary and volcanic rocks of the Mississippian to Middle Jurassic Bridge River Group. The exotic blocks in the melange represent a structural sampling of this footwall, introduced into the serpentinite melange by subcretion.

Palinspastic reconstruction taken together with geochemical data indicate that the Shulaps Complex and associated rocks were formed in a back arc basin environment during the Upper Palaeozoic to Lower Mesozoic and subsequently accreted onto the North American continental margin during the ensuing Cordilleran orogeny in late Mesozoic time.

INTRODUCTION

The Shulaps Complex occupies an area of approximately 180 square kilometres and forms one of the largest ultramafic/mafic bodies in the Cordillera of British Columbia. It lies approximately 200 km north of Vancouver and is one of a number of discrete, fault-bounded microterranes of oceanic and volcanic arc affinity that form part of the Coastal Plutonic Belt, a highly tectonized zone separating the Intermontane superterrane to its east and the Insular superterrane to its west (Monger et al., 1990; Monger, 1986; Journeay, 1990; Schiarriza et al., in press). The eastern boundary of the portion of the Coast Belt dealt with here is the Yalakom Fault and the western margin, the Upper Cretaceous Dickson-McClure batholith.

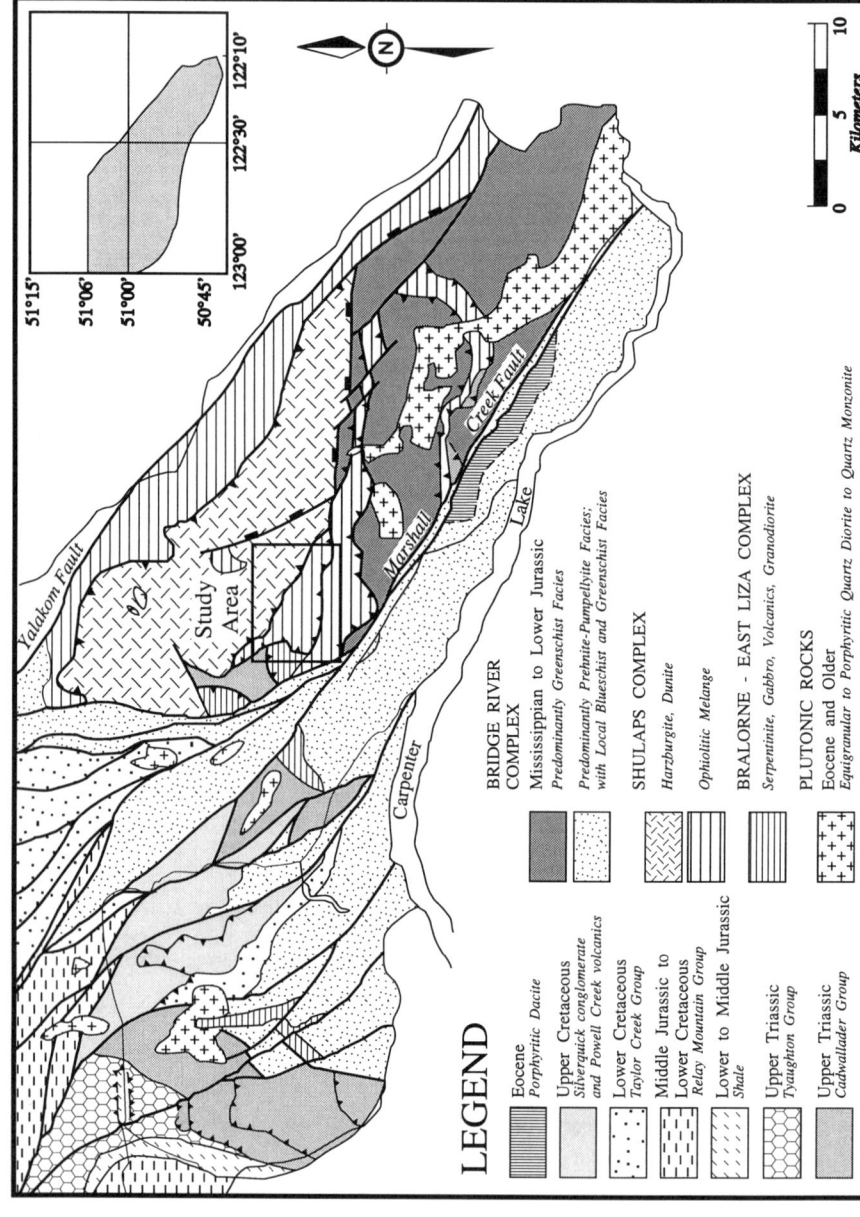

Figure 1. Regional geological setting of the Shulaps Complex in South - Central British Columbia (after Schiarizza et al., 1990)

The Shulaps Complex is an example of those Cordilleran ultramafic/mafic bodies identified as dismembered oceanic lithosphere or ophiolite (Wright et al., 1982; Calon et al., 1990). It occurs as a series of tectonic slivers interleaved with strongly deformed crustal and supracrustal rocks of the Bridge River Complex, the Cadwallader Group, and the Tyaughton Group (Figure 1). In these structural sequences, rocks are juxtaposed along complex systems of Cretaceous to Tertiary faults that reflect a protracted history of contractional, strike-slip and extensional deformation (Schiarriza et al., 1989, 1990).

Wright et al., (1982) interpreted the ultramafic rocks of the Shulaps Complex as peridotite tectonites representing depleted upper mantle, and recognized their ophiolitic origin. They established the western (basal) portion of the complex as a serpentinite melange containing exotic blocks of sedimentary and volcanic rocks. These were tentatively correlated with the supracrustal rocks of the oceanic Bridge River Complex, which structurally underlie the southern portion of the melange (e.g. Potter, 1986; Figure 1). The melange also contains blocks of ultramafic and mafic plutonic rocks, which thus may represent fragments of layer 3 of the oceanic crust (Calon et al., 1990). The Bridge River Complex itself contains cherts that range from at least Mississippian to late Middle Jurassic in age (Cordey and Schiarizza, 1993), and blueschists produced by high pressure, presumably subduction-related metamorphism in the Middle to Upper Triassic (Archibald et al., 1991). The Cadwallader Group includes volcanic rocks (Pioneer Formation) and overlying clastic sedimentary rocks (Hurley Formation) of Upper Triassic age. These units are everywhere imbricated with units of the Bralorne Complex comprising ultramafic and mafic plutonic rocks, volcanic rocks, and granodioritic intrusions of Permian age (Leitch et al., 1991).

LITHOLOGIES OF THE SHULAPS COMPLEX AND RELATED ROCKS

In order to provide a modern account of the structural and petrogenetic evolution of the Shulaps Complex and its Mesozoic accretionary history, geological mapping and geochemical analyses were carried out along the southwestern edge of the complex, covering an area of approximately 20 square kilometres centred around the upper courses of Jim Creek and East-Liza Creek (Figures 2 and 3). This area comprises the critical transition from coherent thrust sheets of residual mantle peridotite of the Shulaps Complex to the underlying ophiolitic melange. It displays the relationships between three major ophiolitic units which are, from structural top to bottom:
i) The Shulaps peridotite suite, exposed to the north (subunit 1)
ii) The Shulaps ophiolitic melange, along the southwestern margin of the peridotite in the south and central portions of the map area (subunits 2-8)
iii) The East Liza plutonic and volcanic rocks found in the southwestern portion of the study area (subunits 9-10).

In addition, Hurley Formation sediments of the Cadwallader Group are exposed on the west flank of the Shulaps Complex and East Liza suite in the map area (subunit 11). Together, these four units comprise a complicated, southwesterly verging linked thrust system.

The Shulaps peridotite suite

The peridotites form part of a coherent basal thrust sheet of mantle rocks of approximately 300 metres thickness. The sheet is bounded at its top and bottom by shear zones up to 400 metres thick (Figures 2 and 3). These boundaries are sharp structural contacts which are parallel to the schistosity in the serpentinite matrix of the shear zones. Lithologies are

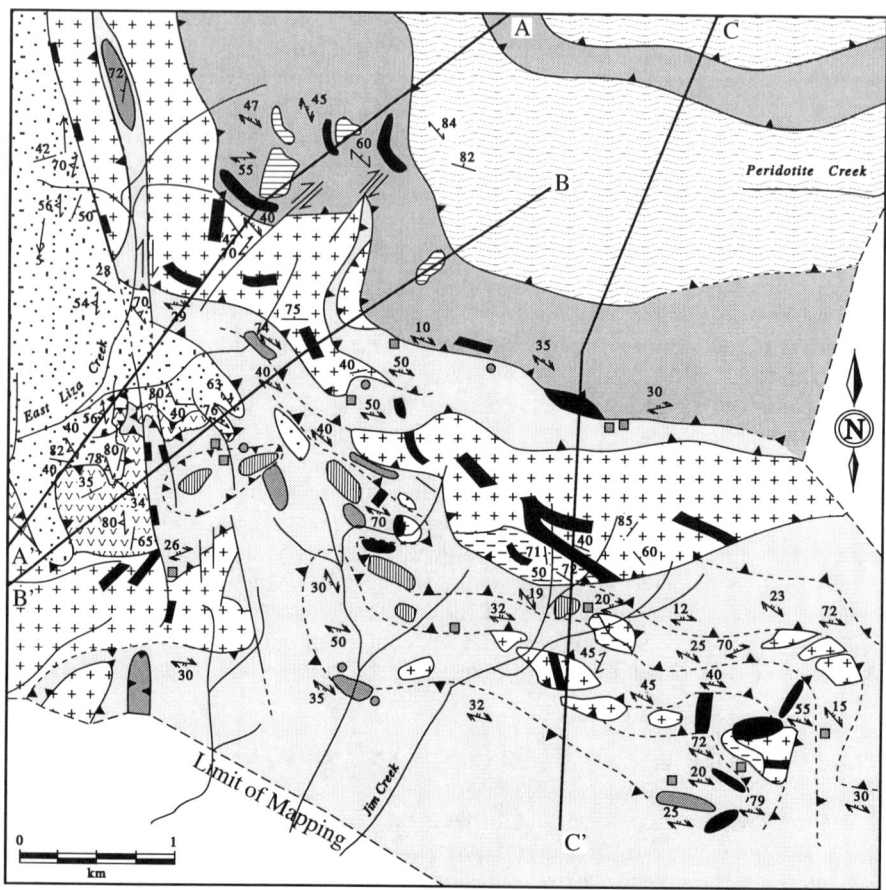

Figure 2a. Geological Map of the Western part of the Shulaps Complex in the area of East Liza Creek and Jim Creek. See Figure 1 for location of study area.

dominated by layered and massive harzburgite, with subordinate dunite and orthopyroxenite, generally as pods within the harzburgite. Pods are variable in size, from less than a metre to tens of metres in diameter and cut the peridotite tectonite fabric in an irregular manner. Most dunitic bodies contain abundant disseminated chromite and thin chromite stringers with variable orientations. Chromite grains range up to 1 centimetre in size and are generally euhedral.

The harzburgite shows a penetrative mineral foliation and lineation, generally parallel to the compositional layering which is defined on a centimetre to metre scale by modal variations in orthopyroxene and olivine. The linear aspect of the structural fabric is outlined by chromite pull-apart textures and is subvertical in the foliation plane which dips steeply to the north-northeast or south-southwest. This orientation reflects the regional fabric attitude observed throughout the Shulaps Complex (Wright et al., 1982).

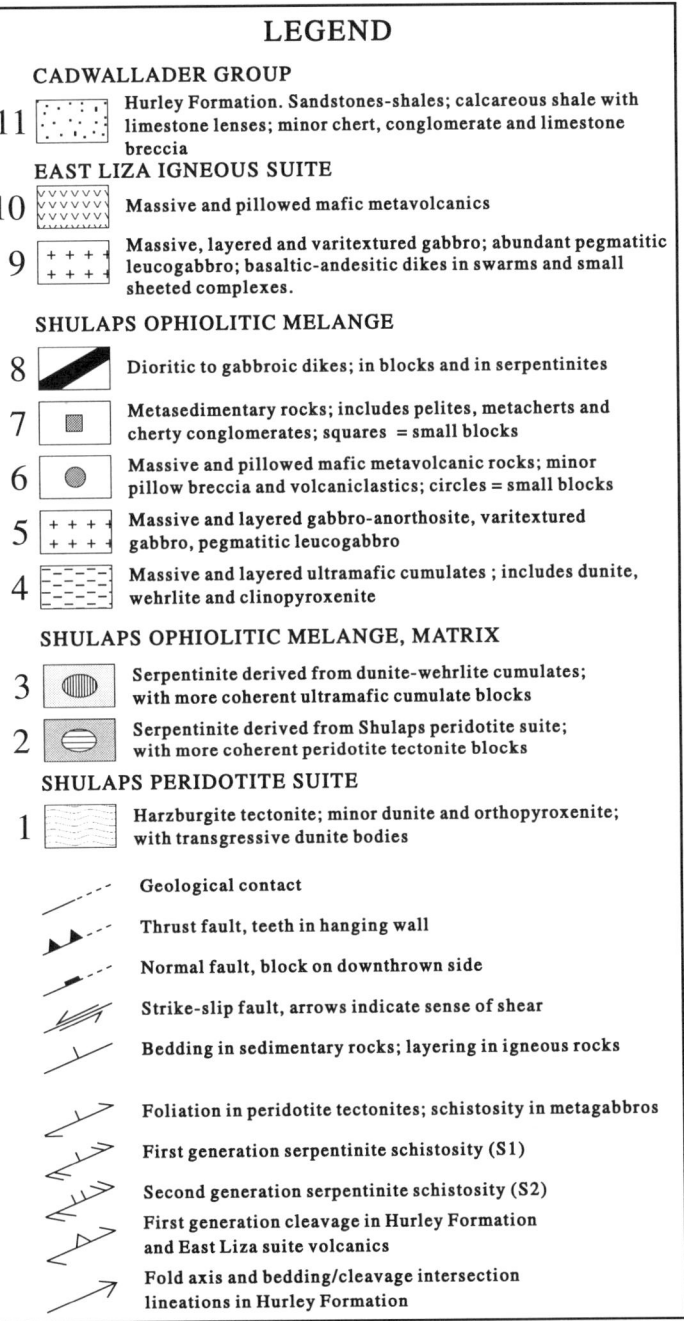

Figure 2b. Legend to detailed geological map of Shulaps Complex in East Liza Creek - Jim Creek area.

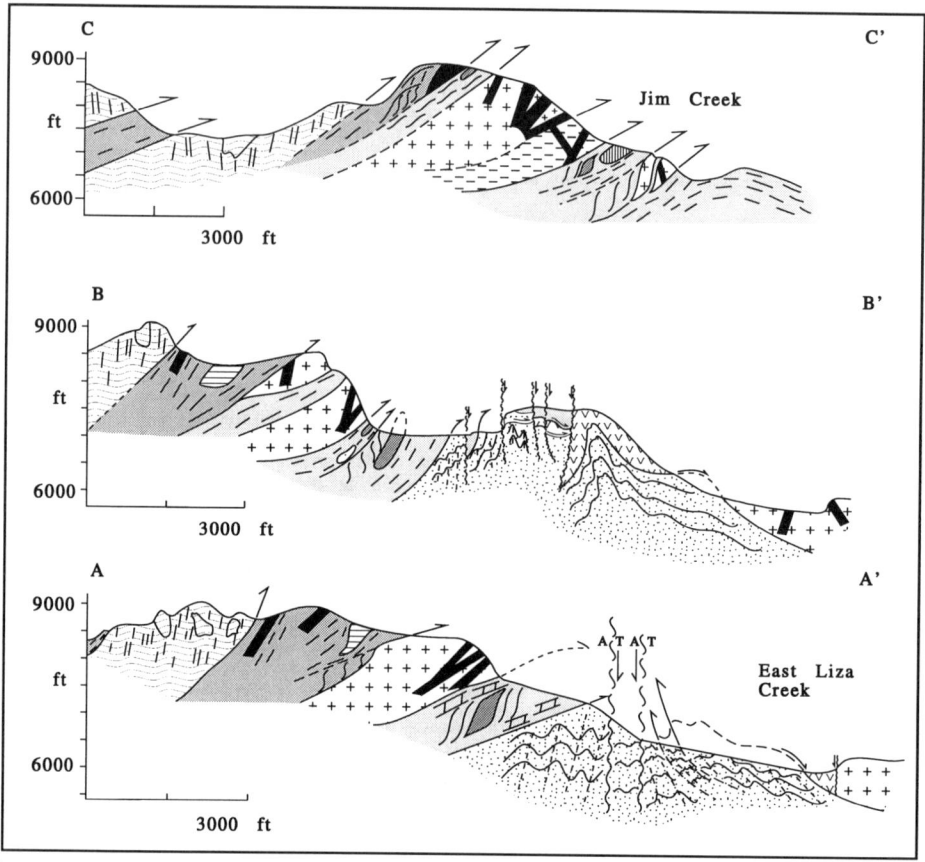

Figure 3. Vertical sections of the Western thrust front of the Shulaps Complex. See figure 2 for legend and location of the section lines.

The Shulaps ophiolitic melange

This serpentinite melange is a complex of lithologies including: mantle peridotite-derived melange at the base of the Shulaps peridotite suite; ultramafic cumulate-derived melange beneath the peridotite-derived melange and occupying most of the eastern and southern parts of the map area; coherent blocks of the ultramafic-mafic plutonic complex; blocks of sedimentary and volcanic rocks; small blocks of amphibolite; dikes intruding the melange matrix as well as the blocks (Figures 2 and 3).

The mantle peridotite-derived melange forms a zone some 400 metres thick which dips to the north and northeast beneath the basal thrust of the Shulaps peridotite. Previous workers (e.g. Nagel, 1979) have assumed that the complete ophiolitic melange was derived from the overriding sheet, but this is true only for its upper part. The matrix of the lower part of the melange consists entirely of serpentinite derived from protoliths such as wehrlite, dunite and lesser clinopyroxenite, which are identical in all aspects to ultramafic cumulates found as coherent sequences in two of the larger blocks within the melange. Notably, the mantle peridotite derived serpentinite melange is devoid of blocks of exotic material (see below).

The main group of blocks within the lower serpentinite melange comprises a variety of intrusive rocks ranging from olivine-rich ultramafic cumulates to varitextured plagioclase-rich gabbros. They display complex multiple intrusive relationships and local evidence for heterogeneous high-temperature plastic deformation synchronous with intrusion. The ultramafic cumulates are mainly found in two large blocks near the present structural base of the melange and comprise dominantly wehrlite and clinopyroxenite with subordinate chromitiferous dunite. The rocks are generally massive, poorly layered and have isotropic texture with anhedral olivines, randomly oriented stubby diopsides and altered plagioclase. Where phase layering is observed, it is usually non-planar and discontinuous even over short distances. Gabbroic lithologies constitute the most voluminous component of plutonic rocks as melange blocks and exhibit intrusive relationships between various phases, e.g. cross-cutting phase domains, xenolith-charged margins of late intrusive stocks, structurally controlled phase boundaries and late intrusive dike swarms. All of these features suggest that the gabbros did not evolve by crystallisation in a large, simple magma chamber, but rather through spatially and temporally complex intrusive processes in a number of small chambers. Compositionally, the gabbroic suite ranges from clinopyroxenite and rare websterite to two-pyroxene gabbro and anorthosite. Varitextured gabbros are the dominant component in most of the blocks but massive, layered and, in some places, foliated gabbros also occur. Fine-scale compositional layering with grain-size graded or phase graded aspect is common in a number of localities but layering is bounded by crosscutting varitextured gabbros or by shear zones and faults. The deformed gabbros are generally affected by metamorphism to amphibolite and greenschist facies assemblages.

The lower serpentinite melange also contains a number of blocks of sedimentary, volcanic and volcaniclastic rocks. Whereas the plutonic rock blocks can reasonably be considered as indigenous to the belt, these blocks of supracrustal rocks represent a truly exotic element, justifying use of the term "melange". The sedimentary blocks comprise mainly bedded and massive chert, and thin- to medium-bedded turbiditic siltstone and sandstone. In one block, bedded chert is interlayered with a unit of strongly silicified and mineralized volcanic rocks of 10 metres thickness. Another small block of coarse pyroclastic rock was found near a block containing an upward-facing 20-metre sequence of pyritiferous laminated shale-siltstone, white bedded chert with shale partings, and massive greywacke with siltstone rip-up clasts. This sequence appears correlative with a more extensive unit of siliciclastic rocks with interbedded chert and rare volcaniclastic rocks which is exposed on the lower slopes west of Jim Creek (Figure 2). Deformation in the form of mesoscopic folding and cleavage development is prominent in the larger sedimentary blocks.

Volcanic blocks are less abundant than sedimentary blocks in the melange. They comprise massive and pillowed lava and pillow breccia. In some localities, the lavas show variolitic and/or vesicular texture; some contain feldspar phenocrysts and chlorite pseudomorphs presumably after primary pyroxene or amphibole. Pillow breccias locally contain lenses of chert and limestone up to several metres thick. The volcanic rocks range from basaltic to dacitic in composition, although they are generally strongly altered due to silicification and low greenschist facies metamophism, and show heterogeneous deformation in the form of flattening of pillows and cleavage in the matrix of pillow breccias.

Besides the exotic blocks of supracrustal rocks, the lower serpentinite melange also contains rare small blocks of foliated and lineated amphibolite and amphibolite breccia. These rocks are correlated with the deformed gabbros of the plutnoic blocks in the melange. Preliminary dating has yielded Permian ages for the metamorphic amphibole (Archibald et al., 1991).

Numerous disrupted fragments of dikes occur within both parts of the ophiolitic melange. These dikes range in composition from gabbroic to dioritic; hornblende-porphyritic quartz diorite is a partcularly abundant component of the dike suite. Some large dikes in the eastern part of the area are multiple intrusions, ranging in composition from pyroxenite to gabbro and flow-banded feldsparphyric diorite. Some gabbroic dikes are strongly altered to either rodingite, greenschist or talc schist. On the other hand, many dioritic dikes are remarkably fresh and preserve well developed chilled margins.

Straight dike fragments of the dioritic suite generally preserve chilled margins against the serpentinite matrix. Their contacts are, however, invariably sheared, and the fine-grained chill zones often show a foliation related to post-intrusion deformation of the matrix. Such field relationships suggest that both pre-deformation and post-deformation dikes are present in the melange. Gabbroic dikes are both early and late, whereas dioritic dikes are almost all post-deformation of the melange matrix. Dikes in the mantle peridotite-derived serpentinite melange are relatively rare compared to the abundant occurrences in the ultramafic cumulate-derived melange and appear to be mostly of the late dioritic suite.

The East Liza plutonic and volcanic rocks

The ultramafic cumulate-derived serpentinite melange overlies the East Liza suite, a series of gabbroic and volcanic rocks, with clear thrust contact along its western margin (Figures 2 and 3). These rocks, in turn, structurally overlie sedimentary rocks of the Cadwallader Group.

The series comprises mafic to intermediate intrusive and extrusive rocks displaying complicated igneous relationships. Leech (1953) and Nagel (1979) have noted that a transitional contact exists between the gabbros and volcanic rocks and although the contact zone is poorly exposed, local field relationships suggest that it dips gently north-northeast. The intrusive sequence consists mainly of finely layered two-pyroxene gabbros with minor interlayered websterite, clinopyroxenite and anorthosite. These rocks show a well-developed tectonic foliation sub-parallel to layering, as well as discrete plastic shear zones overprinting the foliation. The deformed gabbros are cut by small, irregularly shaped stocks of isotropic, fine-grained to pegmatitic gabbros, by varitextured gabbroic veins, and by abundant fine-grained gabbroic to dioritic dikes which have highly variable orientation. This intrusive sequence resembles the varitextured gabbros in the plutonic blocks of the melange in many respects.

The contact zone with the volcanic rocks is characterized by an increase in the occurrence of dikes, by frequent microgabbroic stocks, and narrow screens of intensely sheared pillowed and massive lavas between intrusive phases. Locally, dike swarms appear to have coalesced into small sheeted dike sections, but a sheeted dike complex is certainly not well developed along the contact zone. The volcanic rocks comprise mainly pillow lava with subordinate massive flows and pillow breccia. The pillows are fine-grained to aphanitic and locally vesicular and porphyritic. The rocks are strongly altered to low greenschist facies assemblages with abundant quartz, epidote and chlorite. Compositionally the lavas appear to range from basalt to dacite.

The thrust contact between the volcanic rocks and underlying sedimentary rocks of the Hurley Formation is well exposed along the upper eastern slopes of East Liza Creek (Figure 2 and section B-B' in Figure 3). In the lavas it is a zone of silicic, banded mylonite and phyllonite with well-developed C-S fabrics up to 1 metre thick. At one locality, the thrust is clearly cut by a diorite dike that can be traced over some distance into the underlying sedimentary rocks. The presumed thrust contact between gabbros and Hurley Formation is nowhere exposed in the study area.

Cadwallader Group, Hurley Formation

This unit comprises a variety of siliciclastic and calcareous sedimentary rocks which, on the basis of lithological correlation, are assigned to the Upper Triassic Hurley Formation (Russmore, 1987). The most prominent sequence in the unit consists of thin- to medium-bedded grey sandstones and laminated grey to black siltstones, which are turbiditic in nature, displaying grading as well as convolute bedding and cross-lamination. The unit contains interbedded limestone, chert and pebble conglomerates, which are more abundant towards its stratigraphic base. The turbidite sequence becomes more calcareous towards its stratigraphic top, there consisting of medium to thick-bedded, graded calcarenites, calcareous shales and rare, thin, discontinuous limestone beds. The unit is cut by a few dikes ranging in composition from basalt to quartz diorite.

STRUCTURE OF THE SHULAPS COMPLEX

Three broad categories of deformational events, which represent a relative time sequence, are recognized in the study area: deformation related to the construction of the ophiolite complex; deformation related to the accretion of the Shulaps Complex, the East Liza Suite, and the Cadwallader Group; regional, post-accretion deformation in the ensuing orogenic belt. No individual unit records all events, and the younger events related to accretion and post-accretion deformation are variably represented in the different units in terms of style, orientation patterns and distribution of superposed structural elements.

Deformation that can unambiguously be related to the construction of the Shulaps ophiolite complex is recorded in a variety of ways in the harzburgites of the Shulaps peridotite suite (Wright et al., 1982), the ultramafic to mafic plutonic blocks in the Shulaps melange and the gabbro sequence of the East Liza igneous suite (Calon et al., 1990). The development of penetrative high-temperature foliations and lineations of orthopyroxene and spinel in the coherent sheets of peridotite, the coarse mildly porphyroclastic textures associated with these fabrics, and the truncation of the fabrics by intrusive or metasomatic dunite bodies, are all typical features attributed to hypersolidus flow in the residual upper mantle section of an ophiolite succession (Nicolas and Prinzhofer, 1983). Construction-related deformation features observed in the blocks of ultramafic-mafic plutonic rocks in the melange and the gabbroic section of the East Liza igneous suite are identical in style and in the temporal and spatial relationships that they show with respect to magmatism. They include: synmagmatic, high-temperature, plano-linear tectonite fabrics; discrete plastic and brittle-plastic shear zones; and block-faulted domains between dikes cutting the plutonic complexes.

Nicolas and Violette (1982) have recognized two distinct mantle flow patterns at oceanic spreading centres on the basis of geometric relationships observed in ophiolite complexes; these are the diapiric pattern, where mantle foliations and lineations are at a high angle to paleohorizontal, and the spreading limb pattern, where mantle fabrics are aligned parallel to the attitude of the petrological Moho. In the case of the Shulaps ophiolite, the type of mantle flow pattern according to this classification cannot be directly established; however, the high-angle truncation of the mantle fabrics by gently dipping, obduction-related thrust zones suggests that the complex may preserve a diapiric mantle flow pattern, provided that no large rigid body rotations occurred during the emplacement of the mantle section along shallow dipping thrusts.

Calon et al. (1990) have provided the first detailed descriptions of the accretion-related macroscopic structure of the Shulaps Complex (Figure 3). The belt as a whole is interpreted as a southwest verging thrust stack, probably more than three kilometres thick. It comprises a telescoped section of a dismembered ophiolite suite displaying a structurally inverted disposition of its original lithological sequence. The upper plate of the stack consists of a shingled array of thrust sheets of coherent, upper mantle peridotite, separated from one another by up to 400 metre-thick shear zones consisting of serpentinite derived from the mantle peridotites.

This unit overlies the serpentinite melange, up to 2 kilometres thick, which is interpreted as a huge, composite hinterland-dipping duplex structure that is deeply exhumed along its southern side. Within the duplex, the serpentinite schistosity generally dips steeply to the north and northeast and curves into the duplex boundaries. The schistosity wraps around lozenge-shaped blocks of the plutonic rock and creates the appearance that originally much larger coherent sections of the plutonic complex were telescoped along shear zones injected by the serpentinite. The overall attitude of the duplex structures is flatlying in the southern part of the belt but steepens to northerly ~50° dips at the contact with the overlying mantle peridotite section. Most of this change in attitude is focussed on a belt of large plutonic rock blocks which appears to have acted as a footwall ramp to the overlying thrust system.

The internal geometry of the duplex structure is quite variable along strike and in cross section even on the scale of the map area (Figures 2 and 3) and appears to be largely controlled by the distribution, size and shape of the blocks of plutonic rocks. Characteristic shear zone structures in the serpentinite matrix of the melange include duplexes centred around shingled small blocks, serpentinite cleavage duplexes, C-S fabrics, shear bands, asymmetric kink-style fold trains and rotated tailed clasts and blocks, and all provide evidence for southwest-directed tectonic transport.

The relatively small "exotic" blocks in the melange made up of metasedimentary and metavolcanic rocks which display lithostratigraphic affinities with rocks of both the Bridge River Complex and Cadwallader Group are now interpreted as subcreted fragments (Karig, 1982) incorporated into the duplex structure of the melange by out-of-sequence thrusting from the footwall of the obducting ophiolite slab. Their capture in the duplex structure appears related to the ramp configuration of the floor thrust. It occurred mainly in the area where the floor thrust climbed from a level within the mantle peridotites to the basal ultramafic portion of the lower crustal layer 3 of the ophiolite succession. The development of folds and cleavage in the diverse exotic blocks attests to penetrative footwall deformation of the evolving thrust wedge, presumably under sizeable lithostatic load.

The East Liza Intrusive suite in the western portion of the map area constitutes a large horse along the base of the ophiolitic thrust stack. It is the only lithotectonic element in the belt that preserves the transition from the lower to upper crustal units of the ophiolite succession. In this area, the basal part of the ophiolitic thrust stack is intensely folded in large upright structures together with the underlying Hurley Formation (Figures 2 and 3). At present, it is not clear whether these macroscopic folds represent late-stage accretionary structures or developed during regional, post-accretion deformation.

The telescoped ophiolite complex forms a large, lozenge-shaped block with an outcrop area of approximately 180 square kilometres within the collage of northwesterly trending, fault-bounded tectonostratigraphic parcels that defines the southwestern margin of the Intermontane superterrane. The complex is truncated on its northeastern side by the Yalakom Fault and on its southern side by the Marshall Creek Fault; a zone of transfer faults between the Yalakom and Marshall Creek Fault systems terminates the ophiolite complex and the adjacent fragment of the Cadwallader Group on their western sides (Figure 1). The kinematic evolution of this complicated array of fault systems has been discussed in some detail by Schiarizza et al. (1990). The most apparent movement on the fault systems is dextral strike slip, creating a large transtensional structure to the west of the Shulaps Complex, and is, at least in part, post-Eocene in age. It is followed by a phase of extensional faulting on the Marshall Creek Fault, characterized by southwest-side down displacement. Regionally, dextral strike slip may have been preceeded by a phase of early Upper Cretaceous sinistral transpression, inferred by Schiarizza et al. (1990) for fault systems to the west of the Shulaps Complex.

GEOCHEMISTRY OF SHULAPS COMPLEX BASIC ROCKS AND BASIC ROCKS FROM ASSOCIATED COMPLEXES

We here present new geochemical data comprising major and trace element analyses of 22 samples of basic igneous rocks collected during 1990 from the East Liza Igneous Suite, the Bridge River Complex and the Pioneer Formation of the Cadwallader Group (Table I). In addition, a number of analyses of volcanic rocks published by Russmore (1985) and Potter (1983) are used in Figure 5. These geochemical plots are essentially discriminant function diagrams which theoretically allow distinction of tectonic setting of basaltic rocks on the basis of immobile element concentrations. The level of metamorphism of the basic rocks, for the most part low grade greenschist facies, is such that these diagrams should be valid in their assumption of the immobility of the elements used.

Three distinct volcanic suites are recognized in the data set based on trace element chemistry (Figures 4 and 5). The Pioneer volcanic rocks consistently plot as island arc tholeiites in Ti vs V, Zr/Y vs Zr and Cr vs Y discrimination diagrams. The Bridge River and East Liza volcanic rocks plot as ocean floor basalts in the Ti vs V plot and overlap the MORB and within plate basalt fields in the Zr/Y vs Zr, and Cr vs Y plots. Volcanic rocks sampled from blocks within the serpentinite melange of the Shulaps Complex show variable geochemical affinities and overlap the compositional fields of both the East Liza and Bridge River suites (Figure 5A).

Ti vs V System
Shervais (1982) used variation in Ti vs V ratios with changing oxygen fugacity to distinguish between arc related tholeiites, Mid-Ocean Ridge basalts (MORB), and alkali basalts. Data from his work suggest that rocks with Ti/V ratios <20 are formed in volcanic arc environments, Ti/V ratios between 20 and 50 are MORB and ratios >50 have alkalic compositions. The Ti vs V plot in Figure 4A clearly distinguishes between the Pioneer volcanics which for most samples plot in the island arc tholeiite field, and the Bridge River and East Liza volcanic rocks which plot primarily as ocean floor basalts. Within the ocean floor basalt field, there is clear separation between rocks from Bridge River which have a MORB or perhaps even more enriched signature and the East Liza volcanic rocks which straddle the volcanic arc and ocean floor basalt fields, overlapping with both the Bridge River and Pioneer fields. Three samples from the Bridge River volcanics are fairly alkalic with Ti/V values approaching 50. One East Liza sample plots within the Pioneer field.

Table 1.
Major and trace element analyses of volcanic rocks from the Shulaps area
(Major element oxides in wt. %, trace elements in ppm. nd = not detected)

	Pioneer 1	Pioneer 2	Pioneer 3	Pioneer 4	Bridge River 1	Bridge River 2	Bridge River 3	Bridge River 4	Bridge River 5	East Liza 1	East Liza 2
SiO_2	61.5	54.8	53.33	54.8	46.2	47.2	44.1	46.4	51.9	48.5	49.2
TiO_2	0.92	0.96	1.04	0.84	1.8	1.12	1.8	1.2	1.64	1.4	1.4
Al_2O_3	13.8	15.5	14.4	15.3	16.4	14.39	15.1	14.9	13.1	15	14.5
Fe_2O_3	7.35	8.27	10.3	8.34	9.97	9.72	11.72	9.34	9.6	10.49	10.47
MnO	0.11	0.12	0.22	0.11	0.15	0.16	0.17	0.13	0.13	0.17	0.17
MgO	4.19	4.98	6.45	4.98	6.31	6.22	6.01	6.36	8.46	7.01	7.1
CaO	3.1	5.4	4.7	6.3	7.98	10.12	10.84	10.6	3.1	9.98	8.88
Na_2O	5.28	5.24	3.74	4.43	4.1	4.29	3.04	3.15	8.99	3.75	3.96
K_2O	0.05	0.05	0.37	0.05	0.42	0.24	0.54	1.8	1.26	0.11	0.09
P_2O_5	0.07	0.08	0.05	0.07	0.18	0.1	0.21	0.22	0.36	0.1	0.1
LOI	2.22	2.78	3.9	3.54	4.91	5.56	5.54	5.51	2.41	2.58	2.38
Cr	22	18	53	21	261	186	416	664	670	103	202
Ni	18	15	26	18	124	62	256	433	187	51	68
V	324	405	376	351	263	294	274	249	252	376	328
Cu	88	100	25	45	52	27	70	83	61	62	66
Zn	53	64	80	62	84	104	96	84	115	86	87
Rb	nd	nd	nd	nd	15	nd	13	32	25	nd	5
Ba	9	nd	124	17	117	30	155	137	374	22	29
Sr	44	64	217	56	319	186	261	249	280	212	129
Ge	14	16	16	17	20	19	18	18	22	17	17
Nb	nd	nd	nd	nd	19	nd	10	nd	8	nd	nd
Zr	50	52	41	44	135	80	171	62	168	75	83
Ti	5515	5755	6235	5036	10791	6714	10791	7194	9832	8393	8393
Y	14	15	15	14	22	19	22	17	19	23	21

	East Liza 3	East Liza 4	East Liza 5	East Liza 6	East Liza 7	East Liza 8	East Liza 9	East Liza 10	East Liza 11	East Liza 12	East Liza 13
SiO_2	49.8	50.4	52.4	47.8	48.8	46.7	51.7	46.9	49.3	48.3	50.7
TiO_2	1.84	1.48	0.96	1.2	1.2	1.56	1.28	1.36	1.16	1.64	1.68
Al_2O_3	15.1	14.1	14.7	15.2	14.8	14.1	14.2	15.4	14.5	14.4	13.9
Fe_2O_3	10.33	10.51	9.86	9.37	9.93	12.42	9.34	9.64	8.57	11.26	11.48
MnO	0.16	0.17	0.15	0.14	0.17	0.2	0.15	0.16	0.15	0.19	0.18
MgO	7.05	6.58	6.04	7.34	4.97	8.12	5.74	7.89	8.84	7.23	5.6
CaO	6.68	8.58	7.36	11.56	13.44	8.66	8.88	10.42	9.36	8.16	8.32
Na_2O	4.97	4.42	4.66	2.3	3.16	3.17	4.8	3.27	3.62	4.08	4.32
K_2O	0.4	0.07	0.15	0.76	0.03	0.16	0.12	0.41	0.12	0.12	0.05
P_2O_5	0.17	0.11	0.06	0.08	0.09	0.1	0.11	0.09	0.1	0.18	0.14
LOI	2.67	2.21	2.32	3.02	1.94	2.97	1.99	3.14	2.98	2.79	2.03
Cr	254	97	133	341	228	150	110	174	118	278	209
Ni	107	50	53	90	97	77	41	72	59	106	76
V	358	340	312	302	300	359	300	310	279	271	390
Cu	49	68	61	87	15	90	51	45	95	38	38
Zn	99	90	84	70	82	121	75	77	52	149	99
Rb	12	5	7	17	nd	nd	nd	11	6	-	7
Ba	176	33	nd	133	nd	22	30	880	28	21	13
Sr	151	158	105	301	348	196	167	185	155	122	95
Ge	16	13	11	17	18	18	18	14	13	19	17
Nb	nd	nd	nd	nd	nd	nd	nd	-	-	-	-
Zr	73	72	61	73	72	92	92	80	72	106	103
Ti	11031	8873	5755	7194	7194	9352	7674	8153	6954	9832	10072
Y	23	17	17	23	15	28	23	24	17	24	27

Volcanic blocks from the ophiolitic melange are characteristically high in Ti and tend to cluster in or near the alkalic basalt field. Some overlap with the Bridge River compositional field and others plot within the East Liza field (Figure 5A).

Shervais (1982) examined the evolution of magmas during partial melting and crystal fractionation with systematic fluctuations in fo_2 using the trivalent cation V with respect to an index of fractionation Ti. The variation in Ti/V ratios in primary melts produced by various degrees of partial melting at a set fo_2 are represented by variation in the bulk distribution coefficients for Vanadium (D_v) and is shown in Figure 5A. Subsequent changes through various degrees of fractional crystallization of ol +/- plag +/- cpx mineral assemblages are also shown. It is apparent that the Pioneer and East Liza volcanic suites may have been produced by progressive partial melting of a single source but evolved through a slightly different fractionation history. Alternatively the two suites evolved from

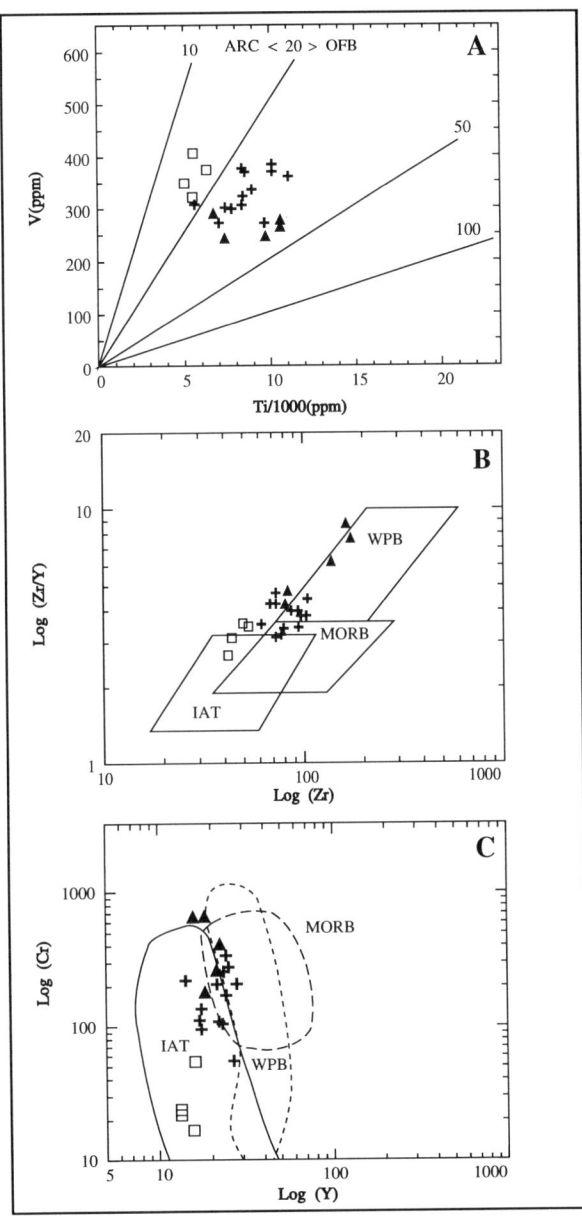

Figure 4. A) V vs. Ti plot after Shervais (1982); B) Zr/Y vs. Zr after Pearce and Norry (1979); C) Cr vs. Y after Wood (1979). □ = Pioneer volcanics, + = East Liza volcanics, ▲ = Bridge River volcanics. MORB = Mid Ocean Ridge Basalt, WPB = Within Plate Basalt, IAT = Island Arc Tholeiite, OFB = Ocean Floor Basalt.

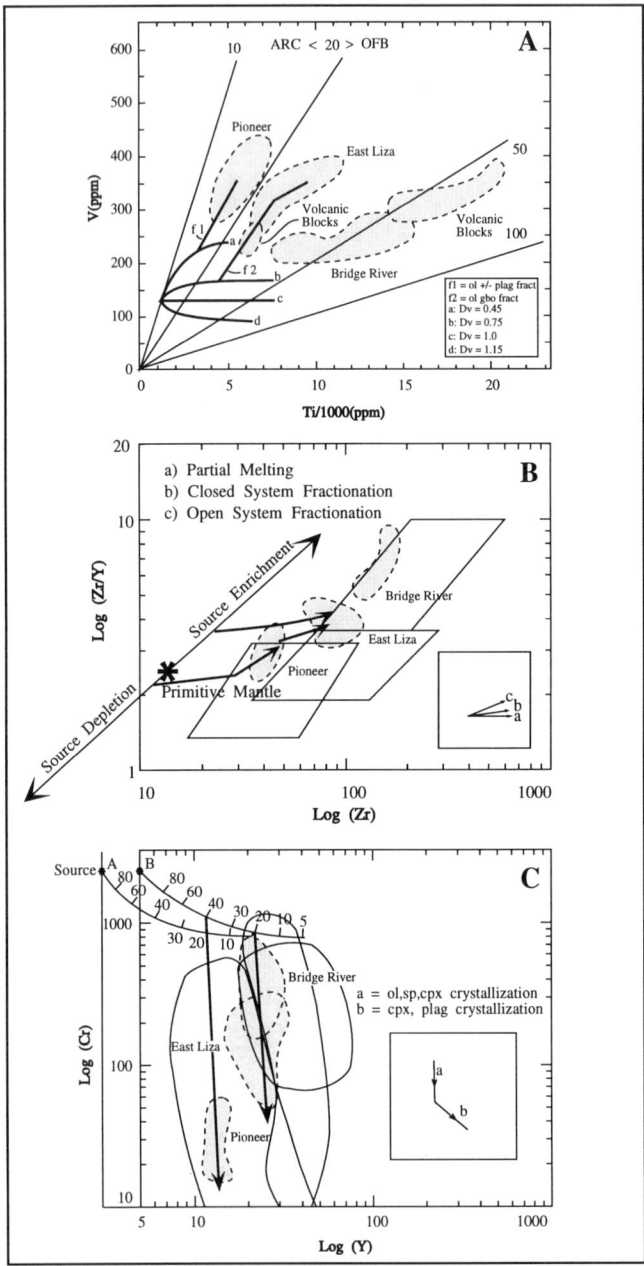

Figure 5. Petrogenetic schemes for the three volcanic suites. Although 5A suggests the Pioneer and East Liza suites are derived from a similar source, this does not appear to be the case when Zr, Y and Cr are used. See text for explanation.

two distinct primary melts produced at slightly different fo_2 in different tectonic settings. The Bridge River volcanics and the volcanic blocks from the ophiolitic melange appear to have formed under more reducing conditions and evolved through the fractional crystallization of ol:plag:cpx mineral assemblages.

Zr/Y vs Zr system
Pearce and Norry (1979) used the systematic variation of the elemental ratio Zr/Y with respect to an index of fractionation Zr, to place constraints on various genetic processes that cause variation in the Zr/Y ratio, and evaluate changes in this ratio in various tectonic settings. In the Zr/Y vs Zr plot (Figure 4B), there are clear distinctions between the Pioneer, Bridge River and East Liza suites that can be attributed to the initial tectonic setting in which the rocks were formed. The Bridge River samples plot as within plate basalts (WPB), the East Liza samples as MORB with some overlap into the other fields and the Pioneer samples as arc basalts with overlap into the MORB field. As in the Shervais plot, the Pioneer and East Liza fields appear to overlap.

Figure 5B, adopted from Pearce and Norry (1979), examines the change in Zr/Y ratios relative to processes involved with magma generation and evolution, including partial melting and open and closed system crystal fractionation. Suggested evolutionary paths for the three volcanic suites are shown. This model suggests that the Pioneer and East Liza volcanic rocks may be related by either progressive partial melting of a somewhat depleted mantle lherzolite source followed by open system crystal fractionation of an ol:cpx:plag mineral assemblage or that they may be derived from the same primary melt and are related through crystal fractionation. The Bridge River rocks and volcanic blocks from the ophiolitic melange are enriched in Zr/Y relative to the Pioneer and East Liza suites. Several of these rocks have sometimes strong alkalic signatures suggesting they are derived from partial melting of a previously enriched source similar to that in present day volcanic ocean island settings.

Cr vs Y system
The discrimination diagram using the immobile incompatible element Y vs an index of fractionation Cr is extensively discussed by Wood (1979) and Pearce (1982). In the Cr vs Y discrimination diagram (Figure 4C), Pioneer basaltic rocks are extremely depleted in Cr and Y, and plot in the island arc field. The East Liza compositional field overlaps with the Bridge River field but on the whole contains less Cr, plotting as MORB and within plate basalt. The Bridge River volcanics are enriched in Cr relative to both the East Liza and Pioneer volcanic rocks and plot as MORB and within plate basalts. Two samples from the suite show extreme enrichment in Cr which may arise from partial melting from a primary magma source enriched in Cr or some fractionation process whereby Cr-bearing minerals (spinel) are retained by the magma.

Figure 5C, from Pearce (1982), models primary melt compositions from a mantle lherzolite. Two crystal fractionation trends can be modelled in this system: a) representing the crystal fractionation of olivine, spinel and clinopyroxene and b) representing the crystal fractionation of olivine, spinel and clinopyroxene and plagioclase. When applied to the data set, this model clearly distinguishes between a magma source for the Pioneer volcanics and a second source that may have given rise to the East Liza and Bridge River volcanic rocks. The model goes further to suggest that the Pioneer rocks were produced from either a large degree of melting ($\sim$ 40%) of a mantle lherzolite source or a lesser degree of melting (20-25%) from a previously depleted mantle source. In either case this source differs from the source(s) which gave rise to the other two volcanic suites.

DISCUSSION

Rocks of the Shulaps Complex, Bridge River Complex, East Liza and Bralorne suites, Cadwallader Group and Tyaughton Group are all part of a long-lived Panthalassic oceanic domain (Cordey and Schiarizza, 1993), that evolved from the late Paleozoic to mid-Mesozoic. The presence of blueschists suggests that the whole assemblage accumulated as an accretionary complex in response to plate subduction. The nature and diversity of the oceanic tract related to this subduction are clearly demonstrated in the chemostratigraphy of the various lithotectonic elements.

The volcanic suites examined have relatively clear, distinguishable geochemical signatures in a variety of discrimination diagrams. The Bridge River volcanics appear to be a chemically diverse suite of MORB, within plate tholeiites and alkali basalts, the latter derived from a somewhat enriched mantle source. They therefore likely represent the volcanic portion of an oceanic crust (e.g. Potter, 1986) with superimposed seamount volcanism. The Pioneer volcanics of the Cadwallader Group consistently plot as island arc tholeiites, and appear to have been derived from a previously depleted mantle source. This, taken together with the composition of associated sedimentary rocks (Hurley Formation), suggests their formation in a volcanic arc complex (e.g. Russmore, 1987). The East Liza volcanics appear intermediate between the Bridge River and Pioneer volcanics and may be distinct from most of these in terms of their petrogenesis. They were derived from somewhat depleted mantle and represent volcanics more akin to MORB than any other in the region, except for some Bridge River samples. Exotic volcanic blocks in the serpentinite melange appear to be a structural sampling of the Bridge River Complex as well as the East Liza suite.

Structural and stratigraphic relationships observed within the Shulaps thrust system allow for a tentative palinspastic restoration of the various litho-tectonic elements incorporated into this part of the accretionary complex (Figure 6). The upper part of the Shulaps thrust system represents a sampling of tectonized residual upper mantle, possibly derived from a position nearby a spreading centre, and plutonic rocks interpreted as layer 3 of the Shulaps oceanic crust. These units occur in a structural stacking order in which the ophiolitic stratigraphy is reversed. This indicates that the upper mantle slice was emplaced over a crustal-scale ramp which exceeded 2-3 kilometres in height and climbed from a position within the upper mantle to a level at least near the top of layer 3 (Figure 6, right-hand side). The floor thrust of the system continued as a flat within the basal, ultramafic portion of layer 3. Emplacement of the thrust system was facilitated by pervasive serpentinization of the ultramafic rocks, particularly along brittle-plastic thrust zones. Subcretion of the exotic supracrustal blocks probably occurred in front of the foreland migrating crustal ramp.

The transition from layer 3 plutonics to layer 2 volcanics of the Shulaps oceanic tract is no where preserved in the upper part of the thrust system. In part, this may be explained by the deep level of exhumation of the western thrust front (Figure 3). The occurrence of plutonic blocks without their contiguous volcanics in several stacked duplexes within the middle part of the thrust system suggests, on the other hand, that the upper and lower sections of the oceanic crust may effectively have been detached from one another (Figure 6, central part).

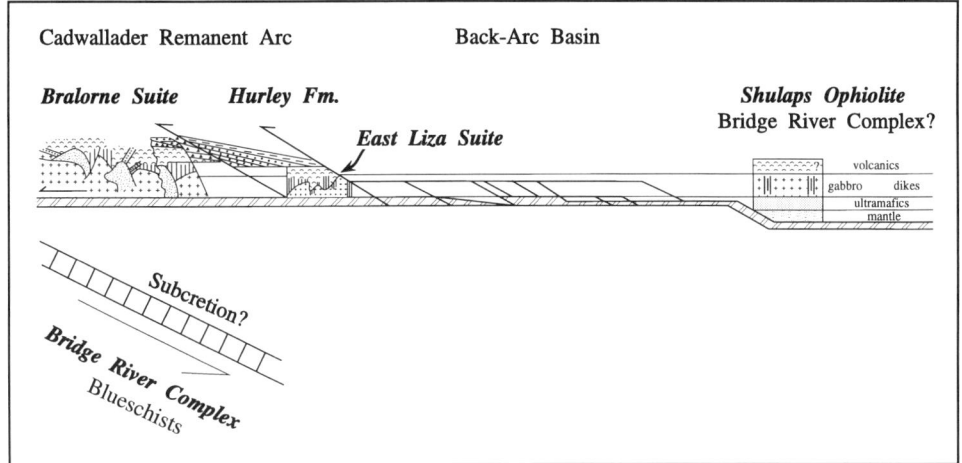

Figure 6. Schematic restored section showing inferred palinspastic relations of lithotectonic units in the study area. Also shown are trajectories of thrust faults (in bold lines) explaining the observed stacking order of units in the Shulaps thrust system. Diagram is not drawn to any scale.

The mantle and lower crustal components of the ophiolite were emplaced directly on top of gabbros and overlying volcanic rocks of the East Liza suite. In many respects, these gabbros correlate well with the mafic plutonic rocks which occur as melange blocks interpreted above as layer 3 of the Shulaps oceanic crust. The East Liza volcanics may therefore represent the upper crustal, layer 2 of the Shulaps oceanic tract. The transition between gabbros and volcanics occurs along a zone of discrete dike swarms and screens, but lacks the development of a well-formed sheeted dike complex.

The East Liza suite is itself directly emplaced onto the Hurley Formation, interpreted by Russmore (1987) as the fringing sedimentary apron of the Upper Triassic Cadwallader arc. The Pioneer volcanics represent the extrusive edifice of this arc. The distinctive geochemical signature of the East Liza volcanics sets this suite apart from the Pioneer volcanics. The proposed palinspastic restoration places the East Liza suite as a contiguous part of the Shulaps oceanic tract to the east of the Cadwallader arc (Figure 6, left-hand side).

The Bralorne suite, exposed to the west of the Shulaps Complex (Figure 1), contains many elements similar to the East Liza suite and Shulaps Complex, including serpentinized ultramafic rocks and variably deformed and metamorphosed gabbroic plutonic rocks. Notably, volcanic rocks associated with the Bralorne suite and previously assigned to the Pioneer Formation carry the distinctive geochemical signature defined here for the East Liza suite (P. Schiarizza, pers. comm.). The ultramafic-mafic plutonic rocks of the Bralorne suite were intruded by hornblende diorite and plagiogranite, possibly in a spreading centre environment, during the early Permian (Leitch et al., 1991). The ages of the Shulaps Complex and East Liza suite remain uncertain to date. Blocks of amphibolite in the melange, correlated with the gabbros of the Shulaps Complex, have yielded Permian ages (Archibald et al., 1991). The various stratigraphic relationships suggest that the Shulaps Complex and Bralorne - East Liza suite that evolved from at least the Permian onwards. The much younger Cadwallader arc may have been built upon this oceanic substrate (Figure 6).

Acknowledgements

We would like to thank Mr. Darryl Williams and Mrs. Nancy Fagan for their invaluable support in the preparation of this work. Financial support was provided by the Mineral Resources Division of the Ministry of Energy, Mines and Petroleum Resources, Province of British Columbia, and is gratefully acknowledged.

References

D.A. Archibald, P. Schiarizza and J.I. Garver. $^{40}Ar/^{39}Ar$ Evidence For the Age of Igneous and Metamorphic Events in the Bridge River and Shulaps Complexes, Southwestern British Coumbia (920/2; 92J/15, 16). In: *Geological Fieldwork, 1990*, B.C.Ministry of Energy, Mines and Petroleum Resources, Paper 1991-1, pp. 75-83 (1991).

T.J. Calon, J.G. Malpas and R. Macdonald. The anatomy of the Shulaps ophiolite. In: *Geological Fieldwork 1989*, B.C. Ministry of Energy, Mines and Petroleum Resources, Paper 1990-1, pp. 375-386 (1990).

F. Cordey and P. Schiarizza. Long-lived Panthalassic remnant: the Blue River accretionary complex, Canadian Cordillera. *Geology*, 21, 263-266 (1993).

J.M. Journeay. Structural and Tectonic Framework of the Southern Coast Belt, British Columbia. In: *Current Research*, Part E, Geological Survey of Canada, Paper 90-1E, pp. 183-197 (1990).

D.E. Karig. Deformation in the forearc: implications for mountain belts. In: *Mountain Building Processes*. K.J. Hsu (Ed.). pp. 59-71. Academic Press, London (1982).

G.B. Leech. Geology and Mineral Deposits of the Shulaps Range. B.C. Ministry of Energy, Mines and Petroleum Resources, *Bulletin 32*, 54 pages (1953).

C.H.B. Leitch, P. Van Der Heyden, C.I. Godwin, R.L. Armstrong and J.E. Harakal. Geochronometry of the Bridge River Camp, southwestern British Columbia. *Canadian Journal of Earth Sciences*, 28, 195-208 (1991).

J.W.H. Monger. Geology Between Harrison Lake and Fraser River, Hope Map Area, Southwest British Columbia. In: *Current Research*, Part B, Geological Survey of Canada, Paper 86-1B, pp. 699-706 (1986).

J.W.H. Monger, J.M. Journeay, C.J. Grieg, and J. Rublee. Structure, Tectonics and Evolution of Coast, Cascade and Southwestern Intermontaine Belts, Southwestern British Columbia; Notes to Accompany a Field trip, *Geological Association of Canada-Mineralogical Association of Canada*, Vancouver Meeting, Field trip B6 (1990).

J.J. Nagel. The Geology of part of the Shulaps Ultramafite near Jim Creek, Southwestern British Columbia. Unpublished M.Sc. thesis, The University of British Columbia, 74 pages (1979).

A. Nicolas and A. Prinzhofer. Cumulative or residual origin for the transition zone in ophiolites: structural evidence. *Journal of Petrology*, 24, 188-206 (1983).

A. Nicolas and J.F. Violette. mantle flow at oceanic spreading centres: models derived from ophiolites. *Tectonophysics*, 81, 319-339 (1982).

J.A. Pearce. Trace Element Characteristics of Lavas from Destructive Plate Boundaries. In: *Andesites*. R.S. Thorpe (Ed.). pp. 525-548, John Wiley and Sons, London (1982).

J.A. Pearce and M.J. Norry. Petrogenetic Implications of Ti, Zr, Y, and Nb Variations in Volcanic rocks. *Contributions to Mineralogy and Petrology*, 69, 33-47 (1979).

C.J. Potter. Geology of the Bridge River Complex, Southern Shulaps Range, British Columbia: A Record of Mesozoic convergent tectonics. Unpublished PhD thesis, University of Washington, Seattle (1983).

C.J. Potter. Origin, Accretion and Postaccretionary Evolution of the Bridge River Terrane, Southwest British Columbia. *Tectonics*, 5, 1027-1041 (1986).

M.E. Russmore. Geology and Tectonic Significance of the Upper Triassic Cadwallader Group and its Bounding Faults, Southwest British Columbia. Unpublished PhD thesis, University of Washington, 174 pages (1985).

M.E. Russmore. Geology of the Cadwallader Group and the Intermontane-Insular Superterrane Boundary, Southwestern British Columbia. *Canadian Journal of Earth Sciences*, 24, 2279-2291 (1987).

A.P. Schiarizza, R.G. Gaba, J.I. Glover and J.I. Garner. Geology and Mineral Occurrences of the Tyaughton Creek area (920/2, 92J/15, 16). In: *Geological Fieldwork, 1989*, B.C. Ministry of Energy, Mines and Petroleum Resources, Paper 1989-1, pp. 115-130 (1989).

A.P. Schiarizza, R.G. Gaba, M. Coleman, J.I. Garver and J.K. Glover. Geology and Mineral Occurrences of the Yalakom River Area. (920/1,2, 92J/15, 16). In: *Geological Fieldwork 1990*, B.C. Ministry of Energy, Mines and Petroleum Resources, Paper 1990-1, pp. 53-72 (1990).

P. Schiarizza, R.G. Gaba, J.K. Glover, J.I. Garver and P.J. Umhoefer. Geology and Mineral Occurrences of the Taseko Bridge River Map Area. *B.C. Ministry of Energy, Mines and Petroleum Resources Bulletin* (in press).

J.W. Shervais. Ti-V Plots and the Petrogenesis of Modern and Ophiolitic Lavas. *Earth and Planetary Science Letters*, **59**, 101-108 (1982).

D.A. Wood. A Variably Veined Suboceanic Upper Mantle-Genetic Significant for Mid-ocean Ridge Basalts from Geochemical Evidence. *Geology*, **7**, 499-503 (1979).

R.L. Wright, J.I. Nagel and K.C. McTaggart. Alpine Ultramafic rocks of Southwestern British Columbia. *Canadian Journal of Earth Sciences*, **19**, 1156-1173 (1982).

The Diversity of Accreted Oceanic Lithosphere in the Brooks Range, Alaska

K.R. WIRTH[1], J.M. BIRD[2] and J.N. WESSELS[1]
[1]Geology Department, Macalester College, Saint Paul, Minnesota, USA, 55105
[2]Department of Geological Sciences, Cornell University, Ithaca, New York, USA, 14853

Abstract. The Late Jurassic - Early Cretaceous Brooks Range fold and thrust belt of northern Alaska hosts a diverse assemblage of oceanic rocks. Field and petrographic studies indicate that Middle Jurassic mafic and ultramafic rocks in the western Brooks Range are characterized by: relatively thick sections of cumulus rocks, early crystallization of clinopyroxene, depleted residual mantle peridotites (harzburgite and dunite), and common magmatic amphibole and magnetite in upper cumulus sequences. Although there are significant variations among the exposed plutonic sections, most contain evidence of an origin within or near supra-subduction zones. Mafic volcanic rocks (Mississippian to Middle Jurassic) structurally underlie the plutonic rocks and are interpreted to be fragments of upper portions of oceanic lithosphere. Lithostratigraphic, petrographic, and geochemical characteristics of the volcanic rocks are extremely variable and are indicative of formation in diverse tectonic settings, including back-arc basin, volcanic arc, and within-plate. Evidently, most of the oceanic rocks in the Brooks Range are fragments of anomalous oceanic lithosphere that originated in diverse tectonic settings within a large ocean basin; few of the rocks are interpreted as being typical oceanic lithosphere or as having originated along mid-oceanic spreading ridges. The diverse features and long transport history of the Brooks Range oceanic rocks contrast with many other ophiolites (e.g. Tethyan) and might be characteristic of oceanic lithosphere obducted along the margins of large ocean basins (e.g. Pacific).

Key words: Alaska, basalts, Brooks Range, emplacement, genesis, geochemistry, island-arcs, Mesozoic, obduction, ophiolite, petrology, seamounts, tectonics, ultramafics

INTRODUCTION

Mafic and ultramafic rocks are exposed over a large region that extends more than 600 km across northern Alaska. Although many of the rocks were structurally disrupted during Middle Jurassic - Early Cretaceous detachment and emplacement, all of the rock units that characterize ophiolites are present within the Brooks Range. In this paper, we summarize existing data and present new field, petrographic, and geochemical evidence indicating that the oceanic rocks of the western Brooks Range originated in diverse tectonic environments. The gabbroic and ultramafic rocks have petrographic and geochemical features similar to other supra-subduction zone ophiolites, whereas the mafic volcanic rocks have geochemical characteristics indicating they originated in various oceanic and convergent margin settings. The diversity of oceanic rocks in the Brooks Range exceeds that recognized in most other orogenic belts; the region may be unique for studying the character of oceanic lithosphere and the processes of crustal evolution.

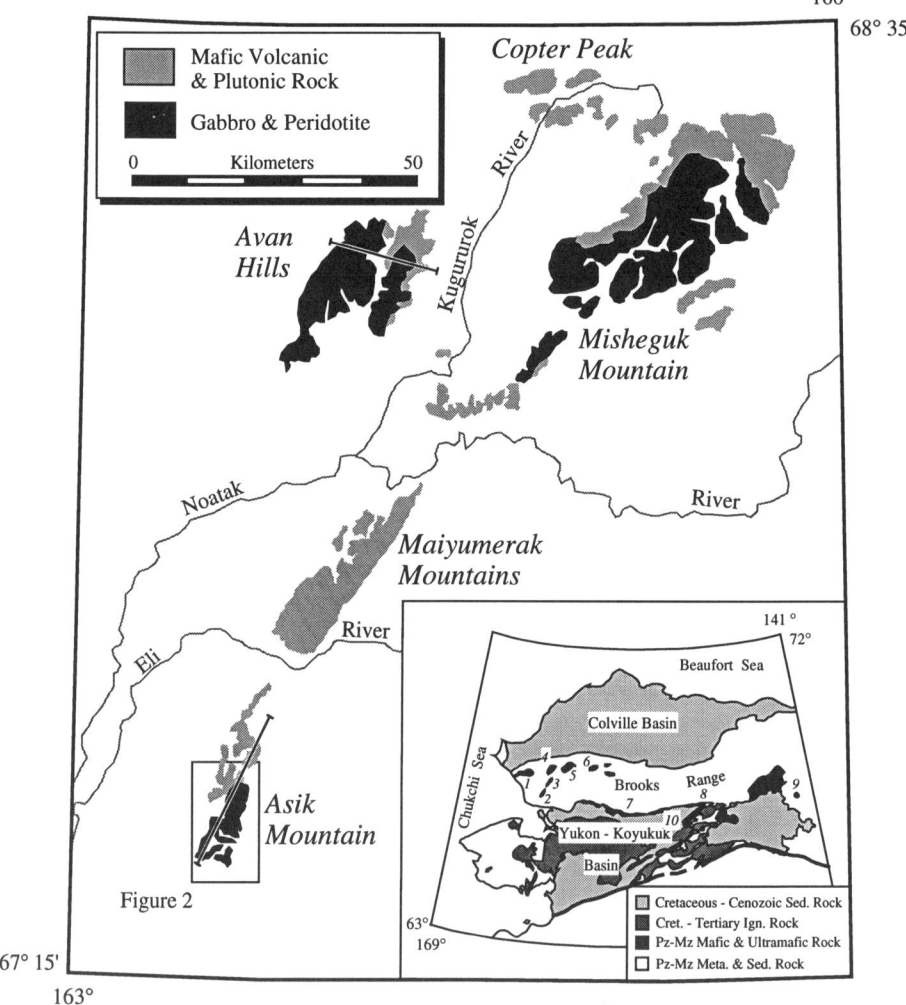

Figure 1. Map of ophiolitic rocks in the western Brooks Range. The regions shown in Figure 2 and in the cross-sections of Figure 3 are also indicated. Inset is a generalized geologic map of northern Alaska showing the principal exposures of ophiolitic rocks: (1) Iyikrok region, (2) Asik Mountain, (3) Maiyumerak Mountains, (4) Avan Hills, (5) Misheguk Mountain-Copter Peak, (6) Siniktanneyak Mountain, (7) Angayucham Terrane, (8) Cathedral Mountains, (9) Porcupine River region, and (10) Kanuti. Modified from Wirth and Bird [1].

Regional Geology

The Brooks Range (Figure 1) is situated at the northwestern limit of the North American Cordillera and consists of Paleozoic through Early Cretaceous sedimentary and minor igneous rocks that were folded and thrust during Late Jurassic through Early Cretaceous time [2, 3, 4, 5, 6, 7]. Uplift and erosion of these rocks during the Cretaceous and early Tertiary resulted in the deposition of thick sequences of sediment in the Colville and Yukon-Koyukuk Basins north and south, respectively, of the Brooks Range. In the Yukon-Koyukuk Basin, Middle to Late Jurassic (175 - 154 Ma) tonalite and trondhjemite plutons [8, 9] are unconformably overlain by Neocomian calc-alkaline and alkaline volcanic rocks and sediments [10, 11] and are interpreted to be an intra-oceanic volcanic arc that formed in response to southward-directed subduction of the Arctic Alaska continental margin during the Middle Jurassic - Early Cretaceous [5, 6, 7].

In the western Brooks Range (Figure 1), mafic and ultramafic oceanic rocks are exposed in the structurally highest thrust sheets [2, 12, 13, 14, 15, 5]. The uppermost thrust sheet consists of gabbroic and ultramafic rock, and a lower thrust sheet consists of volcanic rock. Similar mafic and ultramafic rocks (Angayucham terrane) occur along the southern margin of the Brooks Range [16] and dip southward beneath the Yukon-Koyukuk Basin (Figure 1). These rocks have been interpreted to be in the "root zone" for the western Brooks Range ophiolites [14, 17, 18, 19, 7] or possibly the result of post-contractional extension [20]. Similar rocks dip northward along the southeast margin of the Yukon-Koyukuk Basin [18].

Previous Work

The oceanic character of the western Brooks Range ophiolites was initially recognized during reconnaissance mapping of the Brooks Range [2, 13, 14, 15]. Recent studies of the composition, stratigraphy, and structure of several mafic and ultramafic complexes in the western Brooks Range confirm that they are similar to other ophiolites and to modern oceanic lithosphere. Initially, K-Ar studies of hornblende- and biotite-bearing lithologies suggested that the ophiolites were Middle to Late Jurassic [14, 21, 22]. Recently, however, $^{40}Ar/^{39}Ar$ studies of several of the western Brooks Range ophiolites indicate that the gabbros crystallized during the early Middle Jurassic (~185 Ma) and were detached at approximately 165 Ma [23, 24, 25]. Stratigraphic and geochronologic relationships indicate that the ophiolites were not emplaced on the continental margin until about 145 Ma, approximately 40 m.y. after the time of crystallization [1].

The general lithostratigraphy and structure of Avan Hills was mapped by Curtis et al. [26] and Curtis et al. [27]. The geology and rock types in the central region of the Avan Hills were described by Zimmerman et al. [28] and Frank and Zimmerman [29]. Zimmerman and Soustek [30] recognized a transition zone between mafic cumulates and deformed and recrystallized peridotites. Nelson and Nelson [21] conducted detailed mapping and petrographic studies of the rocks at Siniktanneyak Mountain (Figure 1) and described ultramafic rocks consisting mostly of medium-grained dunite and wehrlite with rare harzburgite and olivine pyroxenite. The peridotites are overlain by medium- to coarse-grained gabbro, olivine gabbro, and leucocratic hornblende-pyroxene gabbro. They also describe small stocks in the northern (upper?) part of the complex that consist of medium-grained rocks that range from hornblende diorite to biotite-hornblende alkali granite. The presence of these small, intermediate to felsic rock bodies at Siniktanneyak Mountain, and elsewhere in the Brooks Range, led Hamilton [31] to suggest that the mafic-ultramafic complexes in the western Brooks Range are not oceanic in origin, but are the roots of continental volcanic complexes.

The ophiolite at Misheguk Mountain (Figure 1) has been studied by Boak et al. [22] and by Harris [24]. Ultramafic tectonites at Misheguk Mountain consist predominantly of harzburgite, and clinopyroxene generally follows olivine in the crystallization sequence (olivine --> clinopyroxene --> plagioclase). Harris [24] suggested that the ophiolite at Misheguk Mountain is compositionally transitional between mid-ocean ridge- and arc-type ophiolites on the basis of the depleted character of the peridotites (low Ti, Al, and Ca) and the compositions of olivine (Fo_{73-87}) and plagioclase (An_{75-94}) in the layered gabbros.

Intensely metamorphosed amphibolite, schist, pelite, marble, or chert occur locally along the fault surface that separates the overlying ultramafic and gabbroic rocks from the underlying mafic volcanic rocks [32, 33, 27, 34, 22, 35, 36, 24]. Peak metamorphic conditions within the metamorphic sole ranged from hornblende- to pyroxene-hornfels facies in the Avan Hills [33], to amphibolite facies near Iyikrok Mountain (Figure 1) [22].

Mafic volcanic rocks throughout the western and southern regions of the Brooks Range occur structurally beneath the tectonites and magmatic rocks. The field relationships of the volcanic rocks have been described by Nelson and Nelson [21], Mayfield et al. [5], Gottschalk [37], Moore [38], Barker et al. [39], Harris [24], Karl and Dickey [40], Pallister et al. [41], Wirth et al. [42, 43], Karl and Long [44], and Wirth [45]. Hitzman et al. [17, 46] divided basalts in the Angayucham Mountains into two different units on the basis of age and metamorphism. Additional mapping, geochemical, and paleontological work by Pallister et al. [41] confirmed the division of volcanic units. Radiolaria from interflow and interpillow chert within the volcanic units indicate that the basalts are Jurassic (south) and Triassic (north) [41]. A third thrust sheet, consisting of Paleozoic (Devonian and Mississippian) basalt, occurs north of the Triassic basalt and contains interlayered and interpillow chert, tuff, and limestone. Major element abundances suggest that the basalts are hypersthene-normative, olivine tholeiites [41]. Although the two basalt units appear to be similar in the field, geochemical data indicate that they are different. Trace element analyses indicate that the rocks are most similar to recent oceanic island and oceanic plateau basalts [39, 41]. Mafic volcanic rocks of the Cathedral and Twelve-Mile Mountains, in the central part of the southern Brooks Range, contain chert layers that yield Carboniferous radiolaria [47, 37] and are similar to present-day mid-ocean ridge basalt (MORB) [37]. Geochemical studies of mafic volcanic rocks in the western Brooks Range have identified volcanic rocks that have enriched-MORB (E-MORB) or within-plate characteristics. Moore [38] compared the tholeiitic and enriched light rare-earth element (LREE) character of basalts from Copter Peak and those underlying the gabbros at Siniktanneyak Mountain, to the basalts of the Angayucham Terrane. Similarly, Harris [24] concluded that basalts beneath the gabbros and peridotites at Misheguk Mountain are transitional between mid-ocean ridge and within-plate basalt. Basalt and gabbro in the Maiyumerak Mountains have MORB and volcanic arc characteristics [42, 40, 45].

GEOLOGY AND PETROGRAPHY OF WESTERN BROOKS RANGE OPHIOLITES

Regional studies indicate that ophiolitic rocks in the Brooks Range are characterized by similar lithostratigraphic sections; by the presence of metamorphic soles; by the absence of sheeted dikes, volcanic flows, and sedimentary rocks; and by the common occurrence of subordinate leucocratic intrusive rocks. We studied the lithostratigraphy and composition of two ophiolite exposures, Asik Mountain and Avan Hills, in the western Brooks Range (Figure 1). In most regions of the Brooks Range, the lithostratigraphy can be inferred from the nearly continuous exposures of locally-derived, frost-heaved rock. However, because bedrock exposures are limited in many regions, detailed structural relationships are often not well constrained. Igneous rock names given below are based on the IUGS classification system [48].

Lithostratigraphy of Asik Mountain and Avan Hills Ophiolites
Rocks of the Asik Mountain ophiolite (Figure 2) are grouped into four units from southwest to northeast: layered peridotite; layered gabbro; massive gabbro; and gabbro and diorite,

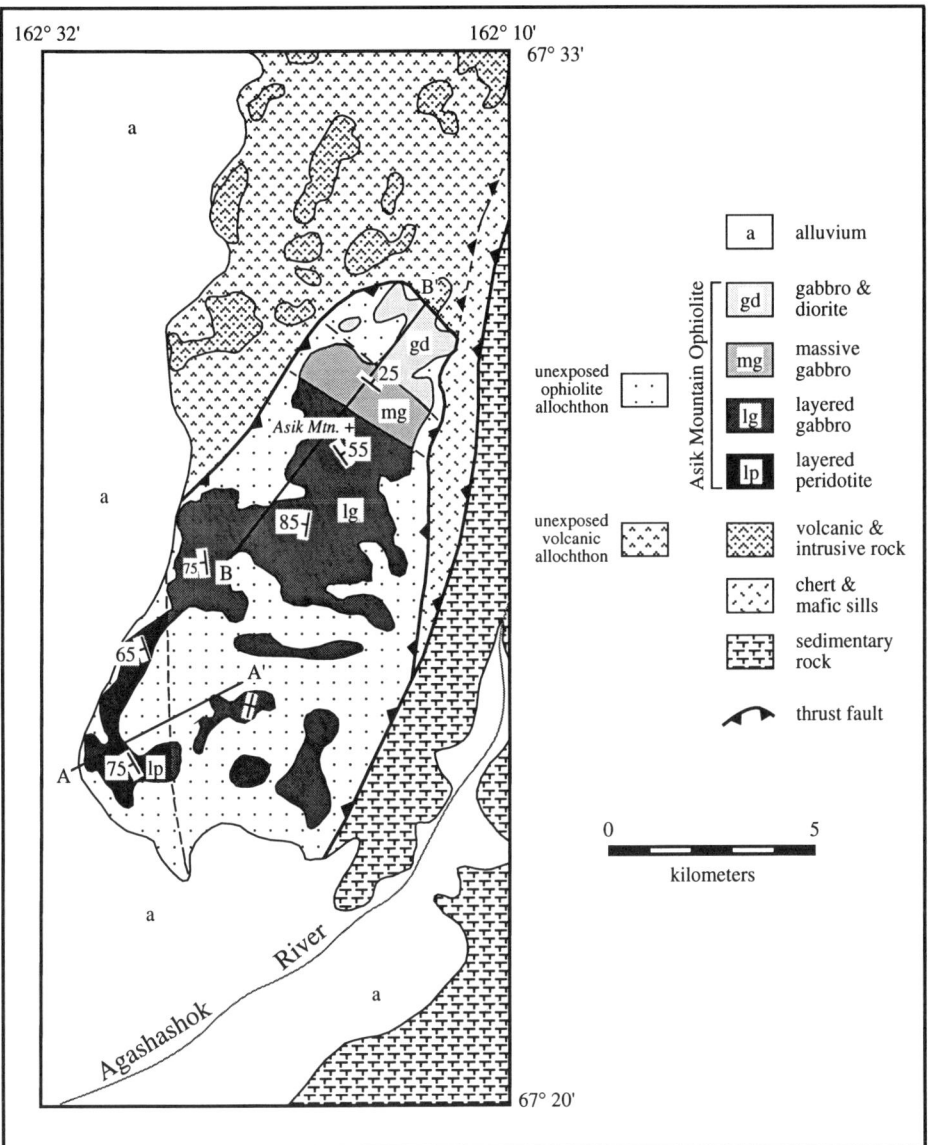

Figure 2. Geologic map of the Asik Mountain region.

respectively. These rocks structurally overlie mafic volcanic flows and minor plutonic rocks that have been variably metamorphosed and deformed (Figure 3). Compositional layering trends northwest and dips steeply to the southwest in the region of southwestern Asik Mountain. To the north and east of Asik Mountain, the orientation of the layering is progressively rotated through vertical and dips northeast.

The layered peridotite unit (> 2.5 km thick) consists of dunite, wehrlite, and olivine clinopyroxenite with minor clinopyroxenite and gabbro (Figures 4 and 5). Contacts between interlayered dunite, wehrlite, and olivine clinopyroxenite range from gradational to sharp. The

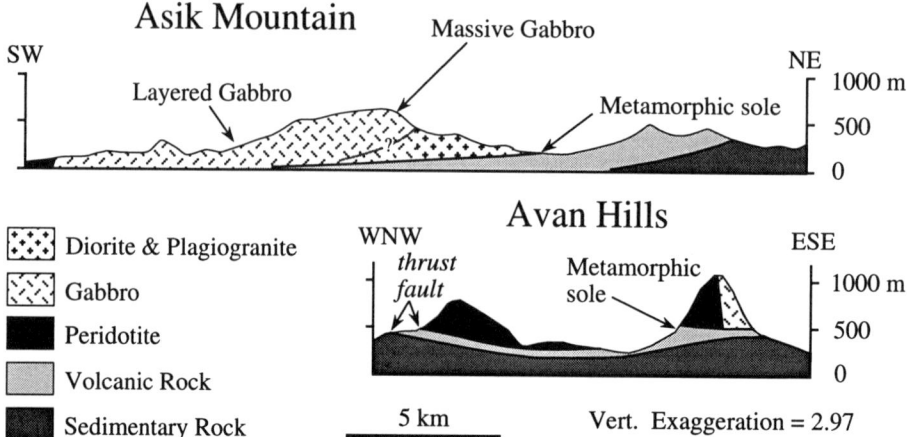

Figure 3. Cross-sections through the Asik Mountain and Avan Hills ophiolites.

stratigraphically lowest olivine-rich rocks exposed in the section (e.g. samples 130, 145, 152) have a scaly weathering habit that roughly parallels the trend of compositional layering in the region and is interpreted to be foliation. In thin-section, the presence of bent and pinching twin lamellae in plagioclase and of deformation bands in pyroxene suggest that portions of these rocks have undergone high-temperature solid-state (plastic) deformation. Higher in the section, the rocks are undeformed and clinopyroxene is predominant.

The layered gabbro unit (~4.5 km thick) consists mostly of gabbro with minor interlayered olivine gabbro, gabbronorite, and anorthosite. Layering within the gabbro unit is defined by changes in grain-size and the relative proportions of pyroxene and plagioclase. Grain-size and modal changes are gradational within each layer, but change abruptly between layers. Layers range from approximately 15 cm to nearly one meter thick and are rarely deformed or folded. In thin-section, the layered rocks exhibit planar lamination (magmatic foliation) of elongate minerals and consist of adcumulus cpx + plag ± ol with intercumulus orthopyroxene or hornblende. Olivine is more abundant in the lower part of the layered gabbro unit, and a thin zone of dunite in the lower part of the layered gabbro might be a small out-of-sequence intrusive. The relatively uniform composition of the layered sequences suggests that the parent magma compositions remained relatively constant throughout formation. Exceptions to this include several layers (?) of wehrlite and olivine-bearing gabbro (e.g. samples 11, 15, 17, 24, 40, 45) that occur relatively high in the section on the east side of Asik Mountain. If these rocks are part of the layered sequence, then they indicate late-stage changes (injection of primitive magma?) in the composition of the magma chamber. Alternatively, they might be late-stage sills intruded into the upper parts of the layered sequence. Because of poor outcrop exposure in this region, it is difficult to interpret the relationship of these olivine-rich gabbros with the surrounding layered gabbros.

The massive gabbro unit (> 2 km thick) at Asik Mountain consists primarily of noncumulus gabbro, pyroxene-hornblende gabbro, minor hornblende gabbro, diorite, and rare pegmatitic and crescumulus hornblendite. Variations in grain size and composition are common and occur within a single outcrop. In thin-section, plagioclase commonly exhibits compositional zoning. Structurally above the massive gabbro unit the rocks consist mostly of quartz diorite, tonalite, and plagiogranite with screens of massive gabbro. Many of these rocks exhibit

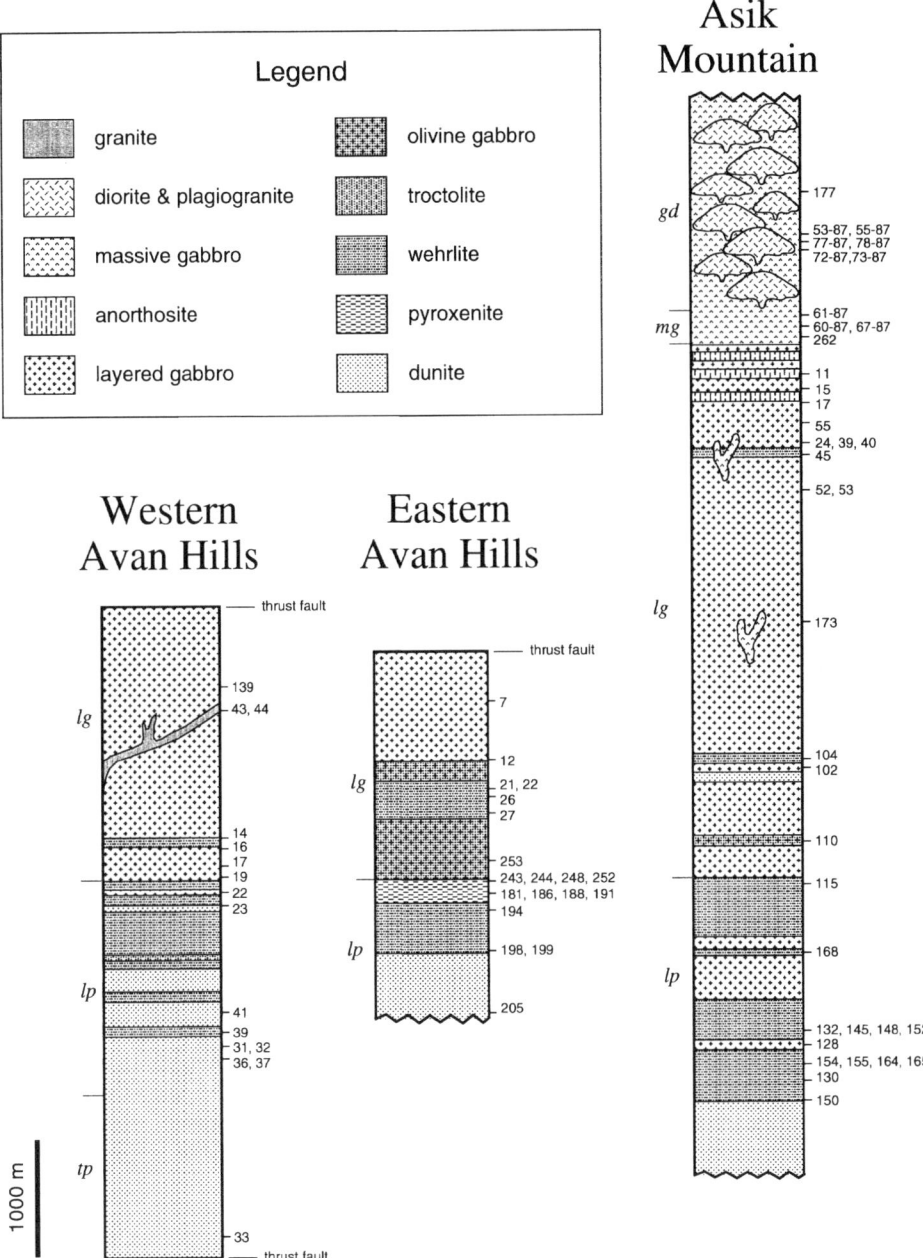

Figure 4. Generalized lithostratigraphic sections through Avan Hills and Asik Mountain ophiolites. The Asik Mountain section was reconstructed along A-A' and B-B' in Figure 2. Sections through the Avan Hills ophiolite were reconstructed along the line shown in Figure 1. Samples studied are shown along the right margins of the sections. Map units (see Figure 2) are indicated along the left margins of the columns; *tp* in the western Avan Hills section is tectonized peridotite.

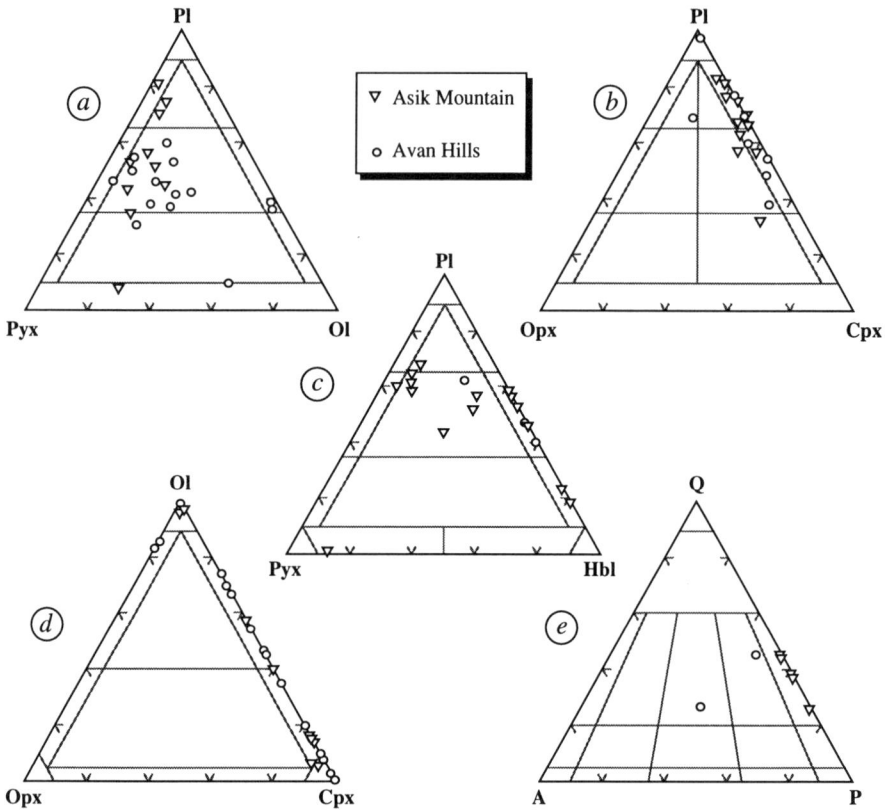

Figure 5. Ternary plots of modal compositions of Asik Mountain and Avan Hills ophiolite a) olivine gabbros, b) gabbros, c) hornblende gabbros, d) ultramafic rocks, and e) granitic rocks. Each sample is plotted on only one diagram. See text for discussion. Diagrams and field boundaries are from Streckeisen [48].

evidence of weak deformation (elongate and polycrystalline quartz and a weak hornblende lineation). Gabbro that occurs in the screens between the other intrusive rocks is nearly identical in character to the gabbro in the the massive gabbro unit suggesting that these rocks are continuous with the underlying massive gabbros, but that they have been invaded by numerous late-stage, mafic to intermediate intrusions. A sheeted dike complex is not exposed at Asik Mountain, perhaps due to removal by tectonic or erosional processes. Alternatively, sheeted dikes may not have been originally present in the section.

Ultramafic and mafic rocks exposed in the Avan Hills are similar to those at Asik Mountain, although there are several notable differences. Compared with Asik Mountain, there are extensive exposures of mantle peridotite in the Avan Hills consisting of mostly tectonized dunite and minor harzburgite. Harris [24] also reports the presence of minor lherzolite in the Avan Hills. Another difference is that rare plagioclase occurs sporadically with olivine in troctolites in the middle of the layered peridotite sequence (Figure 4), orthopyroxene is less abundant, and there is a general lack of amphibole in the gabbroic rocks. Furthermore, plagioclase in the gabbroic rocks of Avan Hills is characterized by greater anorthite contents (relative to Fo content in coexisting olivine) than at Asik Mountain. A small dike of granite (samples 43 and 44) intrudes the layered gabbro in the northern part of the Avan Hills and has been interpreted as a younger, thrust-related intrusion [25].

Petrogenesis of Western Brooks Range Ophiolites

The modal compositions of the magmatic rocks at Asik Mountain and the Avan Hills exhibit a uniform progression from olivine-rich, to olivine- and pyroxene-rich, and finally to pyroxene- and plagioclase-rich lithologies (Figure 4). These overall lithostratigraphic changes parallel the sequence of crystallization inferred from petrographic relationships. Compositional data from olivine, pyroxene, and plagioclase in the layered sequences exhibit relatively little compositional variation (Figure 6), and zoning is generally absent. Similar features were observed in the Semail ophiolite and were interpreted as evidence for crystallization from a melt of relatively uniform composition with periodic replenishment by primitive magma [49]. The upper, massive gabbroic rocks exhibit greater modal and compositional variations indicating that they crystallized from relatively closed magmatic systems that underwent more extensive fractionation.

The lithostratigraphy and petrography of the magmatic rocks at Asik Mountain and Avan Hills are similar, but there are significant differences between them and the other magmatic sequences in the western Brooks Range. Orthopyroxene is more abundant in the gabbroic rocks at Asik Mountain, as are cumulus amphibole and magnetite. Plagioclase appears fairly late in the crystallization sequence in the Asik Mountain and Avan Hills ophiolites and in most of the other ophiolite complexes in the western Brooks Range. However, in the Avan Hills, thin layers of troctolite are interlayered with dunite and olivine-clinopyroxenite in the lower part of the layered peridotite section.

Only two other ophiolite massifs, Misheguk Mountain and Siniktanneyak, in the western Brooks Range have been studied in detail. Many of the features of the ophiolite at Misheguk Mountain are similar to those of Asik Mountain and the Avan Hills. At Misheguk Mountain the residual mantle peridotites are mostly harzburgite and dunite with rare lherzolite; hornblende-bearing gabbros are common in the upper portions of the magmatic sequence [24]; orthopyroxene is rarely present at Misheguk Mountain, similar to Asik Mountain and Avan Hills. Coexisting olivine and plagioclase in the Misheguk Mountain ophiolite [24] have relatively low Fo (<85) and high An (>85) contents similar to gabbroic rocks from volcanic arcs [50]. Gabbroic rocks from the Avan Hills and Asik Mountain ophiolites have olivine and plagioclase compositions that are transitional between MORB and volcanic arcs. Harzburgite and dunite are the most abundant mantle peridotites in the ophiolite at Siniktanneyak Mountain, and hornblende is common in many of the gabbroic rocks [21].

Many aspects of the Kanuti ophiolite, located on the southern margin of the Yukon-Koyukuk Basin (Figure 1), are similar to the ophiolites of the western Brooks Range. The ophiolite overlies a thrust sheet of volcanic flows and consists of a lower, refractory, mantle suite of harzburgite and dunite and an upper magmatic suite of layered ultramafic and magmatic rocks [18]. The order of crystallization in the Kanuti ophiolite parallels those of the western Brooks Range ophiolites (ol-->cpx-->plag-->opx), and the compositions of olivine and clinopyroxene in the residual harzburgites and the cumulus ultramafic rocks are similar in the Brooks Range and Kanuti ophiolites. Furthermore, the gabbroic suites in both regions are relatively thick and orthopyroxene is relatively abundant. Loney and Himmelberg [18] suggest that these features in the Kanuti ophiolite, in conjunction with the compositions of spinels and cumulus

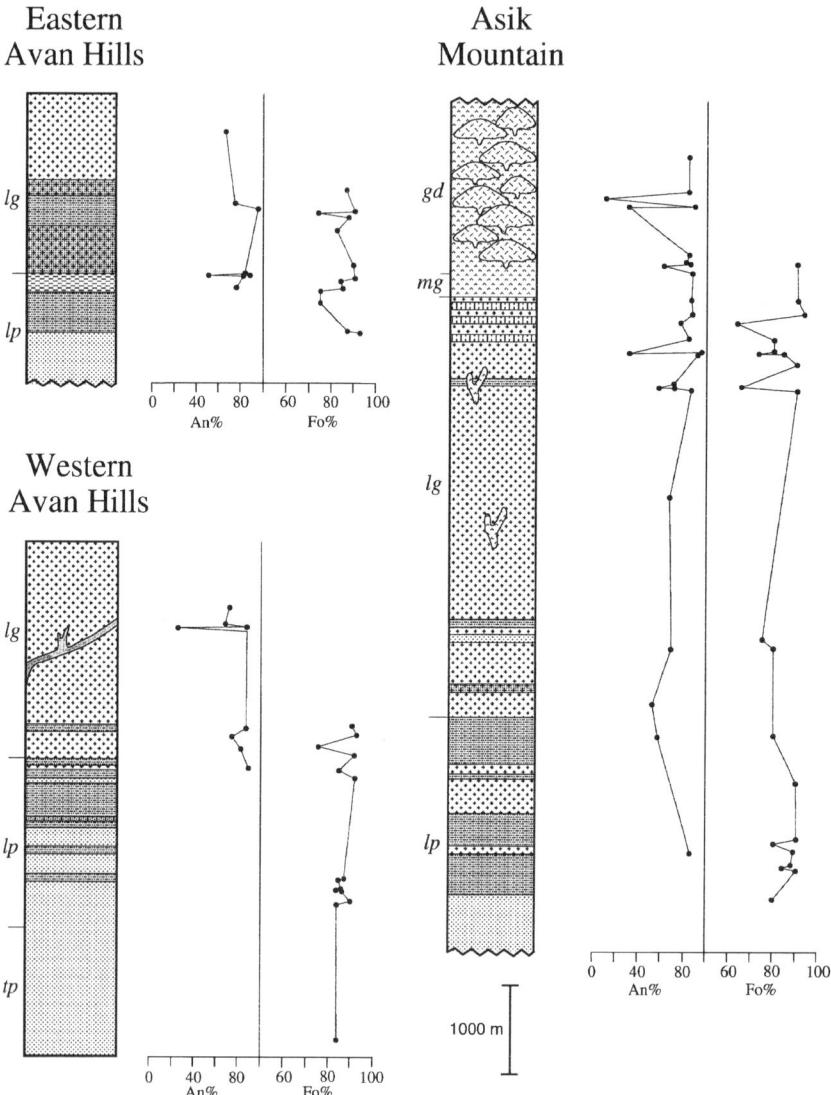

Figure 6. Olivine (Fo) and plagioclase (An) compositions along sections through the Asik Mountain and Avan Hills ophiolites. Patterns as in Figure 4.

olivine, clinopyroxene, plagioclase, and amphibole are best modelled by formation of the ophiolite in a volcanic arc environment.

The sequence of crystallization in the cumulus rocks of the Brooks Range ophiolites, and in the composition of the associated residual mantle peridotites, are most similar to the Yakuno ophiolite (Figure 7) which is considered to be intermediate in character between the Poroshiri (Liguria-type) and Horokanai (Papua-type) ophiolites in Japan [51]. These three ophiolite types are interpreted as the petrologic associations that result from progressive melting of a lherzolite source [52].

Western Brooks Range		Japan		
Asik, Avan (& Misheguk)	Ophiolite	Poroshiri (Liguria-type)	Yakuno	Horokanai (Papua-type)
not exposed	Sediment Cover	Cret.? Mudstone	Permian Mudstone	Jurassic Chert
not exposed	Volcanic Cover	Evolved MORB	Primitive MORB	Primitive MORB
not exposed	Sheeted Dikes	Absent	Absent	Absent
Clinopyroxene	Cumulate Type	Plagioclase	Clinopyroxene	Orthopyroxene
(0-0.2 wt. %)	TiO_2 in Cpx	0.8 wt. %	0.4 wt. %	0.1 wt. %
Ol Pl Cpx Opx	Crystallization Sequence	Ol Pl Cpx Opx	Ol Pl Cpx Opx	Ol Pl Cpx Opx
	Mafic Cumulates			
	Seismic Moho			
	Ultramafic Cumulates			
Cpx-bearing Harzburgite	Residual Mantle Peridotite	Lherzolite and cpx-rich Harzburgite	Cpx-bearing Harzburgite	Cpx-free Harzburgite
(0.5-1.9 wt. %)	Bulk Al_2O_3 + CaO	3-5 wt. %	1-2 wt. %	0-1 wt. %
~75 (32-75)	Cr# of Spinel	30-50	50-70	70-90
(0.46-2.7 wt. %)	Al_2O_3 of Opx	2-4 wt. %	1-2 wt. %	0-1 wt. %
(90-91)	Olivine Fo	90	90-91	92
	Other Examples	Alps Trinity	Bay of Islands Samail (Oman) / Troodos Miyamori / Vourinos Mariana	Khan Taishir Adamsfield

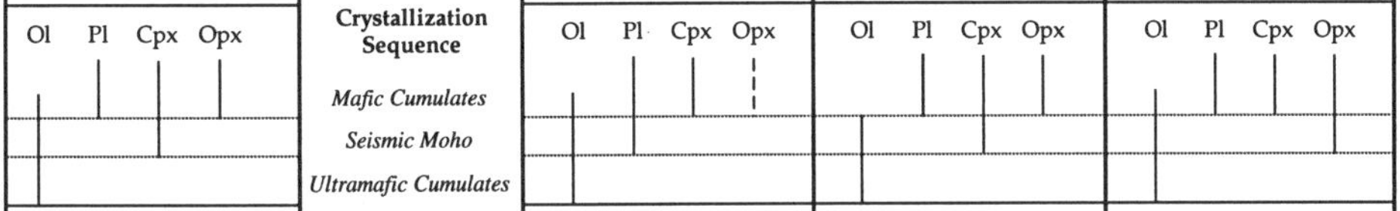

Figure 7. Comparison of Brooks Range ophiolites with three ophiolite types in Japan. Modified from Ishiwatari [51]. Data for Misheguk Mountain are from Harris [24].

The lithostratigraphic and petrographic features of the Brooks Range ophiolite complexes are interpreted to indicate crystallization from relatively large, steady-state, high-budget magma chambers, perhaps analogous to those described from the East Pacific Rise (e.g. summarized by Detrick [53]). The relatively refractory character (dunite and harzburgite) of the mantle peridotites of the Brooks Range ophiolites is similar to the composition of the mantle sequences of harzburgite ophiolite types (HOT) which are interpreted to originate along fast-spreading centers with high degrees of melting [54]. These features contrast with those observed in ophiolites interpreted to have formed along slow-spreading ridges with low degrees of melting and low magma budgets (e.g. Josephine ophiolite [55]). The sequence of crystallization (ol-->cpx-->plag) is fairly typical of ophiolites interpreted to have formed in a volcanic arc setting (e.g. Acoje block, Zambales Range [56]); the common occurrence of primary magmatic amphibole and magnetite in the cumulus sequence implies formation in a high f_{H_2O} and f_{O_2} environment such as has been inferred for supra-subduction zones [57].

MAFIC VOLCANIC ROCKS

Apparently, all of the ultramafic-mafic complexes in the western Brooks Range structurally overlie mafic volcanic flows. There are several additional exposures of mafic volcanic rocks in the Brooks Range (Figure 1) that are not overlain by ultramafic-mafic complexes (e.g. Maiyumerak Mountains, Copter Peak, Angayucham Mountains, Cathedral Mountain, and along the Porcupine River). It is not known if those mafic volcanics not overlain by ultramafic-mafic complexes are more deeply eroded or if they had different structural histories. Although not complete ophiolites by definition [58], the volcanic rocks of the Brooks Range clearly originated within oceanic settings and are considered to be detached fragments of the upper portions of oceanic lithosphere.

The ages of the mafic volcanic rocks in the western Brooks Range are not well constrained; attempts to date the basalts using Sm-Nd and Rb-Sr isotopes have been unsuccessful (Wirth, unpublished data). Structural relationships require that the volcanics are older than ~165 Ma, the time of thrusting and metamorphism. Mesozoic radiolaria have been observed in chert within mafic rocks in the eastern Avan Hills [27]. Similarly, chert layers within the mafic volcanic rocks at Copter Peak [38] and Misheguk Mountain [34] contain Triassic radiolaria. Blocks of limestone in volcanic flows beneath the mafic and ultramafic rocks in the Siniktanneyak Hills contain Frasnian (lower Upper Devonian) stromatoporoids and corals [21], and Devonian to Mesozoic ages were reported for the mafic volcanic rocks in the Maiyumerak Mountains [40].

Field Relations
Mafic volcanic rocks in the western Brooks Range (Figure 1) are typically variably metamorphosed and deformed. Pillow structures are common and flows typically consist of plagioclase- and clinopyroxene-bearing basalt. Olivine phenocrysts are not common, but have been observed in some regions (e.g. Avan Hills). Equilibrium metamorphic mineral assemblages vary locally and indicate that the basalts attained peak metamorphic conditions ranging from prehnite-pumpellyite to upper greenschist facies. Disequilibrium mineral assemblages indicate that some metamorphic reactions are locally incomplete. Zones of intense hydrothermal alteration, silicification, and sulfide mineralization also occur locally. Layer 1 sediments have not been recognized in association with the volcanic rocks, but some volcanic sequences contain minor amounts of inter-flow sediment. Inter-pillow chert, thin layers of inter-flow chert, and volcaniclastic sediment have been observed in the Maiyumerak Mountains, the Avan Hills, and Asik Mountain. Minor carbonate, chert, and shale have been observed near Copter Peak and in the Siniktanneyak Hills. In regions where the volcanic rocks are structurally overlain by ultramafic-mafic complexes, metamorphic rocks are commonly present along the fault surfaces and are interpreted to be metamorphosed sediments. Intermediate to mafic intrusive rocks occur locally within the volcanic sequence [27]. These rocks have textures that range from medium-grained and porphyritic to fine-grained and equigranular and are classified as gabbro and diorite.

Mafic dikes are not generally exposed in most of the plutonic and volcanic rocks in the Brooks Range. The only exposure of sheeted dikes is in the southwestern Maiyumerak Mountains where porphyritic basaltic dikes intrude flows. The dikes are generally tabular and range from 0.5 meters to over 10 meters in width. Evidence for multiple intrusion of the dikes comes from the relations of chilled zones on the margins of the dikes. Basaltic dikes also occur within the gabbro sequence at Siniktanneyak Mountain. Locally these dikes comprise as much as 10-20% of the total rock. The dikes have thin (~1-5 mm thick) chill-margins and have not appreciably metamorphosed the surrounding gabbro host-rock. Roeder and Mull [15] suggested that the dikes might be analogous to the lower portion of a sheeted-dike unit of an ophiolite. However, Nelson and Nelson [21] later concluded that the dikes lack asymmetric chilling and are younger than the main part of the ophiolite.

Basalt, basaltic tuff, and radiolarian chert [14] are exposed along the southern margin of the Brooks Range in a nearly continuous belt, the Angayucham terrane. Principal exposures of these rocks (Figure 1) are along the Porcupine River [59, 60], in the Cathedral Mountains [37], and in the Angayucham Mountains [17, 46, 41]. The basalts along the Porcupine River are unique because they are apparently interlayered with sedimentary rock. At one locality, cobble-size fragments of basalt, chert, and limestone occur in a matrix of immature sandstone [59]; cobbles of limestone have also been reported to occur in the basalt. In thin-section, the basalts consist of primary phenocrysts of pyroxene and plagioclase and are remarkably undeformed and unmetamorphosed. A sample of massive basalt yielded a Late Triassic date (225.7 ± 0.6 Ma; Wirth and Bird, unpublished $^{40}Ar/^{39}Ar$ whole-rock date).

Geochemistry
In this section, the primary (magmatic) geochemical features of the Brooks Range volcanics are summarized and compared with those of rocks from various tectonic settings. A detailed discussion of the analytical methods and geochemistry of the volcanic rocks is given by Wirth [45] and Wirth et al. [in prep.]. Summarized below are the results of more than 120 major and trace element whole-rock analyses from ten exposures throughout the Brooks Range. The results of these studies are also compared with those from previous geochemical studies of mafic volcanic rocks in the Angayucham Terrane by Barker et al. [39] and Pallister et al. [41].

Data from the Brooks Range volcanic rocks were plotted on numerous geochemical variation diagrams, the results of which are illustrated (Figure 8) on a diagram of Th versus Ta and Hf [61]. Undoubtedly, some of the scatter of the data on this and other trace element plots is due to post-magmatic alteration and metamorphism. However, the relatively tight clusters of most of the data on these diagrams, coupled with the consistent tectonic classifications of the volcanic rocks on plots of different elements, suggest that the abundances of the less-mobile trace elements have remained largely unaffected by post-magmatic processes. The rare-earth element patterns of the volcanic rocks also differ significantly from each other (Figure 9). These features suggest that the observed trace element abundances are largely due to differences in source region composition and therefore provide clues to the tectonic origins of the rocks.

The variety of geologic and geochemical characteristics exhibited by the volcanic rocks in the western Brooks Range indicate that they originated in diverse tectonic settings. Of the rocks we studied, the basalts from Avan Hills, Copter Peak, Cathedral Mountain, Iyikrok Mountain, and Misheguk Mountain are characterized by relatively high LREE and large-ion lithophile element (LILE) abundances. Samples from Avan Hills, Copter Peak, Iyikrok Mountain, and Misheguk Mountain also have high TiO_2 (>1.70 wt. %) and P_2O_5 (>0.20 wt. %) similar to E-MORB and within-plate basalt (Figures 8 and 9). Of these, the Avan Hills, Iyikrok Mountain, and Copter Peak basalts are most similar geochemically to the Jurassic basalts of the Angayucham Terrane. The presence of interlayered chert and basaltic tuff (Avan Hills and Copter Peak) and the tholeiitic to alkaline compositions of basalts from these regions are similar to those observed in Jurassic volcanics of the Angayucham Terrane [39, 41]. Basaltic flows at Misheguk Mountain exhibit trace element ratios (e.g. Figure 9) similar to those of the Jurassic volcanics of the Angayucham terrane [41], but have stronger within-plate enrichment (e.g. Th, Ta, and LREE).

Rare earth (REE) and trace element data for the Cathedral Mountains volcanics are similar to those further east in the Angayucham terrane [39] and to those of the Triassic basalts of the Angayucham Mountains [41]. The slight enrichments in the LREE, Ta, Nb, Th, and LILE in

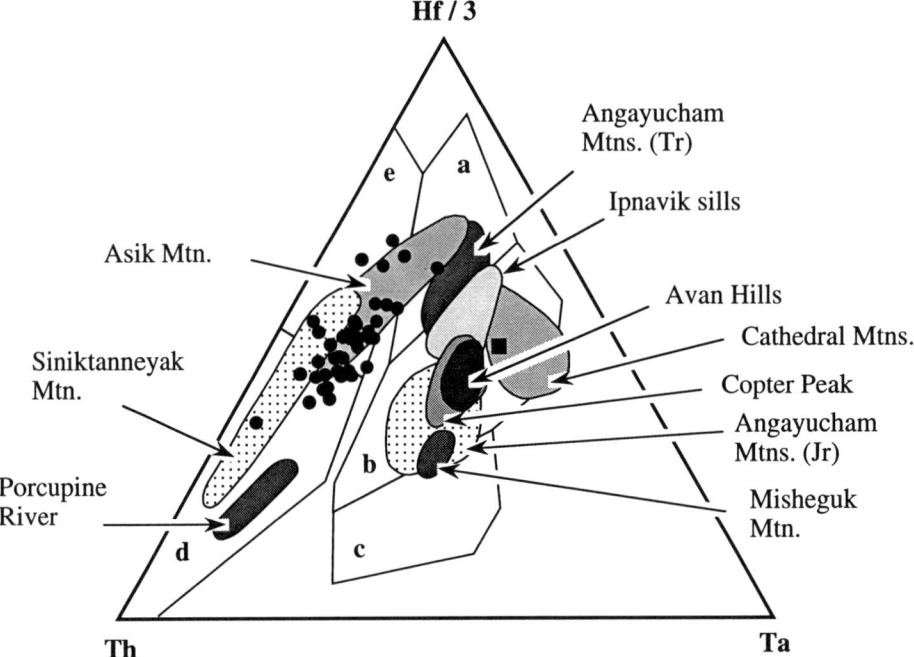

Figure 8. Th-Ta-Hf diagram of Brooks Range volcanic rocks [61]. Tectonic field boundaries are: (a) N-MORB, (b) E-MORB, (c) within-plate basalt, (d) calc-alkaline basalt from convergent margins, and (e) tholeiitic basalt from convergent margins. The compositions of Maiyumerak Mountain flows and dikes are shown with filled circles; flows from Iyikrok Mountain are shown with a filled square. Data from the Angayucham terrane are from Pallister et al. [41].

the Cathedral flows relative to MORB (Figures 8 and 9) suggest that they are transitional between N-MORB and E-MORB.

Mafic flows of the Maiyumerak Mountains are different from those described in the Angayucham terrane. The thick section of sheeted dikes within the section of volcanic flows is unique among the volcanic rocks exposed in northern Alaska. Furthermore, the relative Ta, Nb, Hf, Zr, and Ti depletions that characterize the Maiyumerak basalts have not previously been observed in the Brooks Range. The Maiyumerak volcanics are characteristically enriched in the LILE and are depleted in the high field-strength elements (HFSE), features that typify magmas erupted along convergent plate margins (e.g. Pearce et al. [57]). The trace element data, in combination with the presence of sheeted dikes and interlayered basaltic tuffs, indicate that the most likely tectonic setting for the formation of the Maiyumerak Mountain volcanic rocks is a back-arc basin or volcanic arc. The common occurrence of pillowed flows with minor interlayered chert, and the lack of continentally-derived sediment, indicate formation in an intra-oceanic setting relatively distant from a continental margin.

The mafic dikes that intrude the layered gabbros at Siniktanneyak Mountain have trace element abundances (Figures 8 and 9) that are like those of some of the Maiyumerak Mountains basalts, suggesting that they originated from an analogous mantle source. The supra-subduction character of the mafic dikes contrasts sharply with the E-MORB character of the basalt flows that structurally underlie the gabbros at Siniktanneyak Mountain [38]. The dikes were interpreted by Nelson and Nelson [21] to be much younger than the layered gabbro

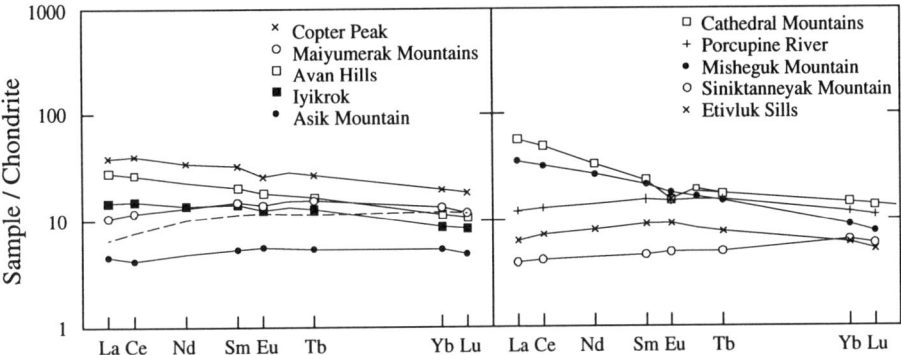

Figure 9. Representative rare earth element patterns of Brooks Range volcanics compared with N-MORB (dashed line; values are from Sun and McDonough [62]). All samples are from volcanic flows except those from Siniktanneyak Mountain dikes and Etivluk sills. Data are normalized to chondrite values of Masuda et al. [63].

portion of the ophiolite, and may have originated as the previously-formed ophiolite was transported through a convergent margin system.

Mafic volcanic flows that occur structurally beneath the gabbros and peridotites at Asik Mountain are variably enriched in the LREE (Figure 9) and LILE, and are relatively depleted in the HFSE. On tectonic discriminant diagrams, the basalts plot mostly in the fields of N-MORB and island arc basalt (Figure 8).

Of the rocks that we studied, basaltic rocks along the Porcupine River (southeastern Brooks Range) exhibit the most unequivocal evidence of formation in a volcanic arc. The common occurrence of pillow basalt indicates eruption in a subaqueous environment and the association of chert, argillite, graywacke, conglomerate, and limestone [64] suggest these basalts are mostly oceanic in origin. The tholeiitic to calc-alkaline composition of the basalts is typical of magmas erupted along convergent plate margins. The supra-subduction zone composition of these lavas is distinct from the within-plate compositions of the similar age Angayucham Terrane basalts in the Cathedral Mountains, and elsewhere in the Brooks Range [37, 39, 41].

Post-Carboniferous mafic sills and dikes that intruded an underlying thrust sheet of chert and shale [65] have REE-patterns similar to MORB, and trace element patterns transitional between N-MORB and E-MORB. On trace element diagrams these rocks consistently plot along with Triassic Angayucham basalt in the fields of N-MORB and E-MORB, or within-plate basalt. The geochemistry of the dikes and sills indicates that the magmas that were intruded into the Arctic Alaska passive margin during the Late Carboniferous through Early Jurassic and were probably derived from E-MORB type mantle.

DISCUSSION

Features that are characteristic of the Brooks Range oceanic rocks include: relatively thick and coherent gabbro-peridotite sequences; a general lack of sheeted dikes associated with volcanic flows or with the upper portions of gabbroic complexes; mafic volcanic rocks with highly variable lithostratigraphies and compositions; and ophiolitic fragments that have undergone long histories (>40 m.y.) of intra-oceanic transport. Several of these aspects support the model of circum-Pacific ophiolites [66] which typically lack sheeted dikes, are highly dismembered, and have several ophiolitic nappes of widely varying age juxtaposed within a single orogenic belt. However, unlike ophiolites elsewhere around the Pacific, the ages of the underlying volcanic rocks in the Brooks Range are generally known or are interpreted to be older than the ages of the overlying ophiolites.

Ideally, the reconstruction of the original tectonic environment of an ophiolite terrane would include lithostratigraphic, structural, and geochemical information. Although limited, the existing data for northern Alaska clearly indicate considerable compositional and lithostratigraphic differences among the Brooks Range ophiolites. Regardless of the specific origins attributed to each of the oceanic rocks in the Brooks Range, new geochemical and lithostratigraphic data indicate that these oceanic rocks originated within very dissimilar tectonic settings. The long transport history and the diversity of crustal compositions preserved in the Brooks Range ophiolites are both interpreted to indicate that the fragments of oceanic lithosphere originated in a variety of tectonic settings within a large ocean basin (e.g. Pacific Ocean). This model of ophiolite genesis contrasts with the more common models of ophiolite formation in marginal basins (e.g. Josephine ophiolite) or in relatively small ocean basins (e.g. Semail ophiolite).

Many ophiolites overlie allochthonous volcanic rocks that are interpreted to have originated in N-MORB, E-MORB, and within-plate settings. In the case of ophiolites emplaced along "dip-slip-dominated" margins, volcanic and sedimentary sequences are often fault-bounded or occur in mélange below the metamorphic sole [67]. Blocks of within-plate lava have been recognized structurally beneath the Ballantrae ophiolite [68] and in mélange units structurally beneath the Troodos ophiolite [69], the Semail ophiolite [70, 71], and the Newfoundland ophiolites (see references cited in Searle and Stevens [72]). Similar rocks also occur in the Franciscan Complex, a strike-slip-dominated setting, and were interpreted by Shervais and Kimbrough [73] as tectonic inclusions of within-plate volcanics that were preferentially preserved during subduction.

Similarly, most of the allochthonous volcanic rocks underlying the Brooks Range ophiolites have stratigraphic, structural, or geochemical features suggesting that they originated in intraplate or supra-subduction zone settings. Harris [74] suggested that these Brooks Range volcanics might be analogous to those in mélange complexes beneath Tethyan-type ophiolites, which have largely been interpreted to be of continental margin origin. However, the allochthonous volcanic rocks in the Brooks Range occur as large (>10 km in extent), coherent thrust sheets and may not be strictly analogous with those in the Tethyan ophiolites which occur as relatively small blocks in mélange. Small blocks (<1 km) of volcanic rock of unknown origin are present in mélange units in the Brooks Range thrust belt [75], but these occur largely in the northern foothills and are structurally distinct from the volcanic thrust sheets underlying the ophiolites. There is also insufficient evidence of continentally-derived sediment associated with the Brooks Range volcanics or in the underlying distally-derived thrust sheets. Lastly, many Tethyan ophiolites contain evidence of relatively short transport histories (<10 m.y.) while geochronologic data for the Brooks Range ophiolites strongly suggest a much longer duration (~40 m.y.), and presumably a much greater distance, of transport.

Western Brooks Range gabbro-peridotite complexes originated primarily within the upper plate of convergent plate-margin systems, whereas the associated underlying mafic volcanic rocks originated as fragments of anomalous (both topographically and compositionally) oceanic lithosphere that were preferentially detached and emplaced during Jurassic - Early Cretaceous subduction along the southern margin of Arctic Alaska. Similar relationships are described by Malpas et al. [this volume] for ophiolitic rocks in New Zealand. The geochemical, lithostratigraphic, structural, and temporal relationships of the Brooks Range volcanics imply that they originated within a large region of oceanic lithosphere. Apparently, however, much of the normal MORB-type lithosphere that was generated along accreting plate margins within the inferred ocean basin was not preserved, but was subducted during convergence.

Acknowledgements
This study was supported by grants from the National Aeronautics and Space Administration (NAS5-28739 and NAGW-1287) and by the Wallace Fund at Macalester College. We are grateful for reviews of the manuscript and improvements suggested by A. Ishiwatari and A. Robertson. The authors also wish to acknowledge the U.S. Park Service for their permission to collect samples within the Noatak National Preserve.

References
1. K.R. Wirth and J.M. Bird. Chronology of ophiolite crystallization, detachment, and emplacement: Evidence from the Brooks Range, Alaska, *Geology.* **20**, 75-78 (1992).
2. I.L. Tailleur and W.P. Brosgé. Tectonic history of northern Alaska. In: *Proceedings of the Geological Seminar on the North Slope of Alaska.* W.L. Adkison and W.P. Brosgé (Eds). pp. E1-E19, American Association of Petroleum Geologists Pacific Section, Menlo Park, CA (1970).
3. M. Churkin, Jr., W.J. Nokleberg and J. Huie. Collision deformed Paleozoic continental margin, western Brooks Range, Alaska, *Geology.* **7**, 379-383 (1979).
4. C.G. Mull. The tectonic evolution and structural style of the Brooks Range, Alaska: An illustrated summary. In: *Geologic studies of the Cordilleran thrust belt.* R.B. Powers (Eds). vol. 1, pp. 1-45. Rocky Mountain Association of Geologists (1982).
5. C.F. Mayfield, I.L. Tailleur and I. Ellersieck. Stratigraphy, structure, and palinspastic synthesis of the western Brooks Range, northwestern Alaska. *U.S. Geol. Surv. Open File Report* OF 83-779 (1983).
6. S.E. Box. Early Cretaceous orogenic belt in northwestern Alaska: Internal organization, lateral extent, and tectonic interpretation. In: *Tectonostratigraphic Terranes of the Circum-Pacific Region.* D.G. Howell (Ed.). Circum-Pacific council for Energy and Mineral Resources, Houston, Texas (1985).
7. W.W. Patton, Jr. and S.E. Box. Tectonic setting of the Yukon-Koyukuk Basin and its borderlands, western Alaska, *J. Geophys. Res.* **94**, 15,807-15,820 (1989).
8. W.W. Patton, Jr. and E.J. Moll. Reconnaissance geology of the northern part of the Unalakleet quadrangle. In: *U. S. Geological Survey in Alaska - Accomplishments During 1981.* W.L. Coonrad and R.L. Elliott (Eds). U.S. Geol. Surv. Circ. 868, pp. 24-27 (1984).
9. W.W. Patton, Jr. and E.J. Moll. Geologic map of northern and central parts of Unalakleet Quadrangle, Alaska. *U.S. Geol. Surv. Misc. Field Stud. Map* MF-1749, scale 1:250,000 (1985).
10. W.W. Patton, Jr. Reconnaissance geology of the northern Yukon-Koyukuk Province, Alaska. *U.S. Geol. Surv. Prof. Pap.* 774-A. pp. 1-17 (1973).
11. S.E. Box and W.W. Patton, Jr. Igneous history of the Koyukuk Terrane, western Alaska: Constraints on the origin, evolution, and ultimate collision of an accreted island arc terrane, *J. Geophys. Res.* **94**, 15,843-15,867 (1989).
12. A.J. Martin. Structure and tectonic history of the western Brooks Range, De Long Mountains and Lisburne Hills, northern Alaska, *Geol. Soc. Am. Bull.* **81**, 3605-3622 (1970).
13. I.L. Tailleur. Possible mantle-derived rocks in western Brooks Range. In: *Geological Survey Research 1973.* U.S. Geol. Surv. Prof. Pap. 850, pp. 64-65 (1973).
14. W.W. Patton, Jr., I.L. Tailleur, W.P. Brosgé and M.A. Lanphere. Preliminary report on the ophiolites of northern and western Alaska. In: *North American Ophiolites.* R.G. Coleman and W. P. Irwin (Eds). Oregon Department Geology and Mining Industry Bulletin 95, pp. 51-57 (1977).
15. D. Roeder and C.G. Mull. Tectonics of the Brooks Range ophiolites, Alaska, *Am. Assoc. Petr. Geol. Bull.* **62**, 1696-1713 (1978).
16. D.L. Jones, N.J. Silberling, P.J. Coney and G. Plafker. Lithotectonic terrane map of Alaska (west of the 141st Meridian). In: *Lithotectonic Terrane Maps of the North American Cordillera.* N.J. Silberling and D.L. Jones (Eds). U.S. Geol. Surv. Map MF-1874-A (1987).
17. M.W. Hitzman, T.E. Smith and J.M. Proffett, Jr. Bedrock geology of the Ambler district, southwestern Brooks Range, Alaska, *Alaska Division of Geological and Geophysical Surveys Geologic Report* 75, scale 1:125,000 (1982).

18. R.A. Loney and G.R. Himmelberg. The Kanuti Ophiolite, Alaska, *J. Geophys. Res.* **94**, 15,869-15,900 (1989).
19. J.W. Cady. Geologic implications of topographic, gravity, and aeromagnetic data in the northern Yukon-Koyukuk province and its borderlands, Alaska, *J. Geophys. Res.* **94**, 15,821-15,841 (1989).
20. E.L. Miller and T.L. Hudson. Mid-Cretaceous extension of a Jurassic-Early Cretaceous compressional orogen, Alaska, *Tectonics.* **10**, 781-796 (1991).
21. S.W. Nelson and W.H. Nelson. Geology of the Siniktanneyak Mountain ophiolite, Howard Pass quadrangle, Alaska. *U.S. Geol. Surv. Misc. Field Stud. Map* MF-1441, scale 1:63,000 (1982).
22. J.M. Boak, D.L. Turner, D.J. Henry, T.E. Moore and W.K. Wallace. Petrology and K-Ar ages of the Misheguk Igneous Sequence - an allochthonous mafic and ultramafic complex - and its metamorphic aureole, western Brooks Range, Alaska. In: *Alaska North Slope Geology.* I.L. Tailleur and P. Weimer (Eds). Pacific Section, Society of Economic Paleontologists and Mineralogists, Bakersfield, California (1987).
23. K.R. Wirth, D.J. Harding, A.E. Blythe and J.M. Bird. Brooks Range ophiolite crystallization and emplacement ages from $^{40}Ar/^{39}Ar$ data, *Geol. Soc. Am. Abstr. Programs.* **18**, 792 (1986).
24. R.A. Harris. *Processes of allochthon emplacement with special reference to the Brooks Range ophiolite, Alaska and Timor, Indonesia.* Unpublished PhD Dissertation. University of London, Great Britain (1989).
25. K.R. Wirth, J.M. Bird, A.E. Blythe, D.J. Harding and M.T Heizler. Age and tectonic evolution of western Brooks Range ophiolites, Alaska: Results from $^{40}Ar/^{39}Ar$ thermochronometry, *Tectonics.* **12**, 410-432 (1993).
26. S.M. Curtis, I. Ellersieck, C.F. Mayfield and I.L. Tailleur. Reconnaissance geologic map of southwestern Misheguk Mountain quadrangle, Alaska. *U.S. Geol. Surv. Open-File Report* OF 82-611, scale 1:63,360 (1982).
27. S.M. Curtis, I. Ellersieck, C.F. Mayfield and I.L. Tailleur. Reconnaissance geologic map of southwestern Misheguk Mountain quadrangle, Alaska. *U.S. Geol. Surv. Misc. Invest. Map* I-1502, scale 1:63,360 (1984).
28. J. Zimmerman, C.O. Frank and S. Bryan. Mafic rocks in the Avan Hills ultramafic complex, De Long Mountains. In: *The U.S. Geological Survey in Alaska: Accomplishments during 1979.* N.R.D. Alber and T. Hudson (Eds). U.S. Geol. Surv. Circ. 823-B, pp. 14-15 (1981).
29. C.O. Frank and J. Zimmerman. Petrography of nonultramafic rocks from the Avan Hills complex, De Long Mountains, Alaska. In: The *United States Geological Survey in Alaska: Accomplishments during 1980.* W.L. Coonrad (Eds). U.S. Geol. Surv. Circ. 844, pp. 22-27 (1982).
30. J. Zimmerman and P.G. Soustek. The Avan Hills ultramafic complex, De Long Mountains, Alaska. In: *The U.S. Geological Survey in Alaska: Accomplishments during 1978.* K.M. Johnson and J.R. Williams (Eds). U.S. Geol. Surv. Circ. 804-B, pp. 8-11 (1979).
31. W. Hamilton. Book Review: Tailleur, I. and Wiemer, P., editors, Alaskan North Slope Geology, *EOS (American Geophysical Union Transactions).* **69**, 869 (1988).
32. I. Ellersieck, S.M. Curtis, C.F. Mayfield and I.L. Tailleur. Reconnaissance geologic map of south-central Misheguk Mountain quadrangle, Alaska. *U.S. Geol. Surv. Open-File Report* OF 82-612, scale 1:63,360 (1982).
33. J. Zimmerman and C.O. Frank. Possible obduction-related metamorphic rocks at the base of the ultramafic zone, Avan Hills complex, De Long Mountains. In: *The U.S. Geological Survey in Alaska: Accomplishments During 1980.* W.L. Coonrad (Ed.). U.S. Geol. Surv. Circ. 844, pp. 27-28 (1982).
34. I. Ellersieck, S.M. Curtis, C.F. Mayfield and I.L. Tailleur. Reconnaissance geologic map of south-central Misheguk Mountain quadrangle, Alaska. *U.S. Geol. Surv. Misc. Invest. Series Map* I-1504, scale 1:63,360 (1984).
35. R.A. Harris. Structural relations of the Misheguk Mountain ophiolite complex, western Brooks Range, Alaska, *Terra Cognita.* **7**, 314 (1987).
36. R.A. Harris. Origin, emplacement, and attenuation of the Misheguk Mountain allochthon, western Brooks Range, Alaska, *Geol. Soc. Am. Abstr. Programs.* **20**, A112 (1988).
37. R.R. Gottschalk, Jr. *Structural and petrologic evolution of the southern Brooks Range near Wiseman, Alaska.* Unpublished PhD dissertation. Rice University, Houston, Texas (1987).
38. T.E. Moore. Geochemical and tectonic affinity of basalts from the Copter Peak and Ipnavik River allochthons, Brooks Range, Alaska, *Geol. Soc. Am. Abstr. Programs.* **19**, 434 (1987).
39. F. Barker, D.L. Jones, J.R. Budahn and P.J. Coney. Ocean plateau-seamount origin of basaltic rocks, Angayucham Terrane, central Alaska, *J. Geol.* **96**, 368-374 (1988).
40. S.M. Karl and C.F. Dickey. Geology and geochemistry indicate belts of both ocean floor and arc basalt and gabbro in the the Maiyumerak Mountains, northwestern Brooks Range, Alaska, *Geol. Soc. Am. Abstr. Programs.* **21**, 100 (1989).
41. J.S. Pallister, J.R. Budahn and B.L. Murchey. Pillow basalts of the Angayucham terrane: Oceanic plateau and island crust accreted to the Brooks Range, *J. Geophys. Res.* **94**, 15,901-15,923 (1989).
42. K.R. Wirth, D.J. Harding and J.M. Bird. Basalt geochemistry, Brooks Range, Alaska, *Geol. Soc. Am. Abstr. Programs.* **19**, 464 (1987).
43. K.R. Wirth, J.M. Bird and M.M. Cheatham. Geochemistry of western Brooks Range basalt, Alaska, *Geol. Soc. Am. Abstr. Programs.* **21**, 160-161 (1989).
44. S.M. Karl and C.L. Long. Folded Brookian thrust faults: Implications of three geological/geophysical transects in the Western Brooks Range, Alaska, *J. Geophys. Res.* **95**, 8,581-8,592 (1990).

45. K.R. Wirth. *Processes of lithosphere evolution: Geochemistry and tectonics of mafic rocks in the Brooks Range and Yukon-Tanana region, Alaska.* Unpublished Ph.D. Dissertation. Cornell University, Ithaca, NY (1991).
46. M.W. Hitzman, J.M. Proffett, Jr., J.M. Schmidt and J.M. Smith. Geology and mineralization of the Ambler District, northwestern Alaska, *Econ. Geol.* **81**, 1592-1618 (1986).
47. K.J. Bird. Late Paleozoic carbonates from the south-central Brooks Range. In: *U.S. Geological Survey in Alaska: Accomplishments During 1976.* K.M Blean (Eds). U.S. Geol. Surv. Circ. 751-B, pp. 19-20 (1977).
48. A. Steckeisen. To each plutonic rock its proper name, *Earth Sci. Rev.* **12**, 1-33 (1976).
49. J.S. Pallister and C.A. Hopson. Samail ophiolite plutonic suite: field relations, phase variation, cryptic variation and layering, and a model of a spreading ridge magma chamber, *J. Geophys. Res.* **86**, p. 2593-2644 (1981).
50. J.S. Beard. Characteristic mineralogy of arc-related cumulate gabbros: Implications for the tectonic setting of gabbroic plutons and for andesite genesis, *Geology.* **14**, 848-851 (1986).
51. A. Ishiwatari. Ophiolites in the Japanese islands: Typical segment of the circum-Pacific multiple ophiolite belts, *Episodes.* **14**, 274-279 (1991).
52. A. Ishiwatari. Igneous petrogenesis of the Yakuno Ophiolite (Japan) in the context of the diversity of ophiolites, *Cont. Min. Petrol.* **89**, 155-167 (1985).
53. R.S. Detrick. Ridge crest magma chambers: A review of results from marine seismic experiments at the East Pacific Rise. In: *Ophiolite Genesis and Evolution of the Oceanic Lithosphere.* T. Peters, A. Nicolas and R.G. Coleman (Eds). Petrology and Structural Geology 5. pp. 7-20. Kluwer Academic Publishers, Dordrecht, Netherlands (1991).
54. A. Nicolas. *Structures of Ophiolites and Dynamics of Oceanic Lithosphere.* Petrology and Structural Geology 4. Kluwer Academic Publishers, Dordrecht, Netherlands (1989).
55. G.D. Harper. Episodic magma chambers and amagmatic extension in the Josephine ophiolite, *Geology.* **16**, 831-834 (1988).
56. J.W. Hawkins and C.A. Evans. Geology of the Zambales Range, Luzon, Philippine Islands: Ophiolite derived from an island arc - back-arc basin pair. In: *The Tectonic and Geologic Evolution of Southwest Asian Seas and Islands.* D.E. Hayes (Ed.). Geophys. Monogr. Ser., vol. 27, pp. 95-123, American Geophysical Union, Washington D.C. (1983).
57. J.A. Pearce, S.J. Lippard and S. Roberts. Characteristics and tectonic significance of supra-subduction zone ophiolites. In: *Marginal Basin Geology.* B.P. Kokelaar and M.F. Howells (Eds). Geol. Soc. of London Spec. Publ. 16, pp. 77-94, Blackwell Scientific Publications, Oxford (1984).
58. Anonymous. Report on the Penrose Field Conference on ophiolites, *Geotimes.* **17**, 24-25 (1972).
59. W.P. Brosgé, H.N. Reiser, J.T. Dutro and M. Churkin. Geologic Map and Stratigraphic Sections, Porcupine River Canyon, Alaska. U.S. Geol. Surv. Open File Map 66-263 (1966).
60. W.P. Brosgé and H.N. Reiser. Preliminary Geologic Map of the Coleen Quadrangle, Alaska. U.S. Geol. Surv. Open File Map 69-370 (1969).
61. D.A. Wood, J.L. Joron and M. Treuil. A re-appraisal of the use of trace elements to classify and discriminate between magma series erupted in different tectonic settings, *Earth Planet. Sci. Lett.* **45**, 326-336 (1979).
62. S.-S. Sun and W.F. McDonough. Chemical and isotopic systematics of oceanic basalts: implications for mantle composition and processes. In: *Magmatism in the Ocean Basins.* A.D. Saunders and M.J. Norry (Eds). Geol. Soc. of London Spec. Publ. 42, pp. 315-345, Blackwell Scientific Publications, Oxford (1989).
63. A. Masuda, N. Nakamura and T. Tanaka. Fine structure of mutually normalized rare-earth patterns of chondrites, *Geochim. Cosmochim. Acta.* **37**, 239-248 (1973).
64. D.L Jones and N.J. Silberling. Lithotectonic terrane map of the North American Cordillera. In: *Lithotectonic Terrane Maps of the North American Cordillera.* D.L. Jones and N.J. Silberling (Eds). U.S. Geol. Surv. Open-File Report 84-523 (1984).
65. C.G. Mull, I.L. Tailleur, C.F. Mayfield, I. Ellersieck and S. Curtis. New upper Paleozoic and lower Mesozoic stratigraphic units, central and western Brooks Range, Alaska, *Am. Assoc. Petr. Geol. Bull.* **66**, 348-362 (1982).
66. A. Ishiwatari. Time-space distribution and petrologic diversity of Japanese ophiolites. In: *Ophiolite Genesis of the Oceanic Lithosphere.* Peters et al. (Eds) Ministry of Petroleum and Minerals, Sultanate of Oman (1991).
67. N.H. Woodcock and A.H.F. Robertson. The structural variety in Tethyan ophiolite terrains. In: *Ophiolites and Oceanic Lithosphere.* I.G. Gass, S.J. Lippard and A.W. Shelton (Eds). Geol. Soc. London Spec. Publ. 13, pp. 321-330. Geological Society, London (1984).
68. M.F. Thirwall and B.J. Bluck. Sr-Nd isotope and chemical evidence that the Ballantrae 'ophiolite', SW Scotland, is polygenetic. In: *Ophiolites and Oceanic Lithosphere.* I.G. Gass, S.J. Lippard, and A.W. Shelton (Eds). Geol. Soc. London Spec. Publ. 13, pp. 215-230. Geological Society, London (1984).
69. J.A. Pearce. Basalt geochemistry used to investigate past tectonic environments on Cyprus, *Tectonophysics.* **25**, 41-67 (1975).
70. M.P. Searle, S.J. Lippard, J.D. Smewing and D.C. Rex. Volcanic rocks beneath the Semail Ophiolite nappe in the northern Oman Mountains and their significance in the Mesozoic evolution of Tethys, *J. geol. Soc. London* **137**, 589-604 (1980).
71. M.P. Searle and G.M. Graham. "Oman exotics" - Oceanic carbonate build-ups associated with the early stages of continental rifting, *Geology.* **10**, 43-49 (1982).

72. M.P. Searle and R.K. Stevens. Obduction processes in ancient, modern and future ophiolites. In: *Ophiolites and Oceanic Lithosphere*. I.G. Gass, S.J. Lippard and A.W. Shelton (Eds). Geol. Soc. London Spec. Publ. 13, pp. 303-319. Geological Society, London (1984).
73. J.W. Shervais and D.L. Kimbrough. Alkaline and transitional subalkaline metabasalts in the Franciscan Complex melange, California. In: *Mantle Metasomatism and Alkaline Magmatism*. E.M. Morris and J.D. Pasteris (Eds). Geol. Soc. Am. Spec. Pap. 215, pp. 165-182. Geological Society of America, Boulder, Colorado (1987).
74. R.A. Harris. Peri-collisional extension and the formation of Oman-type ophiolites in the Banda arc and Brooks Range. In: *Ophiolites and their Modern Oceanic Analogues*. L.M. Parson, B.J. Murton and P. Browning (Eds.). Geol. Soc. of London Spec. Publ. 60, pp. 301-325, Blackwell Scientific Publications, Oxford (1992).
75. R.C. Crane. Cretaceous olistostrome model, Brooks Range, Alaska. In: *Alaska North Slope Geology*. I.L. Tailleur and P. Weimer (Eds). Pacific Section, Society of Economic Paleontologists and Mineralogists, Bakersfield, California (1987).

Tectonic Position of the Mesozoic Ophiolitic and Island Arc Formations in the Koryak Region (Northeastern Russia)

N.FILATOVA and V.VISHNEVSKAYA

Institute of the Lithosphere, Russian Academy of Sciences, Staromonetny per., 22, 109180 Moscow, Russia

Abstract. Our investigations showed the presence of nappe piles composed of different terrane slides in the Koryak region (Northeastern Russia) instead of large monotonous terranes as considered previously. These slides in the nappes consist of the formations varying in age, composition and genesis. Some nappes include the Paleozoic and probably Precambrian rocks as well as ultrabasic, basic and tonalite-plagiogranite complexes. Other nappes involve Lower-Middle Mesozoic formations.

Early-Middle Mesozoic allochthonous formations of the Koryak region derived from different paleogeodynamic environments. These were formed both in convergent and divergent plate boundaries, as well as in intraplate conditions, such as marginal sea basin with spreading centers; segmented island arcs with lateral change in composition of volcanic rocks; oceanic abyssal basins and intraoceanic rises; intraoceanic islands (seamounts) and midoceanic ridges. These Mesozoic tectonic paleostructures originally occupied the vast area (from Boreal to Tethyan realms) and were later tectonically shortened to the width of 300 km of the present Koryak region.

The correlation of the tectonically separated Mesozoic formations is based on the radiolarian, geochemical and lithologic methods. It was established that the nappes comprise ten stratigraphic subdivisions from the Middle Triassic to Hauterivian inclusively, tectonically arranged in various spatial patterns. The nappes involving all these rocks were probably formed as the result of arc-arc and arc-ridge-continent collisions within the Pacific periphery. The main collision took place during the interval between 125 and 105 Ma.

Keywords: NW Pacific continental framing, Koryak region, nappe pile, island arc, marginal basin, MORB, within plate basalt, radiolarian age, Triassic-Cretaceous, paleolatitude, collision.

INTRODUCTION

This paper presents the results of tectonostratigraphic investigations on the Mesozoic allochthonous terranes of the Koryak region (Figure 1), which make part of the continental framing of the Northwestern Pacific. In the north, this region is bounded by the Cretaceous Okhotsko-Chukotsky continental-margin volcanic belt. In the south, this region adjoins the Olyutor zone. Koryak region is part of the Koryak-Kamchatka accretionary domain which is traced into the western part of the Kamchatka Peninsula (see the inset of Figure 1).

This region is underlain by accretionary complexes where allochthonous Paleozoic and Mesozoic (pre-Barremian) formations of various provenance are present. In this paper, the interpretations of the tectonic relations between different allochthonous Mesozoic (Trassic-Neocomian) associations are presented, which commonly were considered as terranes [3, 5, 6, 8, 24, 29, 35]. Some investigators [19, 28] regard each Mesozoic allochthon terrane as undeformed stratigraphic sequences of rocks and undisturbed lateral successions of paleostructures, though others [1, 10-12, 24, 26, 27, 30, 31] restored here complex systems of nappes and imbricated structures. However, some authors, accepting point of view about the nappe structure of the Koryak region, identify the nappe and terrane, so they considered as nappes have monotonous composition (for instance, [24, 26]).

Our tectonostratigraphic studies in the Koryak region included: the deciphering of the tectonic structure in the Early–Midde Mesozoic deposits with help of detail geological mapping and the correlation of lithofacies of the Mesozoic allochthonous siliceous-volcanogenic formations on the base of radiolarian assemblages. The radiolarian analysis has applied to determine age [2], stratigraphic subdivision and correlations of siliceous-volcanic rocks (poor in macrofaunas) between the separated fragments of the different nappes [12, 30]. Besides, the analysis of taxonomic composition of radiolarian assemblages and morphological peculiarities of the radiolarian skeletons has given the opportunity to restore possible depths and initial paleolatitude of sedimentary basins [31]. Petrogeochemical data allowed us to indentify tectonic affinities of volcanic rocks.

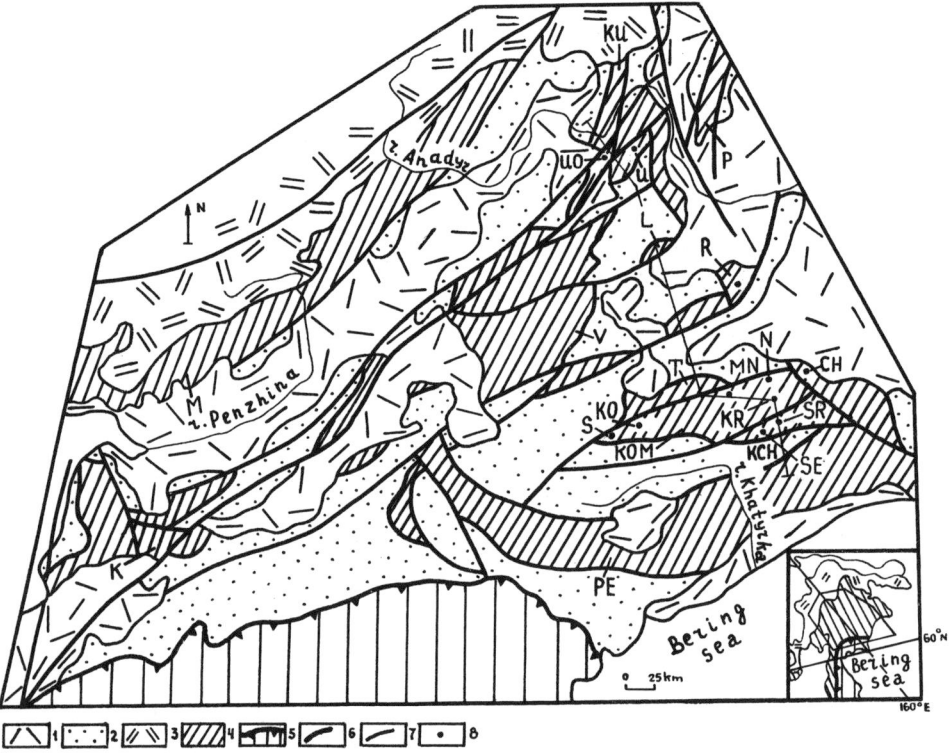

Figure 1. A sketch map of the position of the ophiolites and island arc Mesozoic formations in the Koryak accretionary region (a fragment of the Koryak–Kamchatka suture zone).1. Neoautochthon (KZ).2-3. Intermediate neoautochthon (Albian–Senonian): 2 – terrigenous rocks, 3– volcanic rocks (Ochotsko–Chukotsky volcanic belt). 4. Nappes of the Mesozoic and Paleozoic ophiolites and island arc formations (including basic-ultrabasic and tonalite–plagiogranite complexes). 5. Nappes of the Olyutor zone. 6. Thrusts and strike-slips. 7. Stratigraphic contacts. 8. The locations of simplified columnar section demonstrated in Figure 3. Inset shows location of a sketch map and Koryak–Kamchatka suture zone too (heavily shaded).

Abbreviations indicate the fragments of the nappes: (CH –Chirinay, K – Kuyul, KCH – Kekuro-Chirinay, KU – Kutinskiy, L – Lamutskaya, M – Murgal, KOM – Koyverelan-Maynitz, P – Pekulney, PE–Pikasvayam-Ekonay, R – Rarytkin, T – Tamvatney, U – Utesiky, UO– Ustbelsko–Otrognaya, V – Vaega) and topographic locations in the Koyverelan–Maynitz fragment of the nappe (KO – Koyverelan river, KR – Mt. Krasnaya, MN – Maliy Nauchirinay river, S – Mt.Semiglavaya) and in Kekuro-Chirinay fragment of the nappe (SE – Mt. Seraya, SR – Mt. Srednaya).

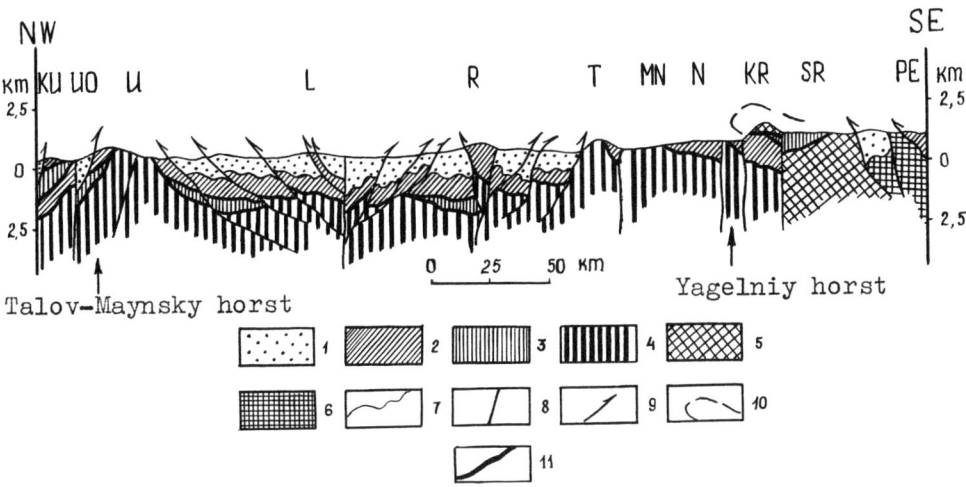

Figure 2. A schematic structural cross-section across northern Koryak region. See Figure 1 for location. Abbreviations are the same as in Figure 1. 1. Intermediate neoautochthon (K_1 al – K_2 sn). 2–6. Nappes (index indicates the age of sequences): 2 – Anadyr-Koryak (J_2 bs – K_1 h), 3 – Elgevayam and its analogues (T_{2-3} –J_1 bj), 4–5 – nappes of the Paleozoic rocks: 4 – Ustbelsko-Nauchirinay (ultramafic rocks consist of lherzolite, dunite and harzburgite), 5 – Kekuro-Chirinay (harzburgites and dunites predominate among ultramafic rocks), 6 – Elgevayam and Kekuro-Chirinay nappes individed. 7. Disconformity. 8. Strike-slip and listric faults. 9. Imbricate thrusts of the Laramian orogeny. 10. Recumbent antiform of the Laramian orogeny supposely. 11. Thrusts of the Pre-Albian orogeny.

We established that the terranes of the Koryak region are presented by group of vast nappes (Figure 2), which include the allochthonous sequences from Paleozoic to Hauterivian in age. Every nappe consists of alternated slides (or slices) varying in age, composition and origin.

Paleozoic rocks form two differently distant nappes: Ustbelsko-Nauchirinay (northern) and Kekuro-Chirinay (southern) (see Figure 2). The blocks of Precambrian rocks are possibly present in these nappes [18, 22].

Ustbelsko-Nauchirinay nappe includes intensively tectonized and metamorphozed (to amphibolite, rarely granulite) complex with lherzolite, dunite and harzburgite in the lower part where lherzolite is dominant in some places and harzburgite is prevailed in some other places. The overlain layered complex consists of the gabbro and ultrabasic rocks as well as amphibolitized gabbro, amphibolite and migmatite. The tectonic slices of tonalite-plagiogranite association also are presented. The K-Ar age of these bodies is ranged from Late Paleozoic to Middle Mesozoic. Sometimes, the fragments of the dike complex occur [29]. It is important to underline the heterogenity of composition of the ultrabasic part of the nappe, which reflected in different ratio of lherzolites and harzburgites in space. Except for basic-ultrabasic and gabbro-plagiogranite complexes, the lower part of the Ustbelsko-Nauchirinay nappe localy may be includes Riphean-Lower Cambrian shales [18] and Middle Paleozoic rocks too.

Kekuro-Chirinay nappe is limited by Elgavayam strike-slip fault from Ustbelsko-Nauchirinay one. The lower part of this nappe mainly made up of harzburgites, rarer dunites. Thus, here more depleted ultramafites is presented in contrast to the Ustbelsko-Nauchirinay nappe. Such difference probably reflects lateral heterogeneity of upper mantle, which had evolutionary changes in the degree of depletion during development of the Paleozoic and Mesozoic island arcs of the Paleo-Pacific ocean.

The lower basic-ultrabasic part contains thick metamorphozed layered gabbro complex with thick "strata" of amphibolites and migmatites. The metamorphic and migmatitic processes, the K-Ar age of which is predominantly Late Paleozoic, here are more intensive than ones in the Ustbelsko-Nauchirinay nappe. The tectonic slices of tonalites and plagiogranites here are thicker too. It is suggested [4] that plagiogranite-plagiorhyolite (with tuffs) association chemically resembles the island-arc rock complex. Its age is probably not younger than Late Paleozoic.

In general, Mesozoic formations are dominant, and Paleozoic rocks are limited in abundance in the allochthons of the Koryak region. The Mesozoic allochthonous formations compose two nappes. The first is not large in volume and consists of the Middle Triassic to Early Bajocian sequences. The isolated slides of this nappe occurs throughout the Koryak region. It is best preserve in the southern area, and is named "Elgavayam nappe" (Figure 2). The second Mesozoic nappe is very large. It is composed of the Late Bajocian to Hauterivian sequences, and is named "Anadyr-Koryak nappe". Both Mesozoic nappes consist of the slides of oceanic, marginal basin and island arc-related formations, which were formed at different paleolatitudinal environments. On the map, the above-mentioned nappes of the Koryak region make up narrow stripes limited by strike-slip and listric faults (see Figure 1). The stripes of the allochthon are separated by the vast area of the intermediate neoautochthon which consists of the Albian-Senonian terrigenous sequences.

The nappes of these stripes form synforms and antiforms. Moreover, the tectono-stratigraphic study showed the presence of recumbent antiforms and synforms. Sometimes these nappes (including Albian-Senonian intermediate neoautochthon) are disturbed by younger (Upper Cretaceous-Cenozoic) faults (see Figures 2), composed the complicated imbricated dislocations [11]. The strongest dislocations are located at the contact between intermediate neoautochthon and allochthon. The West Kamchatka-Koryak volcanic belt represents the Late Eocene-Early Miocene neoautochthon [10].

THE STRUCTURE OF THE NAPPES IN KORYAK REGION

In the Koryak region, above-mentioned nappes consist of the slide piles and dismembered into numerous fragments (see Figure 1). The structure and composition of these fragments are tabulated in Figure 3. In the columns radiolarian assemblages varying in age are presented too. The approximate paleolatitudes of the different radiolarian assemblages are indicated in Figure 3 aside each column.

The Ustbelsko-Otrognaya fragment of the nappes (Figure 3) situated in the northern part of the region and is composed of three nappes dipping to the north-west [1]. The lowermost of the Ustbelsko-Nauchirinay nappe is composed of lherzolite and harzburgite [23] and is covered by the layered complex and migmatites (tonalite-plagiogranitic series). The individual slides of the plagiogranites are disposed above. The maximum oldest of K-Ar ages data are 304 Ma (Upper Paleozoic) [1]. Another data of K-Ar ages (106-183 Ma) [23] correspond to Upper and Middle Mesozoic. So, the age of metamorphism (amphibolites, granitization) is Late Paleozoic and basic-ultrabasic complexes are more ancient. The uppermost thin slides of the Ustbelsko-Nauchirinay nappe is composed of tuffs, cherts, argillites, sandstones and limestones containing Early Carboniferous and Middle to Late Devonian faunas.

The possible presence of Lower Paleozoic and even Precambrian rocks in these areas is confirmed by Ivanov et al. [18], who described the shales with acritarch and algae of the Vendian-Lower Cambrian age.

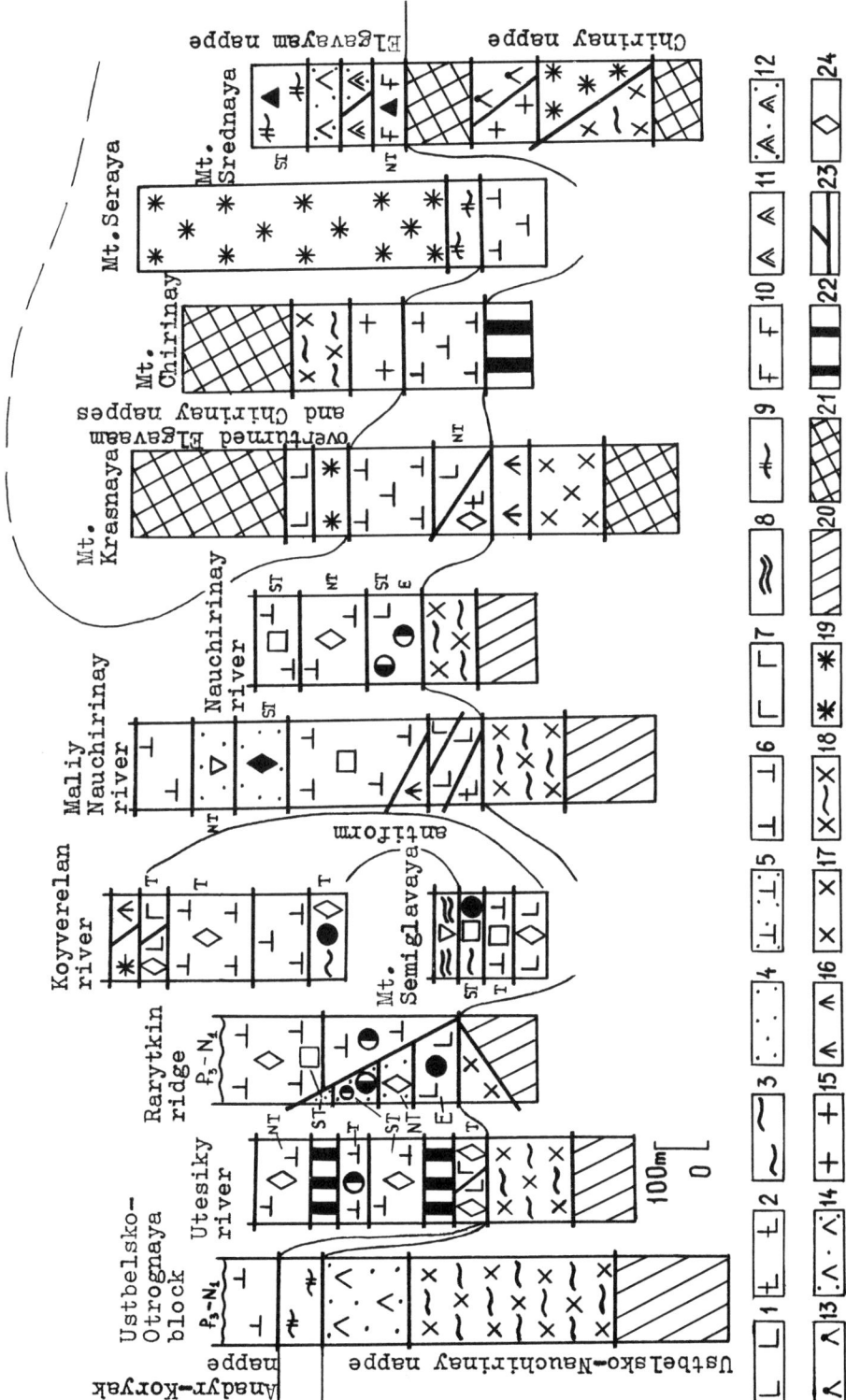

Figure 3. Structure of the fragments of the nappe piles in the Koryak region.
1-7. Middle Jurassic to Lower Cretaceous allochthonous formations: 1- jasper-basaltic MOR, 2- Fe-Ti-basaltic of seamounts, 3- jasperous oceanic abyssal, 4- tuffaceous- jasper-basaltic of backarc basins, 5- tuffaceous-turbiditic of backarc and forearc basins, 6- tuffaceous-volcanic island-arc-related, 7- tholeiitic island-arc-related. 8-14. Another allochthonous formations: 8- Lower to Middle Jurassic chert - jasperous marginal basin – related; 9-10 - Lower Jurassic: 9- chert-terrigenous marginal-basin-related, 10- jasper-alkaline-basaltic of seamounts; 11-12 - Middle to Upper Triassic: 11- volcanic island-arc-related, 12- chert-limestone-terrigenous marginal-basin-related; 13- Upper Paleozoic to Lower Triassic volcanic island-arc-related; 14- Middle Paleozoic island-arc and marginal-basin-related. 15-Plagiogranite (plagiorhyolite) to tonalite assotiation. 16- Diabase dykes and isotropic gabbro. 17- Layered gabbro, gabbro norite, olivine gabbro. 18- Metamorphozed layered gabbro, gabbro norite. 19- Amphibolite, migmatite, layered plagiogranite to tonalite into layered gabbro, gabbro norite, olivine gabbro and ultramafic rocks. 20- Dunite-harzburgite-lherzolite assotiation. 21- Harzburgite. 22- Serpentinite. 23- Thrust. 24-Radiolarian assemblages (age is indicated in Figure 4).

The letters beside columnar sections indicate the paleolatitudes (of Radiolarian complexes): B – Boreal, NT – North-Tethyan, ST – South-Tethyan, T – Tethyan, E – Equatorial. The locations of simplified columnar sections are shown in Figure.1.

The thin nappe of the Middle Jurassic terrigenous rocks is underlain by Ustbelsko-Nauchirinay nappe in the Ustbelsko-Otrognaya fragment. In some other places (for example, in the Vaega fragment) this nappe includes the Triassic volcanogenic-terrigenous sequences. It is necessary to underline that this nappe composed of the Triassic-Bajocian sequences has local distribution. Besides, some slides of this nappe are sometimes alternated with slides of the Ustbelsko-Nauchirinay nappe as a result of the Laramian orogeny.

The uppermost Anadyr-Koryak nappe is represented by essentially tuffaceous island-arc formation containing Late Jurassic-Valanginian Buchias. At present the considered stack of nappes forms the fragment of young Talovsko-Maynskiy horst and is bounded and dissected by series of normal faults.

Southward, in the basin of Algan and Utesiky rivers upper Anadyr-Koryak nappe is represented more completely. In Utesiky fragment of the pile of nappes (Figure 1-3), situated in the northern part of the Koryak region, the lowermost Ustbelsko-Nauchirinay nappe is represented by ultrabasite, gabbro and plagiogranite slides. They are covered by Anadyr-Koryak nappe which consists of alternating of the tectonic slides of MORB-type basalts and jaspers and island-arc tholeiites too.

Radiolarian complexes of the Bathonian-Callovian, Callovian-Tithonian, late Tithonian-Berriassian ages within jasper associated with MORB-type basalts here and also in the adjacent areas (basins of the Konachan, Poperechny Algan, Lamutskaya rivers) were found. IAT-type basalts yield radiolarians of the Bathonian-Callovian and Berriasian-Valanginian ages.

The upper part of this system of slices is composed of two slabs of Bathonian-Callovian island-arc formation, which separated by thin tectonical slice of the ultramafic rocks. It is necessary to emphasize the difference in composition and sedimentary conditions of the coeval rocks of these slabs. The lower of them is composed predominantly of tuffs and tuffaceous jaspers which have been accumulated in peripheric part of deep slope of island arc of South Tethyan Realm according to radiolarian data. The second upper slab contains the tholeiitic basalts regarded to indicate volcanic centers. The radiolarians also show contrast relief which is characterized for the central part of island arc and moreover exhibit oblique the North-Tethyan Realm.

In the more southern exposure of allochthon in the western part of Rarytkin ridge (Figures 1-3) ultramafic and gabbro slides of the Ustbelsko-Nauchirinay nappe is overlain by the Anadyr-Koryak nappe which made up of the slides of the Middle Bajocian-Hauterivian marginal-sea formation. The lowest part is Early Cretaceous in age, the middle part includes Bajocian-Callovian radiolarians and the upper part is Kimmeridgian-Tithonian in age. In accordance with radiolarian data the rocks of these different tectonic slices formed in various paleoclimatic (North-Tethyan and Equatorial) sedimentary environments are put together into single nappe. In the separate areas these slides tectonically amalgamated with slices of the island-arc basalts and thick slides of the island-arc tuff (see Figure 3).

Southward the Paleozoic and Mesozoic allochthonous complexes form the systems of the nappe piles which consist of some fragments: Tamvatney, Koyverelan-Maynitz, Chirinay, Kekuro-Chirinay (see Figures 1-3). Tamvatney and Koyverelan-Maynitz fragments include Ustbelsko-Nauchirinay and Anadyr-Koryak nappes (locally the thin nappe of the Triassic-Middle Jurassic formations may be presented) and Kekuro-Chirinay fragment consists of Chirinay and Elgevayam nappes.

In the western part of the Koyverelan-Maynitz fragment (the basin of the Koyverelan river) the basic-ultrabasic complex as well as MORB-type basalts are very poor exposed. Here, in the Semiglavaya mounatain and on the right bank of Koyverelan river, the overturned slice piles (see Figures 3) occur between the rocks of the intermediate neoautochthon. These slice piles are composed of the allochthonous oceanic abyssal, MORB, island arc and marginal sea formations.

In general, the Tamvatney and Koyverelan-Maynitz fragments form the northern wing and western end of the synform. An axial part of this synform is limited by the Elgavayam strike-slip fault.

In the western centrocline of this synform in basin of the Maliy Nauchirinay river the fragment is represented by the stack nappes inclined to east. The lower nappe (see Figure 3) is composed of the alternation of the wehrlites, clinopyroxenites, rarely dunites and lherzolites [21, 23], which above are overlain by slide of the alternated ultrabasites and gabbro. The slice of banded gabbro norites disposed above. Sometimes these rocks altered into amphibolites and garnet-amphibole metagabbro. In this locality the slides of migmatites are present. Their leucosomes are rocks of the tonalite-plagiogranitic series.

The lower part of the uppermost Anadyr-Koryak nappe consists of slides contaning the jasper and MORB and WPB-types basalts. The individual slices are formed by the island arc boninites [13, 14]. Upper part of the Anadyr-Koryak nappe composed of the facies of the island-arc, fore-arc and back-arc basins (which are represented by rocks ranging from the tuffs to the turbidites) are disposed above. The slides differing in age (Middle Jurassic-Lower Cretaceous, Valanginian) alternate in the section .

The Anadyr-Koryak nappe has the same two-membered structure to east in the valley of the Nauchirinay river (see Figure 3). Here its lower part is composed of slide of the Tithonian to Valanginian jaspers and basalts MORB-type and upper part consists of the Middle-Upper Jurassic island arc tuffs. All slides are composed of the rocks derived from Equatorial to North Tethyan Realm.

In general, such structures of the Anadyr-Koryak nappe is preserved in all localities of the Koyverelan-Maynitz fragment. Both thickness and number of the slides which composed the nappe, are variable. In some areas these slides are interlayered by slices of the serpentinized ultrabasites or serpentinitic melange.
Some investigators [1, 25, 29] point out the Yagelniy melange here. However it is the tectonized and raised block of the pile nappes (see Figure 2) composed of gabbro-peridotite complex, which is covered by slices of the Fe-Ti-rich basalts (WPB-type) and the tholeiitic-basalts (MORB-type).

Toward southeast, near the Elgavaym strike-slip fault, these slides are overlain by the thick tectonic layering island-arc association of the Early Cretaceous age.

Besides, so-called melange zones, which according to numerous publications (for example [24, 26, 32]) are widespread in the Koryak region under detailed survey proved to be well-recognized parts of all above-mentioned nappes. Moreover, isolate slides of nappes in these melange zones do not form chaotic mixture of debris, but are arranged both in space and in the cross-sections of separate nappes.

In the Krasnaya mountain (see Figure 3) the Anadyr-Koryak nappe is overlain by the overturned nappe similar to the Chirinay one (from the Kekuro-Chirinay fragment). This nappe, composed of the crashed slide of the tonalite-plagiogranite complex, is overlain by more crashed thin slide of the MORB-type basalts. On the top of this overturned nappe the thick slide of harzburgite is situated. Such type of the erosion remnants of the overturned nappes was found in many places of the eastern part of the Koyverelan-Maynitz zone. For example, in the Chirinay mountains (Figure 3) the Middle Mesozoic island-arc complex is overthrusted by the nappe, which consists of (from below to up): 1) plagiogranites and tonalites, 2) amphibolized gabbro norites, 3) dunites and harzburgites. We can also observed similar tectonic juxtaposition of two overturned nappes to the southwest, along southern border of the Kekuro-Chirinay nappe, for example in the Seraya mountain. It is important to underline that in this location the Elgevayam nappe is present under the Chirinay nappe (see Figure 3). The latter consists of some slides of the Sinemurian-Pliensbachian marginal sea formation.

The Elgevayam and Chirinay nappe are distinctly exppozed in the Kekuro-Chirinay fragment. This fragment of the pile nappes disposed proximately southeastern side of the Koyverelan-Maynitz fragment. In the Srednaya mountain (see Figure 3) two above-mentioned nappes are present. The lower nappe includes genetically diverse formations which hypothetically are Proterozoic-Paleozoic in the age. The lowest slide is represented by the dunite-harzburgite complex. The upper, thick imbricated slides consist of unequally metamorphozed and migmatized gabbro norites. The gabbro norites are often alternated with websterites, wehrlites, lherzolites and pyroxenites. The gabbro norites are mostly altered to amphibolites. Among metamorphosed rocks epidote-amphibolite, chlorite-amphibolite, and glaucophane schists also occur [1, 2]. Amphibolite and various metamorphozed gabbro usually form individual stripes into the migmatites and are their melanosomes. Leucosome is the successive series of rocks from the tonalite and to the quartz-diorite and plagiorhyolite. Plagiogranites and plagiorhyolites also form independent, elongated tectonic slides of 100-200 meters in thickness. The tonalite-plagiogranitic associations are similar to the island-arc formations due to their chemical characteristics [4]. K-Ar age of the plagiogranites has broad interval: 62-240 Ma [1, 23]. It is more likely that age of migmatization and of tonalite-plagiogranitic series is Late Paleozoic (another analyses are marked as younger).

The age of gabbro-ultrabasic complex is unknown. We suggest that this association has old age (in interval Proterozoic-Early Paleozoic) and multistage development. Pb-Pb termoisochronous ages (2900-900 Ma [22]) of several magmatogenous zircon types from the metamorphic rocks and gabbro norites of the adjacent area (Pekulney ridge) were also taken in account. It is necessary to take into account the finding of Vendian algae and Riphean acritarch in shales of this and adjacent southern area [18].

The lower nappe described above is overlain by the slides of volcanic sequences (basic, intermediate and acidic tuffs, basalts, andesite-basalts, subvolcanic plagiogranites). According to chemical data these rocks belong to the differentiated boninitic series [4]. These sequences do not contain fossils for age dating. But these rocks are cut by the dikes of plagiogranites and plagiorhyolites, and we assume the Late Paleozoic age for this unit.

The second (upper) nappe of the Kekuro-Chirinay fragment has tectonic slide of dunites and harzburgites [23, 25] on the base. The upper part of this nappe is made up of two different slides of Lower Jurassic formations. One of them is the chert-terrigenous marginal-sea formation of South Tethyan province, whereas the jasperous-alkali-basaltic formation of the oceanic seamounts affinity more likely has been derived from North Tethyan province. In some areas these slides alternate with the slices of the Middle-Upper Paleozoic and Middle-Upper Triassic formations.

ALLOCHTHONOUS EARLY-MIDDLE MESOZOIC FORMATIONS OF THE KORYAK REGION

The intensive investigations of the Mesozoic sequences in the above-described nappes allow us to establish several formations and to rank them on age and genesis. Mesozoic allochthonous formations of Koryak region occur in tectonical slices, and their complete stratigraphic sequences and primary contacts are not well preserved. However radiolarian assemblages and geochemic-lithological methods enable us to correlate these sequences as tabulated in Figure 4. Correlation of the separated slice was chiefly realised on the basis of radiolarian stratigraphy: ten radiolarian assemblages were distinguished in the Middle Triassic to Hauterivian interval. The important stratigraphic result of these works is the establishment of broad occurrence of Lower-Middle Jurassic rocks. Previously they were ussually ascribed to Upper Jurassic-Lower Cretaceous (Pekulneiveemskaya or Chirinaiskaya series). Moreover, the rocks of even shorter age interval could not be integrated into a single series (formation), because they contain accumulations of different tectonic regimes.

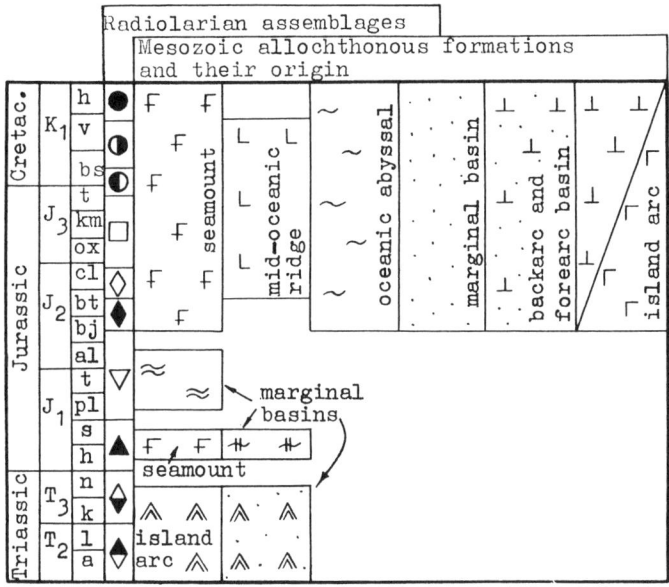

Figure 4. Correlation of Early-Middle Mesozoic allochthonous formations. Legend is shown in Figure 3.

Middle-Upper Triassic
This interval includes two formations: volcanogenic and chert-limestone-terrigenous.

The volcanogenic formation (up to 800 m in thickness) consists of acid and basaltic tuffs and terrigenous rocks. Based on radiolarian determenations, the green and reddish acidic ash tuffs of this formation ranging in radiolarian age from Anisian – Ladinian to Carnian-Norian.

These are typical volcanic rocks of the island arcs. The sharp changing of depths from subaquatic towards abyssal was very characteristic for their bathial conditions of accumulations. However the Triassic island–arc boninites (Figure 5) are known [13, 14] in Ekonay fragment of nappes.

The chert–limestone–terrigenous formation (with maximum exposed thickness about 400 m) is composed of sandstones, siltstones, limestones, cherts and rare acidic tuffs. This is a typical marginal basin formation. Based on radiolarian assemblages, it is subdivided into two intervals: Anisian–Ladinian and Carnian–Norian. The Triassic formations, constructing an island arc–marginal basin system, are widespread in separated slices throughout the Koryak accretionary area [7, 16, 32].

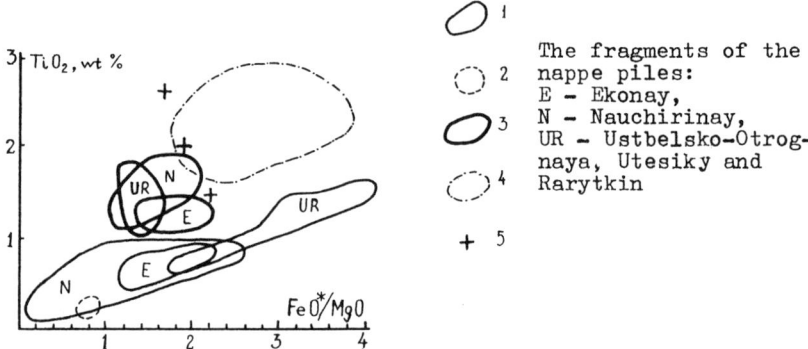

Figure 5. TiO_2 versus FeO*/MgO diagram of Early to Middle Mesozoic rocks (using [32, 33]). 1- Middle Jurassic to Hauterivian island arc tholeiite and boninite fields . 2- Jurassic boninite field of the Ekonay zone [14]. 3- Middle Jurassic to Hauterivian mid-ocean ridge tholeiite fields. 4- Middle Jurassic to Hauterivian within plate to E-MORB field of the Ustbelsko-Rarytkin fragments of the nappes. 5- Hettangian–Sinemurian withinplate basalts and MORB of the Elgevayam nappe.

Lower Jurassic (Hettangian–Sinemurian)
This interval is characterized by two formations: chert–terrigenous and jasper–alkaline–basaltic. The first chert–terrigenous formation (about 300 m in thickness) is made up of chert, rare jasper, alternating with siltstone and fine–grained sandstone. Judging by from lithology this formation probably was formed by turbidity currents. The terrigenous material is arkosic being represented by debris of supposed Paleozoic rhyolites, rhyodacites, plagiogranites, tonalites, plagioclase and quarts.

We suggest that this formation was accumulated in a basin situated far from areas of denudation and island arcs. The fine–grained, well–graded nature of the sediments and the presence of distal ashes support this assumption. The radiolarians also indicate deep slope and deep–water pelagic conditions.

The second, jasper–alkaline–basaltic formation (about 100 m in thickness) is composed of high–titanic basalts and jaspers with a minor amount of limestone. The presence of Ti–riched basalts and some limestone layers may be interpreted to represent condition of midocean ridge or seamount. The data on TiO_2 concentration of these lavas are shown in Figure 5. These basalts include Ti–enriched and low–K basalts (TiO_2=1.90–2.59%, K_2O=0.15–0.29%). This rocks belong to the within–plate basalts and E-MORB. There is single sample of the basalts with TiO_2 intermediate concentration (1.36%) and high–K_2O (1.98%). This rock similar to the shoshonite and its initial tectonic setting is unknown.

Hence, the two Hettangian–Sinemurian formations preserved in the sliced fragments differ in both paleogeodynamic environment (marginal sea or oceanic basin) and paleolatitude

(Central–Tethyan versus North–Tethyan or South–Boreal) of origin.

Lower–Middle Jurassic
The chert–jasper formation of the Pliensbachian–Early Bajocian (about 100 m in thickness) consists of rocks which are typical for a oceanic plateau.These are oceanic represented by alternating of reddish brown jasper and siliceous limestone, greenish–gray chert and limestone with siliceous siltstone. They are often enriched in fine–grained volcanoclastic rocks or acid ash indicating an environments distant from a growing island arcs. More likely it was condition of inner part of a marginal basin.

Upper Bajocian–Hauterivian
Middle Jurassic to Early Cretaceous allochthonous terranes from the Koryak orogenic belt include five formations: jasperous, jasper–basaltic, Fe–Ti–basaltic, volcanogenic–terrigenous and tuffaceous–jasper–basaltic. Radiolarian data indicate the same age for these five formations.

The jasper formation is composed of the bedded jaspers with the siliceous limestone lenses or by the alternation of cherts and siltstone. It is characterized by the Bajocian to Hauterivian radiolarians occurring in coherent sequences. This formation, 80 m in thickness, has been accumulated at a low sedimentation rate environments in an open ocean basin with a minimum terrigenous deposits and pyroclastic input from island arcs. However, the presence of limestone lenses and carbonate admixture in jaspers is evidence of the existence of intraoceanic carbonate reefs with a comparatively shallow–water environment above CCD. The radiolarian assemblage of jaspers indicates an open deep ocean basin environment. The high content of cyrtoid forms and the absence of spongeous shells are indicative of abyssal or transitional depths.

The jasper–basaltic formation (100–150 m in thickness) includes jaspers, cherts and MORB–like tholeiitic basalts.

According to radiolarian analysis, the age of this formation is ranged from Bathonian to Valanginian and it has been accumulated in open ocean basin near the spreading centers.

The comparison between jasper and jasper–basaltic formations shows that they have been accumulated in an open ocean basin. The latter probably tracing areas proximal to a mid–ocean ridges with the tholeiitic basalts eruptions along spreading centers.

Late Bajocian–Hauterivian formations had occupied the vast areas of the ocean. This is supported by the following facts: the jasperous formation has been deposited in the North–Tethyan Realms; the contemporaneous jasper–basaltic formation has been accumulated further south in the Central Tethyan Realm. Presently, they are amalgamated into a single tectonostratigraphic sequences in the narrow Koryak orogenic belt.

The Fe–Ti basaltic formation (up to 150 m in thickness), Late Bajocian–Early Bathonian in age, consists of Fe and Ti basalts with lenses of jaspers. These WPB–like basalts are very characteristic for intraplate oceanic islands. The radiolarian faunas from these jaspers tend to be of special diversity of high–conical Parvicingula which account more than 75% of the assemblage. This implies existence of an upwelling along a deep fault, which could served as a conduit for intraplate basalts.

Terrigenous–volcanic formation (more than 600 m in thickness), Bajocian–Valanginian in age, includes two rock associations: volcanic–tuffaceous, which derived from axial part of a volcanic arc showing occurrence of basalts and wide explosive activity, and tuffo–terrigenous which comprises deposits of the slopes of island arcs and adjoining part of the back arc basin and the deep trough.

This formation yields Bajocian to Valanginian radiolarians in the lower and middle parts and Hauterivian inoceramids in the upper part of sequences [13]. The effusive rocks, although minor in volume, are distinguished by having a very diverse chemical composition. Island-arc tholeiitic basalts with different degrees of depletion are dominant [33,34]. The magmatic components also include a spectrum ranging from calc-alkali basalts, basaltic andesites to boninites.

Considering spatial separation of different magmatic rocks series, segmentation can be inferred for the growing Bajocian to Hauterivian island arcs, supposedly due to lateral heterogenity of mantle sources. The sharp prevalence of the tuffaceous rocks in the volcanogenic-tuffaceous association evidently suggests a high explosive activity during the growth of the island arcs. In places this association is represented by a typical volcanogenic-terrigenous turbidites with abundant "tephra". It may has formed at either the back arc slope and adjacent marginal basin area or the forearc basins or inner trench slope with a ragged topography, as indicated by slumping structures involving non-consolidated fine-grained sedimets. The horizons of terrigenous-siliceous conturites (the thin-bedded turbidites) bearing an admixture of ashes, associated with tuffs and tuffites deposited at lower parts of slopes support this suggestion.

The radiolarian assemblages also indicate the contrasting bathymetric conditions usually encountered in the island-arc basins and at times suggest such oceanographic movements of water masses which may be interpreted as upwelling systems active along the deep-sea trough slope. Toward the marginal basin the volume of chert and jasper is increased. The breccia and conglomerates of Paleozoic to Early Mesozoic ultrabasites and serpentinites, gabbros, plagiogranites, cherts, jaspers and different basalts are olistostromal in origin or are eroded from basement of ragged island arcs.

Tuffaceous-jasper-basaltic formation (about 500 m in thickness), is composed of mid-ocean ridge tholeiitic basalts, related with back arc spreading, jaspers and tuffs.

This formation probably has been deposited in marginal sea basin. The association of tholeiitic basalts of back arc spreading zone and jaspers is completed by rocks, which derived due to growing volcanic arc and its erosion. The above-mentioned segmentation of the Bajocian to Hauterivian broad island-arc system is attested by the difference in the chemical compositions of tholeiitic basalts of separated marginal sea basins. The tholeiitic basalts of these marginal basins and the corresponding segments of island arcs have the same degrees of depletion [34]. The high diversity of all morphological groups of radiolarians is evidence of pelagic environments. Besides, the radiolarian assemblages show coexistence of abyssal and neritic paleoecological niches indicative of an arc-trench systems with intrabasin rises and very likely belonging to the North-Tethyan province.

Geochemical discrimination of the Middle Jurassic to Hauterivian volcanic rocks shows these basalts belong to tholeiitic series in terms of FeO*-FeO*/MgO diagram. Only some basalts belong to calc-alkaline series. The contrast diversity of these tholeiitic basalts showed in the TiO_2 versus FeO*/MgO plot (Fig. 5). The data on low-, middle- and high-Ti volcanic rocks of the Koryak region are plotted as fields. Three levels of TiO_2 concentration versus wide range of FeO*/MgO ratio are clearly distinguished. The low level unites island arc tholeiite from all fragments of the nappes. The low-Ti basalts and boninites from Nauchirinay fragment begin this in more magnesian area of diagram and the boninites discribed in the Ekonay fragment [34] create of high-Mg end of low Ti-field. The Ustbelsko-Rarytkin fragments of nappes follow this trend in a more differentiated direction.

The basalts of MORB affinity have FeO*/MgO ratio 1.0-2.2 and TiO_2 content increases from 1.12 up to 1.52. The part of Nauchirinay and Rarytkin volcanic rocks include within plate basalts (or E-MORB) with TiO_2 content ranging from 1.6 up to 3% (see Fig. 5).

The Ti-enriched Jurassic-Hauterivian volcanic rocks from MORB to within plate basalts have the similar REE distribution. The flat or slightly enriched in LREE patterns are typical for most of these volcanites [33, 34]. In general, chondrite narmalyzed $(La/Yb)_n$ ratio varies in these basalts from 0.78 to 1.20 and $(La/Sm)_n$ is in the range 0.8–1.25.

The low–Ti island–arc volcanic rocks are strongly depleted in LREE ($(La/Yb)_n$=0.8–0.86, $(La/Sm)_n$=0.14–0.8).

The above–considered data about age and composition of the Lower–Middle Mesozoic allochthonous formations from the Koryak region allow us to reconstruct geodynamic paleoenvironments of the formation of Mesozoic ophiolitic and island–arc complexes.

The formations interpreted as generated in divergent and convergent boundaries of lithospheric plates as well as in intraplate environments are recognized among Middle Jurassic–Hauterivian sequences of this region [30, 31]. These formations were accumulated in: 1) the mid–oceanic ridge provinces (jasper–basaltic formation); 2) the oceanic abyssal basin and at the intraoceanic uplifts (the jasper formation, locally enriched in carbonate material); 3) the intraoceanic island formations related with intraplate magmatism (formation of Fe–Ti basalts); 4) the volcanic island–arc (terrigenous–volcanic formation); 5) marginal seas and forearc basins (tuffaceous–jaspers–basaltic formation).

The analysis of radiolarian assemblages from the above–mentioned formations gives evidence of large extent of the Middle Mesozoic ocean: generation of these assemblages spanned from the low paleolatitudes (epiequatorial and South Tethyan regions) to the moderately high paleolatitudes, about 30°N (North Tethyan region and possibly, adjacent part of the South Boreal region). Taxonomic composition of the radiolarian assemblages from the island–arc terrigenous–volcanic formation indicates that the island arcs had mainly in the North Tethyan region, although they might have been partly located in the adjacent South Boreal region.

The development of the Bajocian–Hauterivian island–arc systems was followed by appearance of the marginal basins, the tuffaceous–jasper–basaltic formation being their indicator. The lithological composition of this complex (association of the tephra–terrigenous rocks with the MORB–type tholeiitic basalts) suggests, firstly, a voluminous supply of arc–derived material into the marginal basins and, secondly, the development of the spreading zones with the tholeiitic eruptions in these basins.

The radiolarian assemblages mostly suggest middle–high paleolatitudes for the marginal basin locations (inside of the North Tethyan province, namely in the same paleoclimatic environment as inferred from the radiolarians for the island arcs).

The formations of Early Jurassic (Hettangian–Sinemurian) and Early to Middle Jurassic (Pliensbachian–Bajocian) ages are predominantly represented by deposits of marginal basins (respectively, the chert–terrigenous and chert–jasperous formations). However, the presence of abundant tephra in these formations indicates the existence of active adjacent coeval volcanic arcs. It is important to point out the difference in paleolatitudes of accumulation of the above–mentioned formations. The Hettangian–Sinemurian sequences were originally deposited in the Tethyan province, including the Central–Tethyan or Equatorial areas, whereas the Pliensbachian–Bajocian sequences may have been accumulated in the Boreal and North–Tethyan provinces.

We believe that the Anisian–Norian formations also originated in an island arc–marginal basin system. Its significant distinction from the Bajocian–Hauterivian island–arcs has been established. The Triassic island–arc volcanism is characterized, on the one hand, by a large volume of acidic volcanite and by a very intense explosive activity due to high gas content of the magma, on the other hand. In contrast, the Middle Mesozoic volcanism was predominantly tholeiitic. However, the island arcs with initial boninitic volcanism probably existed in Triassic time [13, 14].

Hence, in the present–day geologic structure of the Koryak orogenic region tectonically juxtaposed are the Early–Middle Mesozoic formation generated in different (oceanic, island–arc, and marginal basin) paleogeodynamic and paleolatitudinal environments.

DISCUSSION

The opinion that the eastern accrecion continental margin of Russia is the gigantic tectonic breccia, which includes large terranes, become at present very common [3, 24, 27, 29, 35]. However, as the result of mapping of the vast territory of Northeastern Russia from the Ochotsko-Chukotsky volcanic belt to the Olyutor zone we do not reveal any chaos in the terrane distribution. On the contrary, we discovered their regular disposition in the vast nappes.

The largest Anadyr-Koryak nappe was observed (from the north to the south) in the basin of the Anadyr river (within the basement of the Ochotsko-Chukotsky volcanic belt) through the basins of the Mayn, Algan, Utesiky river, the Rarytkin ridge and the basins of Koyverelan, Maliy Nauchirinay, Nauchirinay rivers to the basins Pikasvayam and Khatyrka rivers. The fragments of this nappe occur in Northwestern Kamchatka, in the Kuyul and Pekulney mountains. This nappe is composed of alternating tectonic slides represented by Middle Jurassic-Hauterivian rocks of different origin: oceanic (intraplate abyssal and mid-ocean ridge), island-arc and marginal sea. The Mesozoic rocks are not only of different age and genesis but also are from various paleolatitudinal setting (Equatorial, South-Tethyan, North-Tethyan and South-Boreal) and are united in single nappe now. Only a number of the tectonic slides composed of Middle Jurassic-Cretaceous rocks and their thicknesses is varied from place to place in this nappe. The terranes with complete sections are not preserved in the Koryak region.

In the Anadyr-Koryak nappe, the slides of the terranes of the various origin are united, and this nappe is composed by the pile of the different terranes. However, there are the order in the disposition of various formations within the Anadyr-Koryak nappe. The slides made up of MORB, WPB-like basalts and jaspers, usually, are distributed in the low part of the nappe. The age of these ophiolites is younger or equal to the age of island-arc and marginal-basin-related sequences in the upper slides of the Anadyr-Koryak nappe.

Some investigators suggest that in the Koryak region southtowards the Jurassic-Early Cretaceous facies of the island arc substitute by the fore-arc and maginal-basin-related rocks. But we established that the slides of both types of the Jurassic-Early Cretaceous sequences are present in the all parts of the Anadyr-Koryak nappe.

The second nappe with Early Mesozoic (from Triassic to Early Bajocian) rocks has a comparatively little thickness and are locally distributed in the Koryak region.

Both of the above-mentioned nappes overthrust the nappes of the Paleozoic and, possibly, Precambrian ophiolites. The basic-ultrabasic complex is intensively tectonized and metamorphozed (to amphibolite, rarely to the granulite facies) it was shown above. It contains lherzolite, dunite and harzburgite in the lower part where lherzolite in some places is dominant (Ustbelsko-Nauchirunay nappe) and harzburgite is prevailed in some other places (Kekuro-Chirinay nappe).

It is necessary to emphasize that nappes with Early-Middle Mesozoic allochthonous formations overthrust the Paleozoic ophiolites also in the other areas Koryak region (see Figure 1): the Murgal [10], Kuyul [9, 19], Vaegy [32], Ekonay [26, 30], Pekulney [21-23] and Gold [18] ridges. It is in a good agreement with A. Ishiwatari's point of view [17] that Ekonay slice pile and above-mentioned ones are a remnant block of an older accretionary complex but not individual microcontinent [29]. The rocks of the amphybolitic and sometimes granulitic facies are present in the basic-ultrabasic complexes of all mentioned areas. The similar metamorphic rocks were found [15] in the Median ridge of the Kamchatka peninsula.

The Early-Middle Mesozoic formations had overthrusted Paleozoic ophiolites, possibly, as a result of the obduction. This obduction could take place in the Pre-Albian time as far as the Albian-Senonian sequences are the neoautochthon. We consider that the recumbent antiforms (in the Koyverelan river and Krasnaya, Seraya, Chirinay mountains) probably were formed in the Laramian orogeny.

We suggest the preliminary model for the formation of the nappes of the Koryak region. The Koryak region is part of the large Koryak-Kamchatka suture zone, which forms a part of the Northwestern Pacific rim. This zone extend from the Chukotka peninsula through Pekulney ridge and Koryak highland to western part of the Kamchatka peninsula. The northern boundary of this suture zone is the Asian Mesozoic continent. On its south from it there is Olyutor nappe system (see Figure 1). Three stages of collision events may have formed the nappe pik of the Koryak region: (1) Late Paleozoic-Early Triassic (Early Indosinian), (2) Middle Jurassic (Yanshanian) and (3) Early Cretaceous.

The areas of intensive Late Paleozoic metamorphism (to amphibolite, rarely granulite facies) of basic-ultrabasic complexes, accompanied by thick zones of migmatites and bodies of tonalite-plagiogranite association may have formed through Late Paleozoic-Early Triassic (Early Indosinian) orogeny which take place due to collision between large continental blocks in process of closure of the Iapetus and Paleo-Tethyan oceans. The remnants of the Late Paleozoic-Early Triassic collisional volcanic belts occur in the upper part of the Elgevayam nappe, and represent active Early Indosinian plagiogranite-plagiorhyolite magmatism. The intensive collision of Paleozoic island arcs of the Paleo-Pacific, the fragments of which we can see in the Koryak region, perhaps also took place during these Early Indosinian processes. Moreover, oldest (Lower and Upper Proterozoic as well as Lower Paleozoic) structures were involved to this collision. It is confirmed by the findings of the rocks having the Proterozoic age of the magmatic zircons [22], as well as the green shales with relicts of Riphean, Vendian and Early Cambrian acritarch and algae [18].

The other possible manner of transport of the Precambrian-Paleozoic complexes to the Koryak-Kamchatka domain may consists of its removal along the strike-slip faults from the Sino-Korean, Siberian continents and the Mongol-Ochotsky domain. We have revealed numerous such strike-slip faults in the Koryak region.

The clastics of ophiolitic,. Paleozoic metamorphic and tonalites reach a maximum volume in Early Mesozoic allochthonous formations (Middle Triassic-Early Jurassic). In contrast, the debris of the basic-ultrabasic complex as well as jaspers, cherts, volcanic rocks with the minor amount of the arkosic debris from the tonalite-plagiogranite form the bulk of the Mesozoic (Late Bajocian-Hauterivian) allochthonous formation. This suggest different erosion areas during the development of the Early to Middle Mesozoic island arc-back arc basin systems.

Two interval of the Mesozoic island arc growth in the Paleo-Pacific ocean are recognized: 235-182 Ma (Anisian-Early Bajocian) and 180-125 Ma (Late Bajocian-Hauterivian).

The older, Triassic-Early Jurassic island arcs, most likely appeared on the Kula plate near the Equator. Their basement comprised of the oceanic crust with numerous remnant Paleozoic terranes, including island arcs and collision structures with the large bodies of the migmatite and the tonalite-plagiogranite. The abundant arkosic material of these terranes, derived from the vast migmatite and tonalite-plagiogranite bodies, was accumulated into Early-Mesozoic back arc basins.

The Triassic-Early Jurassic island arc formation finished in Middle-Late Jurassic. Probably, the Yanshanian orogeny in the periphery of the Paleo-Pacific ocean followed it. However, the size of the second orogeny was less that which took place in the western Paleo-Pacific ocean (including southeastern Asia).

Hence, accretionary framing, constructed by the collided Early Mesozoic, Paleozoic and older structures of different genesis which are formed the Ustbelsko-Nauchirinay, Chirinay and Elgevayam nappes, probably already existed in Middle-Late Jurassic on periphery of the Paleo-Pacific ocean.

The assemblage of the island arcs and marginal seas far vast as compared with Early Mesozoic ones is reconstructed in Middle Jurassic-Neocomian at Tethyan and Mid-Boreal latitudes of the Paleo-Pacific ocean. The upper mantle suffered a change during Paleozoic and Early Mesozoic was distributed in the basement of these island arcs.

The accelerated drift of the Kula plate at 125-105 Ma [20], toward Asia resulted in intense collision of the Jurassic-Hauterivian island arcs with the continent, on one hand, and with the previously accreted Paleozoic and Early Mesozoic terranes, on the other hand. This Early Cretaceous collision involved the tectonic slides of the Middle Mesozoic island arcs and backarc basins, as well as oceanic structures, including mid-oceanic ridges which were tectonically amalgamated together into single Anadyr-Koryak nappe. This nappe was obducted over the previously accreted Paleozoic and Early Mesozoic terranes composing of a few nappes. Moreover, Anadyr-Koryak nappe was partly obducted over the border of the Asian continent [10]. The intense Cretaceous collision is in a good agreement with the increased spreading rate in the Atlantic ocean. It distinctly affected whole continental framing of the Paleo-Pacific ocean.

The Albian-Senonian olistostrome-molasse deposits were intermediate neoautochthon for all these nappes. Hence, the Koryak-Kamchatka accretional domain is formed in Early Cretaceous orogeny.

The Laramian orogeny (the end of Late Cretaceous and the begining of Paleogene) is the second (after Early Cretaceous) important limit of transform of the Koryak-Kamchatka accretional domain into suture zone. At that time the nappes of Olyutor zone, were obducted over the structures of this zone. The Laramian intensive obduction led to forming of overturned synforms and antiforms in Koryak-Kamchatka zone. Moreover, during this orogeny early formed nappes of the Koryak-Kamchatka zone together with Albian-Senonian cover of neoautochthon were deformed to imbricate structures and were divided in numerous blocks. As result the nappes of Koryak region have got fragmentar character which everyone can see on the geological maps.

CONCLUSION

1. The Mesozoic allochthonous sequences of the Koryak region form two nappes. One of them includes the Middle Triassic to Early Bajocian formations. The second one is more widespread. It involves the Late Bajocian to Hauterivian formations. Besides that, there is the third nappe group in this region. It is composed of the Precambrian and Paleozoic rocks as well as basic-ultrabasic and tonalite-plagiogranite complexes and is distingnished by very complicated structures. The Albian-Senonian sequences are the intermediate neoautochthon of the nappes of the Koryak region and Paleocene-Early Miocene West Kamchatka-Koryak volcanic belt appears to form the neoautochthon.

The nappes of the Mesozoic rocks comprise a slide pile, which consists of the various intervals of Lower Jurassic-Lower Cretaceous. Original stratigraphic relationship between the rock types are not preseved. Hence, the nappe pile composed of the slides of the various terranes are distributed in Koryak region instead a large uniform terranes.

2. The Early-Middle Mesozoic allochthonous rocks of Anadyr-Koryak region dated on radiolarians include the following formations: Middle to Upper Triassic (Anisian-Ladinian), predominantly volcanic; Upper Triassic (Carnian-Norian) chert-limestone-terrigenous; Lower Jurassic (Hettangian-Sinemurian) chert-terrigenous and jasper-alkaline-basaltic; Lower-

Middle Jurassic (Pliensbachian–Bajocian) chert–jasperous; Middle–Lower Cretaceous (Upper Bajocian–Hauterivian) jasper–basaltic (MORB-type) and terrigenous–volcanic (IAB-type).

3. The petrologic and geochemical data integrated with tectonical studies and taxonomic morphologic diversity of radiolarian assemblages, make it possible to establish the tectonical alternation of the Mesozoic allochthonous formations derived from different paleogeodynamic and paleoclimatic environments in the nappes of Koryak region.

a) The Middle Jurassic–Lower Cretaceous (Hauterivian) coeval volcanic–sedimentary formations, being the most widespread in the Koryak orogenic region, were formed at both convergent and divergent plate boundaries, as well as in intraplate environment. They represent a broad lateral succession of structures: marginal sea basins with spreading centers (tuffaceous jasper–basaltic formation), segmented island arcs with laterally variable compositions of volcanic rocks (terrigenous–volcanic formation); oceanic basins and oceanic rises; oceanic islands (Fe–Ti–basaltic formation); mid–ocean ridges (jasper–basaltic formation).

Older Mesozoic formation is represented by following environments: marginal basins with features of adjacent island-arc (Middle–Late Triassic chert–limestone–terrigenous formation, Early Bajocian jasper–chert formations); intraplate magmatism of oceanic floor (Hettangian–Sinemurian jasper–alkaline–basaltic formation) as well as island arcs (Middle–Upper Triassic volcanic formation).

b) The detailed analysis of taxonomic variety of radiolarian assemblages and morphological peculiarities of shells allowed us to establish the bathymetric conditions and topographic peculiarities of sedimentary basins and to determine approximate paleolatitudes of accumulation of the siliceous–volcanogenic sediments. The Late Bajocian–Hauterivian system of island arcs and marginal basins formed at moderately high paleolatitudes (North Tethyan and South Boreal provinces). The coeval oceanic formations (including those from spreading areas) were accumulated in the Tethyan province (North Tethyan and Central Tethyan). The Hettangian–Sinemurian formations were accumulated at different paleolatitudes: the marginal basin–derived jasper–terrigenous formation, in the Central–Tethyan province, and the intraplate jasper–alkaline–basaltic one at the North Tethyan/South Boreal boundary, in contrast to the Pliensbachian–Early Bajocian marginal–basin chert–jasperous formation, which was accumulated in Sub–Boreal province. The Triassic system of island arcs and marginal basins developed in the Tethyan province.

4. The nappes of the Koryak region, accompanied by numerous tectonic slides of the Mesozoic oceanic, island-arc and margin basin formations, are included into Koryak–Kamchatka accretionary orogenic belt now, which was formed in Pre–Albian time and was transformed during Laramian orogeny. The final occurrence of these nappes took place as a result of an arc–arc, arc–ridge and arc–continent collision in the Pacific periphery during Early Cretaceous. By means of this intensive collision the Mesozoic rocks of different genesis (oceanic, island arc, etc.) were obducted over the older (predominantly Paleozoic and, probably, Precambrian) complexes. Before, these complexes were undergone collision of the earliest stages (Indosinian and Yanshanian orogenies) which led to appearance of the metamorphic rocks and tonalite–plagiogranite the magmatism.

The nappes of the Koryak region were underwent additional deformations during Laramian obduction of Olyutor nappes from Pacific ocean side. As a result a separate fragments of the Early Cretaceous nappe pile and antiforms, synforms and imbricated structures too are made up.

Acknowledgements: We thank geologists Drs. A.I. Dvoriankin and A.L. Stavzev for their help during the field studies in the Russian Far East. We express our gratitude to Dr. N. Bogdanov for financial support and for encouragement in the course of this study. Special thanks are due to Professor Akira Ishiwatari for his stimulating comments and criticism on the paper.

References:
1. A.A. Alexandrov. *Imbricated thrusts of the Koryak Highland*. Nauka, Moscow (1978). (in Russian).
2. I.A. Basov and V.S. Vishnevskaya. *Stratigraphy of Upper Mesozoic of Pacific ocean*. Nauka, Moscow (1991) (in Russian).
3. Z.A. Ben-Avraham, D.J. Nor and A. Cox. Continental accretion: from oceanic plateaus to allochthonous terranes. *Science*, **213á**, 47-54 (1981).
4. O.S. Berezner, A.P. Stavsky, and S.K. Zlobin. Early mesozoic volcanic-plutonic association of Northern Koryak ridge. *Izvestia Acad. Sci. USSR*, **3**: 31-42 (1990). (in Russian).
5. N.A. Bogdanov. Tectonic Evolution of Pacific Ocean. In: *Transactions. Third Circum-Pacific Energy and Mineral Resources Conference*. pp.3-7. Honolulu, Hawaii (1992).
6. N.A. Bogdanov, Tilman S.M. Synthesis of tectonics of Northest of USSR. Mobilistic Approach. In: *Abstracts of 28th IGC*. vol.1, p.165, Washington, D.C. USA (1989).
7. N.Yu. Bragin, V.N.Grigoriev, K.A.Krylov, S.D.Sokolov. New finding of Middle-Late Triassic sequences in Koryak Hignland. *Dokl. AN USSR*, **290**, 3, 681-683 (1986). (in Russian).
8. A.K. Cooper, M.S.Marlow, D.W.Scholl. Geological Framework of the Bering sea crust. In: *Geology and resource potential of the continental margin of western North America and adjacent ocean Basins-Beaufort sea to Baja California*, Circum-Pacific council for energy and mineral resources. vol.6, pp. 73-102. Easth Sciences Series, Houston, Texas (1987).
9. N.L. Dobretsov and L.G. Ponomareva. Lawsonite-glaucophanic metamorphic schists of the Pendgin ridge of the NW Kamchatka. *Dokl. AN USSR*, **160**, 1, 196-199 (1965). (in Russian).
10. N.I. Filatova. *Perioceanic volcanic belts*. Nedra, Moscow (1988). (in Russian).
11. N.I. Filatova, A.I. Dvoryankin and A.I. Milekhin. Jurassic-Neocomian sequences and tectonic of Rarytkin zone. *Geology and Exploration*. **12**, 3-17 (1990). (in Russian).
12. N.I. Filatova and V.S. Vishnevskaya. Stratigraphic of the Jurassic-Neocomian sequence of the Anadyr-Koryak region. *Geology and Exploration*. **12**, 3-19 (1991). (in Russian).
13. M.L. Gelman and Ju.M. Bychkov. Triassic volcanites of Kankaran ridge and volcanic zones of Koryak Highland. *Pacific geology*. **1**, 53-62 (1988). (in Russian).
14. M.Z. Gelman, Ju.M.Bychkov and B.S.Levin. Boninites of Koryak Highland. *Isvestia Acad. Sci. USSR, Geol. ser*. **1**, 35-47 (1988). (in Russian).
15. L.L. German. *Oldest crystalline complexes of Kamchatka*. Nedra, Moscow (1978). (in Russian).
16. V.N. Grigoriev, K.A. Krylov and S.D. Sokolov. Main mesozoic formation types of the Koryak Highland and their tectonic setting. In: *Early geosyncline formations and dislocations*. pp.198-244. Nauka, Moscow (1987). (in Russian).
17. A. Ishiwatari. Ophiolites in Japanese islands: typical segment of the circum-Pacific multiple ophiolite belts. *Episodes*, **14**, 274-279 (1991).
18. O.N. Ivanov, A.N. Pertzev and L.N. Ilchenko. *Precambrian metamorphic rocks of the Anadyr-Koryakian region*. SVKNII, Far Eastern Branch, USSR Acad. Sci., Magadan (1989). (in Russian).
19. A.I. Khanchuk, V.N. Grigoriev, V.V. Golozubov, G.I. Govorov, K.A. Krylov, V.B. Kurnosov, I.V. Panchenko, I.E. Pralnikova and O.V. Chudaev. *The Kuyul ophiolite terrane*. Far Eastern Branch, USSR Acad. Sci., Vladivostok (1990). (in Russian).
20. S.Z. Larson and W.C. Pitman. World-wide correlation of Mesozoic magnetic anomalies, and its implications. *Geol. Soc. Amer. Bull*. **83**, 3645-3662 (1972).
21. M.S. Markov, G.E. Nekrasov and S.A. Palandzgyan. Ophiolites and melanocratic foundation of the Koryak Highland. In: *Essay of the Koryakian tectonics*. pp.30-70, Nauka, Moscow (1982). (in Russian).
22. G.E. Nekrasov and L.V. Sumin. Melanocratic foundation of the Pekulney ridge and their Pb-Pb-termoisochronic age. In: *Essay of the NW Pacific tectonic belt geology*. Ju. M. Puscharovsky (Ed.). pp.188-200, Nauka, Moscow (1987). (in Russian).
23. S.A. Palandzhjan. Ophiolite belts of the Koryak Upland; Northeast Asia. *Tectonophysics*, **127**, 341-360 (1986).
24. L.M. Parfenov, L.M. Natapov, S.D. Sokolov and N.V. Tsukanov. Terrane analysis and accretion in North-East Asia. *The Island Arc*, **2**, 35-54 (1993).

25. A.A. Peyve. *Structure and structural position of the Koryak ridge ophiolites.* Nauka, Moscow (1984). (in Russian).
26. S.V. Ruzhentsev, S.G. Byalobjesky, V.N. Grigoriev, K.A. Krylov and S.D. Sokolov. Tectonics of the Koryak ridge. In: *Essay of the Koryakan tectonics.* Yu. Puscharovsky and S.Tylman (Eds). pp. 136–189. Nauka, Moscow (1982). (in Russian).
27. S.D. Sokolov. Exotic Terranes of the Koryak structures. *Geology and Exploration.* **5**, 16–29 (1990). (in Russian).
28. A.P. Stavsky, O.S. Berezner, V.G. Safonov, and S.K. Zlobin. Tectonics of the Maynitz zone of the Koryak Highland. *Pacific Geology,* **3**, 72–80 (1989). (in Russian).
29. A.P. Stavsky, V.D. Chekhovitch, M.V. Kononov and L.P. Zonenshain. Plate tectonics and palinspastic reconstructions of the Anadyr–Koryak region Northeast USSR. *Tectonics.* **9**, 81–101 (1990).
30. V.S. Vishnevskaya, N.I. Filatova and A.I. Dvoriankin. New data about Jurassic deposits of Semiglawaya Mauntain (Koryak Upland). *Isv. áAkad. Sci. USSR, Ser. Geol.* **9**. 21–30 (1990). (in Russian).
31. V.S. Vishnevskaya and N.I. Filatova. The produce environments of the Anadyr-Koryak Middle Mesozoic formation. *Geology and Exploration,* **1**, 29–49 (1992). (in Russian).
32. V.P. Zinkevich. *Formations and stages od tectonic development ofáthe North Koryak Upland.* Nauka, Moscow (1981). (in Russian).
33. S.K. Zlobin, N.I. Filatova, A.I. Dvoriankin and N.I. Kolesov. Composition and origin Jurassic–Lower Cretaceous volcanic rocks of the Koryak Highland, NW Pacific. *Ofioliti,* **17**, 99–115 (1992).
34. S.K. Zlobin, A.P. Stavsky, O.S. Berezner and D.A. Minin. Geochemical peculiarities of the Mainitz paleoisland arc magmatism. *Geochimia,* **1**, 113–124 (1989). (in Russian).
35. L.P. Zonenshain, M.I. Kuzmin and L.M. Natapov. *Plate tectonics of the USSR territory.* Vol. 2. Nedra, Moscow (1990).

Island-arc mafic-ultramafic plutonic complexes of North Kamchatka

V.G. BATANOVA and O.V. ASTRAKHANTSEV

Geological Institute of the Russian Academy of Sciences, Pyzhevsky 7, Moscow, 109017, RUSSIA

Abstract -- Three types of mafic-ultramafic plutonic complexes are recognized within Late Cretaceous to Paleocene island arc terrane in the northern part of Kamchatka. Mineralogy and geochemistry of all plutonic complexes are typical for island-arc magmatism.
Type 1 is represented by linear bodies which are regarded as magma chamber fragments from deep crustal levels. This chambers were dominated by Ol-Cpx-Hb fractionation at P=6-8 kbars. Plutons probably formed in two stages. The first stage - crystal fractionation at depth of 15 to 20 km. In the second stage, crystal cumulates were forcibly expelled from host chambers and emplaced in country rocks coeval with growth of the oceanic arc crust.
Type 2 plutons are concentrically zoned as classic Alaska type intrusions. They are regarded as the intermediate level reservoirs which operated in partially formed arc crust. They were not subjected to the deformation stage.
Type 3 plutons are composed of olivine clinopyroxenite and large volume of gabbroic and dioritic rocks. They represented parts of relatively shallow (2-3km) volcano-plutonic systems related to the evolution of the island arc volcanoes.

INTRODUCTION

The establishing of the original tectonic setting of the mafic-ultramafic cumulate plutonic complexes of constructive continental margins is very important for tectonic and palinspastic reconstructions, as such plutons can either form a part of ophiolite sections or have a non-ophiolitic origin [24]. It was demonstrated that unlike mid-oceanic ridge and within-plate magmas, island arc affinities of pluton-forming magmas are common for both settings [10, 23]. It preclude direct tectonic interpretation of compositional data in the latter case, if geological setting is unclear.
One of the regions where origin of mafic-ultramafic plutons is a subject of controversy is the Olyutor zone of North Kamchatka. For a long time they were considered as ophiolite hypabyssal rocks [1, 2]. Our new geological, mineralogical and geochemical data show that (1) the Olyutor zone plutons were formed within intraoceanic island-arc crust and (2) have distinctive island-arc mineralogical and geochemical features close to those of adjacent arc volcanic suites.

GENERAL GEOLOGY

The Northern Kamchatka region is the southern part of the Kamchatka-Koryakia fold-belt of oceanic and arc-derived tectonostratigraphic terranes of Paleozoic to Cenozoic ages, which were accreted to the continental margin during the Cretaceous to Tertiary times [31]. Studied mafic-ultramafic plutons are exposed in an elongated belt in the frontal part of the Olyutor terrane (Fig. 1). The Olyutor terrane occupies the southern part of the Koryak highlands and has been thrust over the Ukelayat marginal sea deposites of Late Cretaceous to Paleogene age.

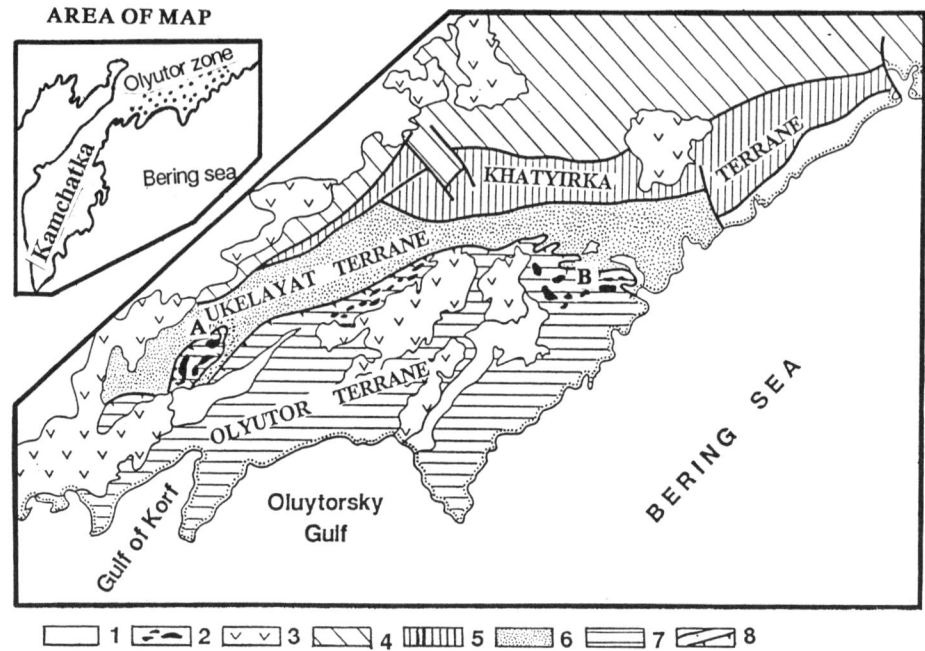

Fig. 1. Map of the Southern Koryak Highlands showing location of mafic-ultramafic plutonic complexes, after [21].
1 - Eocene-Quaternary post-accretionary sedimentary basins, 2 - mafic-ultramafic plutonic complexes, 3 - Oligocene-Quaternary basaltic and andesitic covers, 4 - Mesozoic structures of Koryak Fold Belt, 5 - Khatyirsky terrane of Cretaceous age, 6 - Ukelayat terrane (flysch trough) of Late Cretaceous-Paleogene age, 7 - Olyutor terrane of Late Mesozoic to Early Cenosoic age, 8 - boundaries, a - faults, b - Vatyin thrust, A - southwestern and B - northeastern groups of the Olyutor mafic-ultramafic plutonic complexes.

Three nappes of volcano-sedimentary rocks are included in the Olyutor terrane [21]. The basal unit of each nappe (Vatyin series) is composed of pillow MOR-like basalt flows interbedded with radiolarian cherts, red jaspers and hyaloclastites of Albian to Late Campanian age. The upper unit of nappes (Achayvayam series) is composed of basaltic and picritic volcanics associated with pyroclastics, tuffites, sandstones and siltstones of Late Campanian to Paleocene age. The coal-bearing rocks, formed in shallow waters are observed at the top of Achayvayam suite. The contact between two volcano-sedimentary units is comformable and gradational. The lower suite is interpreted as an ocean crust (ophiolite) assemblage while the upper suite is a product of incipient island arc volcanism on the oceanic crust. The piled nappes were folded and obducted on the Ukelayat marginal-sea flysch at the Paleocene-Eocene boundary. The volcano-sedimentary rocks form as large as 15-20 km recumbent folds plunging northwards.

Two spatial groups of mafic-ultramafic plutons are recognized within the Olyutor terrane (Fig. 1). The first group is located in the souhtwestern part of the Olyutor terrane where the island arc suite is composed predominantly of thin-bedded silisified tuff, tuffaceous sandstones, argillites, black cherts with separate thin volcanic layers. Plutons of the northeastern part of the Olyutor terrane are in close association with island arc basaltic and picritic volcanics, basaltic breccias, and tuff agglomerated lava. As follow from the analysis of volcaniclastic facies, northeastern plutons appear to have marked the active volcanic front of the Achayvayam Late Cretaceous to Paleocene intraoceanic arc [6].

Island-arc Mafic-ultramafic Plutonic Complexes 131

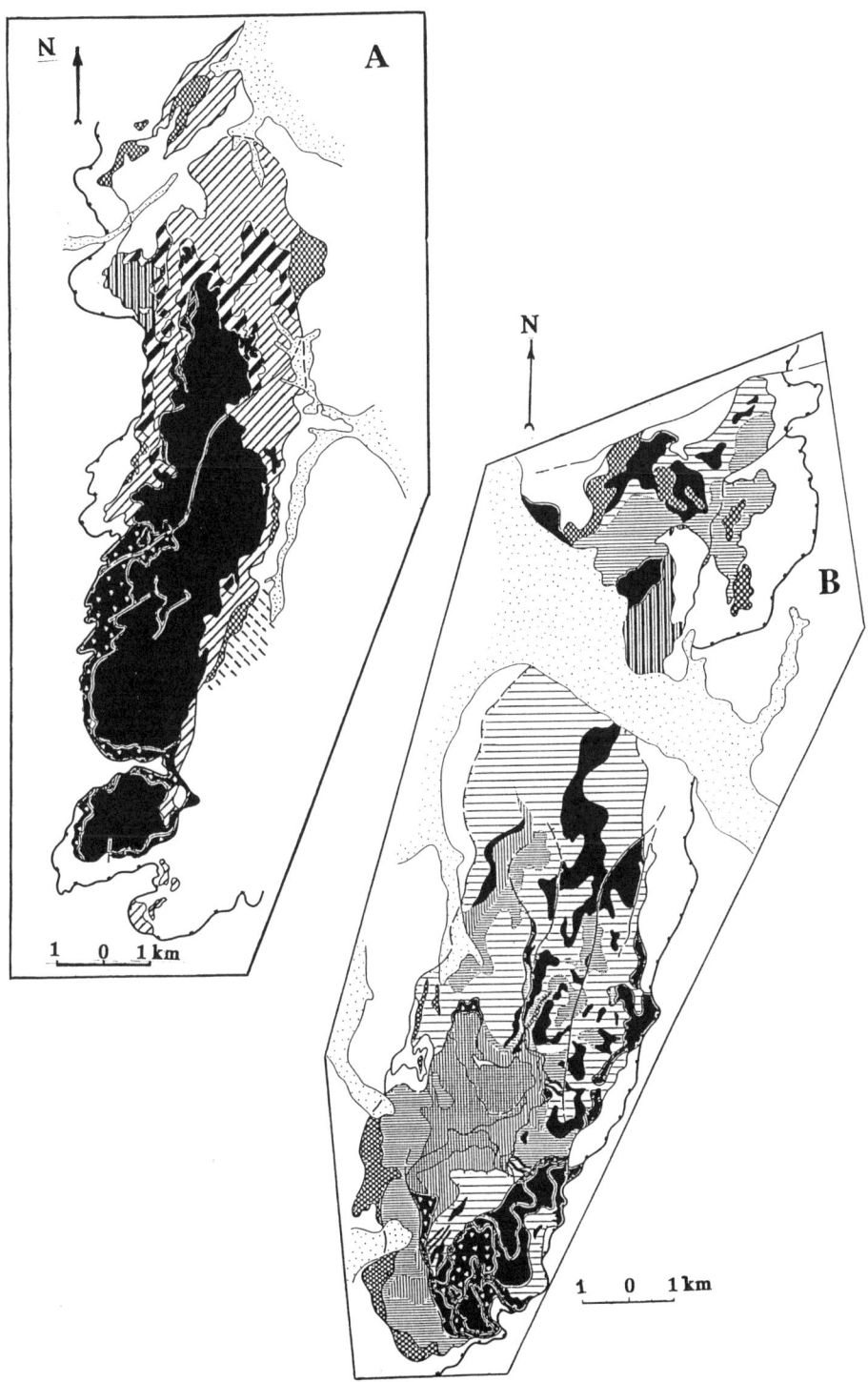

Fig. 2 (A, B).

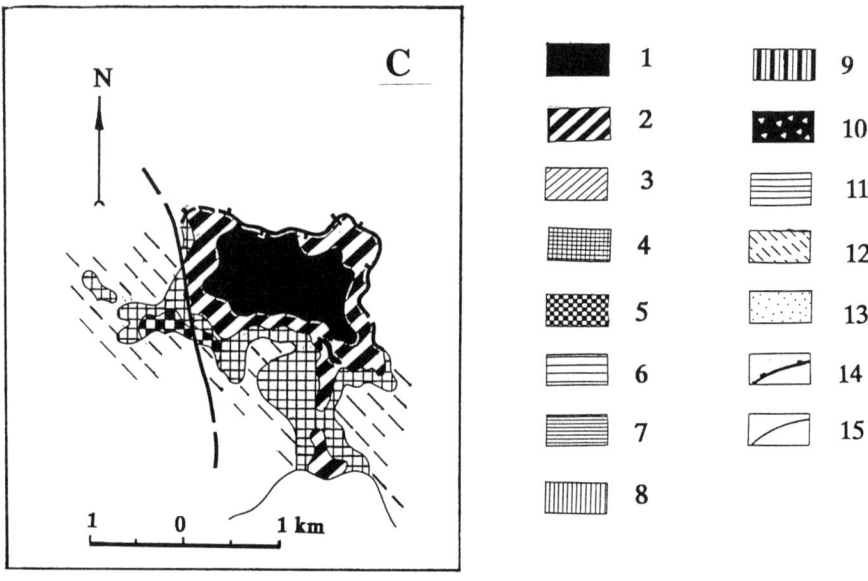

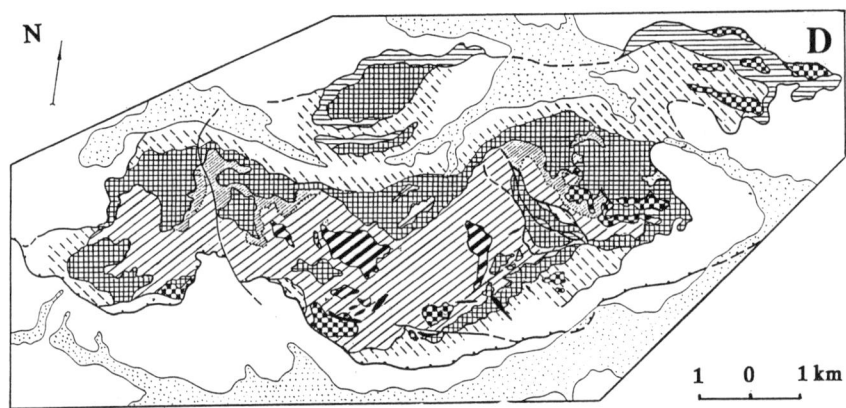

Fig. 2 (C, D). Geological maps of the Olyutor the mafic-ultramafic plutonic complexes. A - Galmoenan, B - Seinav, C - Snegovoy, D - Itchayvayam.
1 - dunite, 2 - wehrlite, 3 - olivine clinopyroxenite, 4 - plagioclase-bearing clinopyroxenite and gabbro, 5 - marginal gabbro (amphibole gabbro, gabbronorite), 6, 7 - dunite-wehrlite-clinopyroxenite layered sequence with: 6 - large scale of layering (thickness of single layers 10-100 m), 7 - small scale of layering (thickness of single layers 0.1-0.5 m), 8 - wehrlite-clinopyroxenite-melanogabbro layered sequence, 9 - hornblendite, 10 - serpentinite melange, 11 - zones of hydrothermal alteration and sulfide mineralization of rocks, 12 - contact aureole, 13 - quaternary rocks, 14 - thrust boundary, 15 - faults.

FIELD AND PETROGRAPHIC FEATURES OF THE PLUTONS

Mafic-ultramafic plutonic complexes are in a tectonic contact with the Ukelayat flysch. There is a serpentinite melange in the contact zone of the Galmoenan massif (Fig. 2A).

Chilled intrusive contacts were revealed between the marginal gabbroic rocks of the Galmoenan and Snegovoy plutons and country grey thin-bedded silicified tuff. The rocks within the contact aureole are fine-grained hornfels composed of biotite, pyroxene and quartz. The country silicified tuff represent a transitional unit between the oceanic and island-arc sequences. Thus plutons intruded all oceanic and lower part of island-arc suites.
The differences of structure, petrology and geochemistry in the plutons of southwestern and northeastern parts of the Olyutor terrane allow to recognize three types of mafic-ultramafic plutonic complexes.
Type one of plutonic bodies is located in the southwestern part of the Olyutor terrane and is represented by large - 3 km in diameter and 20 km long, linear bodies, composed predominantly of ultramafic cumulates (dunites, wehrlites, olivine clinopyroxenites, magnetite-hornblende clinopyroxenites), (Fig. 2A, B). Small amounts of gabbroic rocks are K-feldspar-bearing biotite gabbronorites. The contacts between gabbroic and ultramafic rocks are generally gradual. Magmatic layering is typical for the Seinav massif and also observed in the werhlite-clinopyroxenite rim of the Galmoenan massif. Ultramafic cumulates are characterized by deformational textures produced during high-temperature deformation and recrystallization of rocks. Porphyroclastic textures are typical for dunites and olivine clinopyroxenites.
The second type plutons are concentrically-zoned as the classic Alaskan-type intrusions [18]. That is best illustrated by the Snegovoy pluton (Fig. 2C). The Snegovoy pluton is about 4 km in diameter. The core of the intrusion is dunite, wehrlite, olivine clinopyroxenite, gabbro and gabbro-diorite occur progressively outward. The rocks within gabbro unit range gradationally from olivine-bearing gabbro near the contact with olivine clinopyroxenite to horblende-magnetite gabbro toward the contact with country rocks. Marginal gabbro-diorites are in clearly intrusive contacts with country silicified tuff representing the lower part of island arc sequence. The plutons of this type have not been subjected to high-temperature deformation. There is no evidence of hot plastic flow of the rocks.
Type three is represented by the Itchayvayam massif. The Itchayvayam pluton is composed predominantly of gabbroic and gabbro-dioritic rocks (Fig. 2D). Olivine clinopyroxenites are exposed only in the central, more eroded part of the plutonic body. Intrusive contacts are preserved in the Itchayvayam massif. The marginal micro-gabbros are abundant and overlain by Achayvayam arc volcanic breccias and lavas. Numerous dike swarms are observed in the contact zone of the massif. The dikes are basaltic in composition. They cut marginal gabbroic rocks and country basaltic breccias, but are not traced into the central part of the pluton. Mineralogical and petrochemical data suggest that dikes are probably magmatic feeders for arc volcanics. Estimates of the thickness of the overlying volcanogenic sequence suggest that the pluton was formed at the depth of 2-3 km.

Petrography of ultramafic and mafic rocks
The ultramafic cumulates in the North Kamchatka mafic-ultramafic plutons are dunite, wehrlite, olivine clinopyroxenite, clinopyroxenite. Plagioclase is observed as intercumuluse phase in the ultramafic rocks at the olivine clinopyroxenite - gabbro boundary.
The dunites are olivine - chromite cumulates. Chromite commonly makes up less than 2 percent of the rocks. Clinopyroxene is present in some dunites as a postcumulus phase. The bodies of pargasit-biotite-bearing dunite are present in the Seinav massif. The olivine grains in dunite from Galmoenan and Seinav massifs show the strain defined by deformational kink-bands. The kink-bands parallel to (100) plane as have been reported in olivines from other naturally and experimentally deformed dunites. The petrofabric data suggest the operation {okl} [100] slip system in the olivine during recrystallisation [7].
The werhlites are commonly characterized by cumulus olivine and postcumulus clinopyroxene. Chromium spinel is rare in wehrlites.

Olivine clinopyroxenites, clinopyroxenites are composed of cumulus olivine and clinopyroxene. The magnetite hornblende clinopyroxenites are common. They frequently contain Al-rich spinel.
Mafic rocks are non-cumulates. They are K-feldspar-bearing biotite gabbronorite in the plutons of the southwestern group and hornblende-olivine-bearing gabbros and amphibole magnetite gabbros in the plutons of the northeastern group.

MINERAL CHEMISTRY AND COMPOSITION OF PLUTONS

Mineral chemistry.
 Olivine. In general olivine is fresh. The olivine compositions are Fo_{92}-Fo_{89} in dunites, Fo_{89}-Fo_{79} in wehrlites and clinopyroxenites, Fo_{85}-Fo_{77} in olivine-bearing gabbros. The reaction rims consist of symplectitic intergrowth of orthopyroxene and magnetite are observed in olivine-bearing gabbro of the Seinav massif, and have been interpreted as a subsolidus oxidation features. The NiO content is low in olivine of dunite (Table 1), and is below detection limits in mafic rocks.

Table 1.

Chemical composition of olivines from the Olyutor plutons

Location	G	G	Sv	Sv	S	S	I	I	Sv	Sv	S
Sample	14-16	16-13	90-3	P-S	S28-1	S2-2	I23-6	I23-1	85-19	85-7	S28-9
Rock	d	oc	d	oc	d	d	oc	w	og	og	og
SiO_2	40.45	38.77	40.92	40.48	40.37	39.76	39.59	39.92	38.36	39.14	38.44
FeO	7.43	19.57	9.4	10.88	8.81	13.56	13.1	14.06	20.03	21.77	18.25
MnO	0.15	0.44	0.2	0.18	0.1	0.23	0.24	0.24	0.43	0.00	0.41
MgO	51.5	40.62	49.57	49.11	51.39	45.61	45.53	45.38	40.34	39.94	43.59
CaO	0.05	0.00	0.0	0.02	0.2	0.02	0.07	0.07	0.05	0.05	0.02
NiO	0.00	0.00	0.08	0.00	0.13	0.14	0.2	0.17	0.00	0.00	0.1
Total	100.12	99.4	100.29	100.67	101.0	99.32	98.64	99.84	99.34	100.09	100.8
Fo	92.4	78.7	90.4	89.0	91.2	85.7	86.1	85.2	78.3	76.6	81.0
Fa	7.6	21.3	9.6	11.0	8.8	14.3	13.9	14.8	21.7	23.4	19.0

Notes: Location are follows: G - Galmoenan, Sv - Seinav, S - Snegovoy, I - Itchayvayam. d - dunite, w - wherlite, oc - olivine clinopyroxenite, og - olivine gabbro

Table 2.
Chemical composition of spinels from the Olyutor plutons

Location	G	G	G	Sv	Sv	Sv	Sv	S	S	S	I
Sample	14-4	14-9	14-17	90-3	88-5	273	279	S28-1	S2-4	S2-15	I23-1
Rock	d	d	d	d	d	oc	d	d	d	oc	w
TiO_2	0.47	0.29	0.36	0.74	0.6	0.71	0.65	0.52	0.58	1.46	1.29
Al_2O_3	9.97	6.75	14.77	8.95	11.93	0.30	11.48	8.05	10.38	3.77	14.13
Cr_2O_3	45.86	41.73	44.99	36.41	38.26	9.64	41.97	40.95	40.21	9.09	21.01
FeO	22.00	23.6	21.81	21.65	21.52	22.28	20.16	19.69	22.6	30.25	23.7
Fe_2O_3	12.56	21.38	7.9	25.02	19.2	57.95	13.66	22.58	19.44	53.28	32.76
MnO	0.73	0.66	0.9	0.49	0.62	0.29	0.66	0.52	0.54	0.34	0.52
MgO	6.99	5.81	7.39	7.83	7.89	1.60	8.09	9.21	7.24	1.66	7.14
Total	98.58	100.22	98.12	101.09	100.02	99.77	96.67	101.52	100.99	99.21	100.55
Mg#	36.2	30.5	37.7	39.2	39.5	9.0	41.7	45.5	36.9	8.9	34.9
Cr#	75.5	80.5	67.1	73.2	68.3	95.6	71.8	77.3	72.4	61.8	49.9
Fe^{3+}/R^{3+}	0.16	0.28	0.1	0.32	0.25	0.85	0.18	0.29	0.25	0.78	0.43

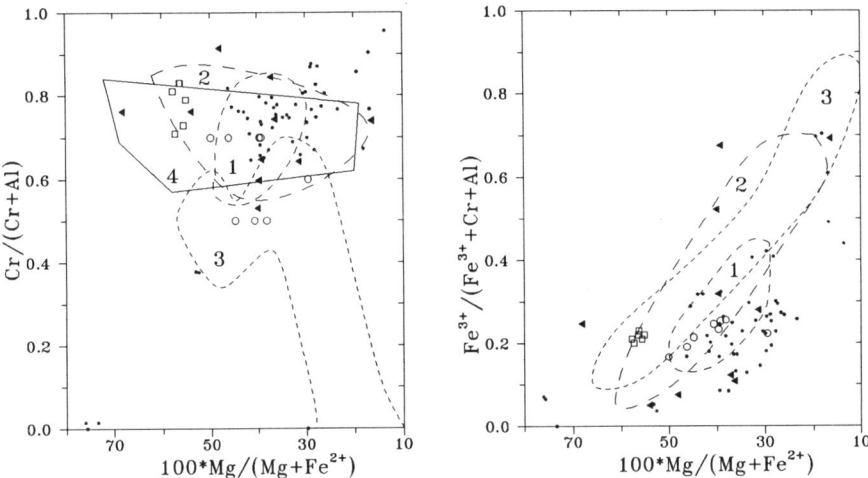

Fig. 3. Spinel compositions of the Olyutor plutonic complexes. 1 - Alaska-type plutons [20], 2 - New Georgia and Aoba island-arc picrites [28], 3 - Grenada island-arc volcanics [28], 4 - Border Range plutonic complexes, South-central Alaska [8], open squares - Olyutor island-arc volcanics [14], open circles - Blashke Island zoned pluton, Alaska [18], fill triangles - Aleutian xenoliths [9], dots - Olyutor plutonic complexes.

Spinel. Spinels vary from Cr-spinel in dunite to Cr-magnetite in olivine clinopyroxenite and to magnetite in hornblende clinopyroxenite and gabbro. Cr-spinel composition is characterized by high Cr/Cr+Al and $Fe^{3+}/(Fe^{3+}+Cr+Al)$ ratios, low Mg# and low TiO_2 content (Table 2, Fig. 3) and are typical to those in arc-related plutonic and volcanic suites [23, 28].

Clinopyroxene. Most of clinopyroxenes are diopsides and show a very limited iron enrichment from ultramafic to mafic rocks (Table 3). Wo-contents of clinopyroxene in the ultramafic rocks are high and range only from 49% to 45%. Clinopyroxenes in gabbro generally have equally high Wo-contents. The Mg# of clinopyroxene range from 90 in ultramafics to 76 in gabbro. Clinopyroxenes are characterized by low TiO_2 and Na_2O contents. The clinopyroxene compositions of cumulates may also be used to determine their geotectonic origin [9, 10, 24]. As follows from Fig. 4 clinopyroxenes from the Olyutor plutonic complexes as well as clinopyroxenes from other arc-related pluonic complexes are clearly characterized by a steeper Al_2O_3 vs. TiO_2 slope than clinopyroxenes from ophiolite cumulates. Accordig to Loucks (1990) this feature is typical for clinopyroxenes of intraoceanic arc cumulates, and can be explained by a considerable water content in arc-related melts.

Hornblende. The hornblende is a ubiquitus mineral in the Olyutor mafic-ultramafic plutonic complexes. It ranges from pargasite to edenite and has a high aluminium content. The hornblende-biotite-bearing dunites from the Seinav massif contain pargasite with high Cr_2O_3 content (Table 4).

Rock geochemistry.
All rocks from the Olyutor mafic-ultramafic plutons follow the calc-alkaline trend in AFM and Miyashiro's diagrams (Fig .5). Trace element geochemistry of the plutonic rocks indicates the arc affinity (Fig. 6). All rocks show depletion in HFSE and high LILE/HFSE ratios. Rare Earth Patterns of plutonic rocks in all types of plutonic

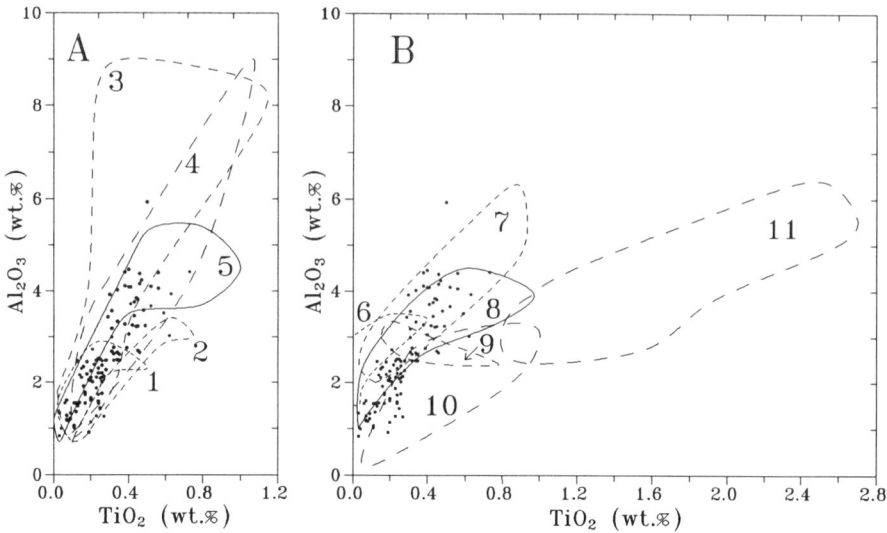

Fig. 4. Clinopyroxene compositions of the Olyutor plutonic complexes compared with arc cumulates (A) and ophiolite and within plates cumulates (B). Dots - clinopyroxene compositions of the Olyutor plutonic complexes. Fields on A: 1 - Blashke Island zoned pluton, Alaska [18], 2 - Bear Mountain pluton, Klamath [30], 3 - Border Range plutonic complexes, South-central Alaska [8, 10], 4 - Aleutian xenoliths [9], 5 - phenocrysts of the Olyutor island-arc volcanics [14]. Fields on B: 6 - North Arm Mountain, Bay of Island [11, 22], 7 - Lewis Hills, Bay of Island [29], 8 - Canyon Mountain, Oregon [17], 9, 10 - Semail, Oman, 9 - intrusive ultramafics [12], 10 - layered complex [26], 11 - plutonic rocks from Piton de la Fornause, Revinuon [3] and Tahiti-Nui caldera, Society Archipelago, French Polynesia [4].

Table 4.

Chemical composition of amphiboles from the Olyutor plutons.

Location	G	G	Sv	Sv	S	S	S	I
Sample	G-1	36-10	51-9	51-11	S2-17	S7-9	S7-10	I12-3
Rock	hc	hc	hd	h	oc	mtg	gd	pc
SiO_2	46.03	42.99	44.08	44.70	46.15	40.54	44.01	43.15
TiO_2	1.83	1.82	1.30	0.87	0.71	1.83	2.22	0.25
Al_2O_3	12.57	13.11	12.62	11.27	11.07	14.12	11.06	11.82
FeO	10.54	6.89	6.01	10.08	7.16	10.40	11.54	10.00
MnO	0.17	0.04	0.09	0.10	0.06	0.12	0.50	0.20
MgO	16.02	16.27	16.72	15.10	17.34	13.94	14.23	15.12
CaO	12.48	12.06	12.48	12.08	12.21	12.10	11.50	11.95
Na_2O	2.31	1.97	2.53	1.67	1.92	2.27	2.21	2.30
K_2O	0.93	1.01	1.19	0.67	0.85	0.91	0.61	11.04
Cr_2O_3	0.00	0.79	1.92	0.63	0.42	0.00	0.00	0.09
H_2O	2.04	2.01	2.10	2.05	2.09	2.01	2.05	2.01
Total	98.82	99.05	101.03	99.22	99.97	98.23	99.96	98.14
Mg#	73.0	80.6	83.2	72.7	81.2	70.5	68.7	72.9

Notes: hc - hornblende clinopyroxenite, hd - hornblende dunite, gd - gabbrodiorite.

Table 3.

Chemical composition of clinopyroxenes from the Olyutor plutons

Location	G	G	G	Sv	Sv	Sv	S	S	S	I	I
Sample	16-2	16-11	36-13	P-D	P-S	282-1	S2-2	S2-7	S2-15	I23-6	I23-1
Rock	w	oc	oc	w	oc	mtc	w	oc	oc	oc	w
SiO_2	53.36	52.86	52.93	53.51	54.04	50.04	53.12	53.93	51.60	52.69	52.84
TiO_2	0.11	0.15	0.12	0.08	0.11	0.46	0.24	0.23	0.21	0.25	0.17
Al_2O_3	1.05	2.1	0.35	1.15	1.54	4.37	1.72	1.36	2.72	2.51	1.73
FeO	0.82	6.03	3.45	3.24	2.89	5.49	4.15	4.7	5.21	4.01	3.76
MnO	0.06	0,2	0.07	0.09	0.10	0.07	0.13	0.10	0.06	0,17	0.12
MgO	17.32	15.23	17.47	17.23	16.71	14.63	16.58	16.74	15.40	16.38	16.57
CaO	23.28	22.56	24.43	23.1	23.96	22.99	24.72	23.22	23.85	23.12	23.11
Na_2O	0.12	0.2	0.11	0.07	0.06	0.17	0.14	0.21	0.05	0.21	0.15
Cr_2O_3	0.24	0.15	0.27	0.94	0.51	0.00	0.42	0.10	0.27	0.52	0.27
Total	99.36	99.50	100.21	99.41	99.92	98.83	101.23	100.49	99.12	99.86	98.27
Mg#	89.0	81.8	90.1	90.5	91.2	82.6	87.7	86.4	84.1	87.9	88.7
En	47.8	43.7	47.3	48.3	49.2	42.7	45.2	46.4	43.7	46.5	46.9
Wo	46.3	46.6	47.5	46.6	50.8	48.3	48.4	46.3	48.1	47.1	47.1
Fs	5.9	9.7	5.2	5.1	4.8	9.0	6.4	7.3	8.3	6.4	6.0

Location	I	G	Sv	S	S	I	I
Sample	I12-3	34-7	85-19	S7-9	S28-9	I6-18	I16-15
Rock	pc	gn	og	mtg	og	g	g
SiO_2	53.00	51.06	51.93	51.74	52.2	50.74	49.09
TiO_2	0.26	0.31	0.24	0.26	0.41	0.41	0.50
Al_2O_3	2.01	3.56	2.59	2.47	3.71	4.08	5.93
FeO	5.45	7.05	4.94	5.71	6.13	6.21	8.03
MnO	0.13	0.31	0.29	0.27	0.10	0.18	0.00
MgO	16.31	14.05	16.53	15.06	15.72	14.58	14.15
CaO	23.20	22.15	22.24	23.32	23.09	22.90	21.90
Na_2O	0.29	0.43	0.36	0.19	0.15	0.36	0.29
Cr_2O_3	0.10	0.00	0.44	0.00	0.00	0.02	0.04
Total	100.79	98.95	99.50	99.03	101.58	99.50	99.99
Mg#	84.2	78.1	85.7	82.5	82.0	80.7	75.9
En	45.2	41.4	46.8	43.0	44.0	42.2	41.2
Wo	46.3	47.0	45.3	47.9	46.4	47.7	45.8
Fs	8.5	11.6	7.9	9.1	9.6	10.1	13.0

Notes: mtc - magnetite clinopyroxenite, pc - plagioclase clinopyroxenite, mtg - magnetite gabbro, g - gabbro, gn - gabbronorite.

complexes are generally flat (Fig. 7). Olivine clinopyroxene cumulates have low total rare earth element abundances and show enrichment in the middle part of rare earth elements spectrum. Olivine clinopyroxenites, gabbros, gabbro-diorites define a continuous series of curves which show a LREE enrichment. According to Gill and others (1981), [16] these patterns are typical for primitive-arc tholeiitic and calc-alkaline series formed at the early stages of the intra-oceanic island-arc evolution.

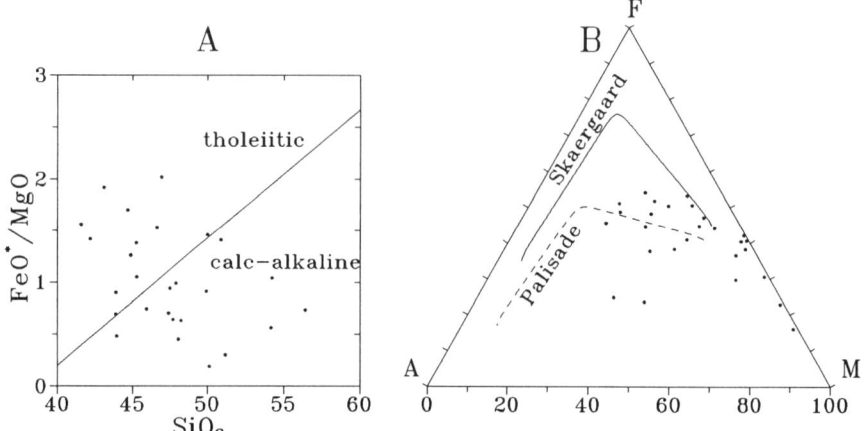

Fig. 5. A. AFM diagram after [25] and B. FeO*/MgO vs. SiO$_2$ diagram after [15]. Dots - rock compositions of the Olyutor plutons.

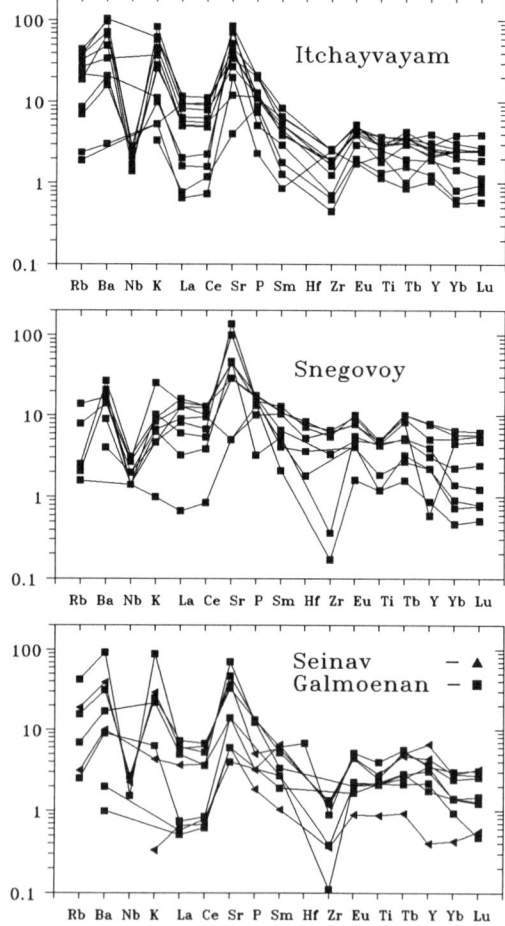

Fig. 6. Trace element contents of the Olyutor plutonic rocks normalized to primitive mantle [32].

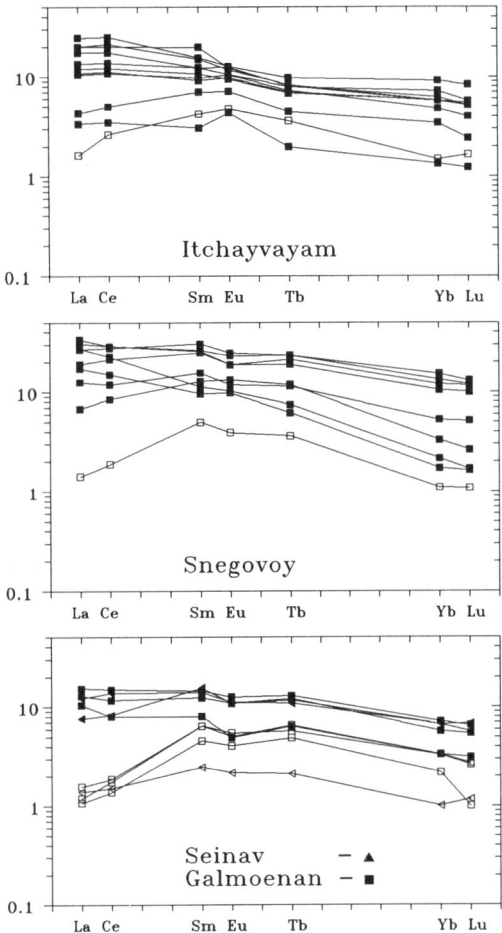

Fig. 7. Chondrite normalized REE patterns of the Olyutor plutonic rocks (open symbols - olivine clinopyroxenites; fill symbols - gabbros and gabbro-diorites). Normalization values from [27].

A comparison of plutonic and Achayvayam arc volcanic suites.
As was mentioned above, the Olyutor plutonic complexes temporally and spatially associate with the Achayvayam suite of Late Cretaceous-Paleocene intraoceanic arc volcanic rocks. These volcanics vary in composition from picrites to basalts. The former are characterized by olivine, chromium spinel and clinopyroxene phenocryst assemblages. A petrological link between plutonic and volcanic rocks is demonstrated by similarities in their mineral compositions (Fig. 3, 4). Moreover, close rare earth element contents of gabbroic and volcanic rocks [14] provide an additional support to the idea of their formation from the same magma types.

DISCUSSION

One of the important results of our study is the establishing of mineralogical and geochemical similarities of North Kamchatka plutonic rocks to island-arc magmatic series, as well as their spatial and genetic association with island-arc volcanics. Revealed

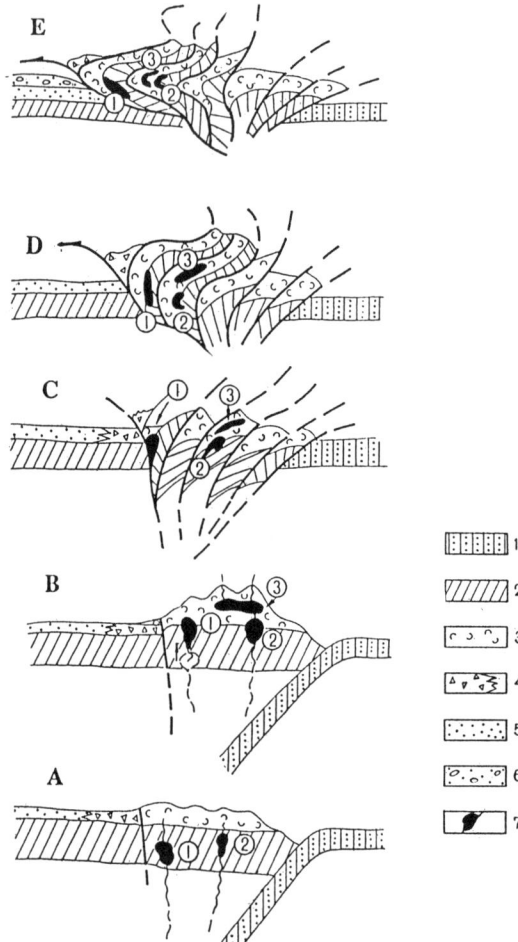

Fig. 8. Evolution of the Achayvayam Late Cretaceous to Paleocene intraoceanic island arc. First stage (A, B) - Emergency of the island arc on the oceanic basement. Second stage (C) - deformation of the island arc and nappe thrusting. Third stage (D, E) - obduction of the pile of nappes on the Ukelayat marginal sea deposits. 1 - undeformed oceanic crust, 2 - partially deformed thickened oceanic crust of Albian - Late Campanian age, 3 - arc related volcanic and sedimentary sequence of Late Campanian - Paleocene age, 4 - olistostrome of Maastrichtian age, 5, 6 - Ukelayat marginal sea flysch of Late Cretaceous - Paleogene age, 7 - location of magmatic chambers of 1, 2 and 3 type plutonic complexes correspondingly.

differences in mineralogy and geochemistry between plutonic complexes may reflect differences in primary melt compositions and magma types involved in their petrogenesis. However, the lack of data at the present preclude determination of the nature of these differences and serial classification of plutonic rocks.

The inferred evolution of the Late Cretaceous - Paleocene intraoceanic Achayvayam island arc, based on the new data, is demonstrated on Fig. 8.

The beginning of the first stage corresponds to the emergence of the arc on an oceanic basement and appearance of island-arc magmatism. This stage was accompanied by formation of magma chambers at different levels in the crust. Mafic-ultramafic cumulates which comprise plutons of the first type appear to be formed in the most deep-seated magma chambers. The early fractionation of olivine-clinopyroxene cumulates took place at pressures 6-8 kbar, according to Hollister's (1987) [19] amphibole geobarometer at depth 15-20 km. High-chromium pargasite from amphibole dunite were used for pressure estimation. This mineral appears to have crystallized close to the liquidus phase. The deformaional structures observed in the ultramafic cumulates provide evidence that the rocks have been recrystallized under the high-temperature conditions. According to Fabrie's olivine-spinel geothermometer (Fabries, 1979) [13], deformation of the dunites occured approximately at 700-900°C [7]. We suggest that deformations of the rocks occured during tectonic emplacement of the intrusive bodies into the upper crust. Prior to total solidification the cumulates were expelled from host magma chambers together with differentiated liquids and emplaced in the country rocks. This emplacement is recorded by the plastic deformations of the cumulates. A significant amount of cumulates could be mobilized as the result of compression of the arc basement.

The second type plutons underwent no plastic deformation. We suppose that they represent shallow magma chambers existed at the beginning of the first stage. The third type plutons were formed later at the first stage, contemporaneously with the main episode of volcanic activity. They appear to represent shallow, only 2-3 km deep, magma chambers of volcano-plutonic systems related to the evolution of island-arc volcanic chains. However, magmas involved in their formation should have undergone some fractionation in deeper chambers.

The second stage of the Achayvayam island arc evolution corresponds to deformation of the arc and nappe thrusting.

During the third stage nappe piles were thrusted over the sequences of the Ukelayat marginal basin. Obtained mineralogical and geochemical data allow us to classify the North Kamchatka mafic-ultramafic plutonic complexes as Alaska-type platinum-bearing ultramafic intrusions. We have predicted [5] that North Kamchatka plutons should also bear platinum and palladium ore deposits, and in fact, platinum places deposites associated with Galmoenan pluton were discovered by a team of geologists of Kamchatka Geological Survey and Insitute of Volcanology, Russian Academy of Science during 1991 field season.

CONCLUSIONS

1. The North Kamchatka mafic-ultramafic plutonic complexes represent magma chambers formed in the crust of the Late Cretaceous - Paleocene Achayvayam island arc.
2. Mineralogical and geochemical data indicate that island-arc magmas were involved in the formation of these plutons.
3. The established differences in mineralogy and geochemistry between plutons may reflect different magma types involved in their petrogenesis, however this point cannot be clarified at the present.
4. The North Kamchatka mafic-ultramafic plutonic complexes correspond to Alaska-type platinum-bearing ultramafic intrusions.

Asknowledgments.
This study was carried out in the Geological Institute of the Russian Academy of Sciences and the paper is a part of Ph. D. dissertation of V. Batanova. Electron microprobe analyses of minerals made in the Institute of Volcanology of the Russian Academy of Sciences. The authors are very grateful to Professors G. Savelieva, A. Perfilyev, S. Sokolov, A. Sharaskin and Dr. L. Danyushevsky for useful disscussion and their critical review of the paper.
We would like to thank G. Polunin for technical assistance in the preparation of the paper.

REFERENCES

1. A.A. Aleksandrov, N.A. Bogdanov, S.A. Palandzhyan and V.D. Chekhovich. Tectonics of the Nothern Part of the Olyutorsk zone of the Koryak Highlands, *Geotectonics* **14**, 242-248 (1980), *(English ed., Amer.Geop.Un. and Geol.Soc.Am.)*.
2. E.S.Alekseyev. Ophiolite complexes of the soustern Koryak Highlands, *Geotectonics* **16**, 313-322 (1982), *(English ed., Amer.Geop.Un. and Geol.Soc.Am)*.
3. T. Auge, P. Lerebour, J.-P. Rancon. The Grand Brule Exploration drilling: new data on the framework of the Piton de la Fornause Volcano, *J. Volcan. Geotherm. Res.* **3**, 139-151 (1989).
4. J.-M. Bardintzeff, H. Bellon, B. Bonin, R. Brousse and A.R. McBirney. Plutonic rocks from Tahiti-Nui caldera (Society archipelago, French Polynesia); a petrological, geochemical and mineralogical study, *J. Volcan. Geothem. Res.* **35**, 31-53 (1988).
5. V.G. Batanova. Mafic-ultramafic plutonic complexes of the southern Koryak Highlands, *Ph.D., Geol.Inst.Rus. Ac. Sci.*, Moscow (1991). *(in Russian, unpubl)*.
6. V.G. Batanova and O.V. Astrachantsev. Tectonic setting and genesis of zoned mafic-ultramafic plutons, northern part of Olyutorskaya zone (Koryak Highlands), *Geotectonics*, **2**, (1992), *(English ed., Amer Geop.Un. and Geol.Soc.Am.)* (in press).
7. V.G. Batanova, O.V. Astrachantsev, E.G. Siderov. Dunite of Galmoenanian basic-ultrabasic massif, Koryak Highlands, *Proceedings of the USSR Ac.Sci.* **1**, 24-35 (1991). *(in Russian)*.
8. L.E. Burns. The Border Ranges ultramafic and mafic complex, south-central Alaska: cumulate fractionates of island-arc volcanics, *Canad.J.Earth.Sci.* **22**, 1020-1038 (1985).
9. W.R. Conrad, R.W. Kay. Ultramafic and mafic inclusion from Adak island: crystallization history and implication for the nature of primary magmas and crustal evolution in the Aleutian arc, *J.Petrol.* **25**, pt. 1, 88-125 (1984).
10. S.M. DeBari, R.G. Coleman. Examination of the deep levels of an island arc: evidence from the Tonsina ultramafic-mafic assemblage, Tonsina, Alaska, *J.Geop.Res.* **94**, NOB4, 4373-4391 (1989).
11. D. Elthon, J.F. Casey, S. Komor. Mineral chemistry of ultramafic cumulates from the North Arm Mountain massif of the Bay of Island ophiolite: evidence for high-pressure crystal fractionation of oceanic basalts, *J.Geop.Res.* **87**, NOB10, 8717-8734 (1982).
12. M. Ernewein, C. Pflumio, H. Whitechurch. The death of the accretion zone as evidenced by the magmatic history of the Semaile ophiolite (Oman), *Tectonophys.* **151**, 247-274 (1988).
13. J. Fabries. Spinel-olivine geothermometry in peridotites from ultramafic complexes, *Contrib.Mineral.Petrol.* **69**, 329-336 (1979).
14. P.I. Federov. Geochemistry and petrology of Late Cretaceous volcanics of the southern of Koryak Highland, *Geochemistry*, **11**, 1583-1594 (1990). *(in Russian)*.
15. D.C. Findlay. Origin of the Tulameen ultramafic-gabbro complex, southern British Columbia, *Can.J.Earth.Sci.* **6**, 399-425 (1969).
16. J.B. Gill. *Orogenic andesites and plate tectonics*. Springer, Berlin (1981).
17. R.G. Himmelberg, R.A. Loney, Petrology of ultramafic and gabbroic rocks of the Canyon Mountain ophiolite, Oregon, *Amer.J.Sci.* **280-A**, pt. 1, 232-268 (1980).
18. R.G. Himmelberg, R.A. Loney, J.T. Craig. Petrogenesis of the ultramafic complex at the Blashke Islands, Southeastern Alaska, *U.S. Geol.Surv.Bull.*, **1662**, 1-14 (1986).
19. L.S. Hollister, G.C. Grisson, E.K. Peters. Confirmation of the empirical correlation of the Al in the hornblende with pressure of solidification of calc-alkaline plutons, *Amer. Miner.* **231**-239 (1987).
20. T.N. Irvin. Chromian spinel as a petrogenetic indicator, *Can.J.Earth.Sci.* **4**, 71-103 (1967).
21. A.D. Kazimirov, O.V. Astrachantsev, A.M. Kheyfets. Tectonics of the Northern part of Olyutor zone. In: *Essay of geology of the Northwestern segment of Pacific tectonic belt*. Y. Pushcharovsky (Ed.), pp.161-183, Nayka, Moscow (1987), *(in Russian)*.
22. S.C. Komor, D. Elthon, J.F. Casey. Mineralogical variationin the layered ultramafic cumulate sequence at the North Arm Mountain massif, Bay of Island ophiolite, Newfoundland, *J. Geop.Res.* **90**, NOB9, 7705-7736 (1985).
23. R. Laurent, R. Hebert. The volcanic and intrusive rocks of the Quebec Appalachian ophiolites (Canada) and their island-arc setting, *Chem.Geol.* **77**, 287-301 (1989).
24. R.R. Loucks. Discrimination of ophiolitic from non-ophiolitic ultramafic-mafic allochtons in orogenic belts by the Al/Ti ratio in clinopyroxene, *Geol.* **18**, 346-349 (1990).
25. A. Miyashiro, F. Shido. Tholeiitic and calc-alkaline series in relation to the behaviors of titanium, vanadium, chromium, and nickel, *Amer. J. Earth. Sci.* **275**, 265-277 (1975).
26. J.S. Pallister, C.A. Hopson. Samail ophiolite plutonic suite: field relations, phase variation, cryptic variation and layering, and a model of a spreading ridge magma chamber, *J. Geop. Res.* **86**, NOB4, 2593-2644 (1981).
27. J.S. Pallister, R.J. Knight. Rare-earth element geochemistry of the Samail ophiolite near Ibra, Oman, *J. Geop. Res.* **86**, NOB4, 2673-2698 (1981).
28. W.R.H. Ramsay, A.J. Crawford, J.D. Foden. Field setting, mineralogy, chemistry and genesis of arc picrites, New Georgia, Solomon Islands, *Contrib.Mineral.Petrol.* **88**, 386-402 (1984).
29. S.E. Smith, D. Elthon. Mineral composition of plutonic rocks from Lewis Hills massif, Bay of Island ophiolite, *J.Geop.Res.* **93**, NOB4, 3450-3468 (1988).
30. A.W. Snoke, J.E. Quick, H.R. Bowman. Bear Mountain Igneous complex Klamath Mountains, California: an ultrabasic to silic calc-alkaline suite. *J.Petrol.* **22**, pt. 4, 501-552 (1981).

31. A.P. Stavsky, V.D. Chechovithc, M.V. Kononov, L.P. Zonenshain. Plate tectonics and palinspastic reconstructions of the Anadyr-Koryak region, Northeast USSR, *Tectonics* **9**, 81-101 (1990).
32. S.-S. Sun, W.F. McDonough, Chemical and isotopic systemetics of oceanic basalts: inplication for mantle composition and proceses. In: *Magmatism in the ocean basins*. A.D. Saunders and M.S. Norry (Eds). *Geol. Soc. Spec. Publ.* **NO 42**, 315-345 (1989).

High and low pressure cumulates of Paleozoic ophiolites in Primorye, eastern Russia.

S.V. VYSOTSKIY

Far East Geological Institute, Vladivostok, 690022, Russia.

Abstract: Two ultramafite-gabbro assemblages were studied in Paleozoic ophiolites of Primorye. They are similar in bulk composition, but sharply differ in mineral paragenesis. The investigation of petrography, mineralogy, and geochemistry of the rocks showed that both assemblages (peridotie - gabbro-norite and peridotite - troctolite) formed from the same-type magmas but at different P-T conditions. The peridotite - gabbro-norite assemblage crystallized at 8-10 kbar within 1200-1000°C temperature interval, while the peridotite - troctolite assemblage crystallized at 5-7 kbar and 1100-900°C. Data obtained suggest that the assemblage with high-pressure mineral parageneses formed under the stress conditions.

Key words: Ophiolites, Primorye, Sikhote-Alin Area, ultramafites, gabbro, mineralogy, petrology, geochemistry.

INTRODUCTION

A chain of outcrops of ultramafites, gabbroides, and diabases associated with Paleozoic basalts and cherts was mapped in the central part of the Sikhote-Alin Area during the long-standing geological investigations [5]. Primarily, researchers suggested basalts associated with cherts to be not originally related to ultramafite and gabbroides. The ultramafite and gabbroides were considered as an independent intrusive complex of the older (up to Proterozoic) age [11, 13, 16]. Further investigations showed that all those rocks represented the tectonically disconnected fragments of a single ophiolite complex [8]. In some ophiolitic allochtons, the relatively complete ophiolite sections were reconstructed [7]. It was suggested that ophiolites formed in the intraoceanic spreading zone and then were dislocated to the foot of the slope of the Late Paleozoic island arc. Other researchers based on some geochemical features of the ophiolites (for example, higher alkali and titanium contents of some complexes of ophiolite association) supposed that the ophiolites formed not in oceanic spreading zone, but in fracture zones inside the continental crust

[13]. The contradictions in the data interpretation suggest that the ophiolite composition of Primorye is non-uniform. It may contain two or more different-type ophiolitic assemblages.

The study of ultramafite-gabbro rocks of ophiolites from two areas in Primorye has shown that the association of different types really exists [14]. The first type, called peridotite-troctolite, is most common in Primorye. It is similar to typical ophiolites of oceans and continents in texture, mineralogy, and geochemistry of the rocks. The second type, called peridotite - gabbro-norite, was revealed as yet only in the north Primorye. Geochemical characteristics of rocks from both types are similar, but their mineral parageneses differ sharply. This allows the specific origin of the peridotite - gabbro-norite assemblage to be suggested. The present paper gives the comparative characteristics of the two types of ultramafite-gabbro assemblages and some ideas of their origin.

GEOLOGICAL CHARACTERISTICS AND AGE OF OPHIOLITES

Paleozoic ophiolites of Primorye are universally dislocated, fragmented, and usually taken part in the sheet-thrusted structures [6,8]. The broken chain of ultramafite-gabbro fragments associated with diabases, basalts, and cherts (Sebucharskaya suite) is traced from the north-eastern coast of the Japan Sea to the Bikin River. Most ultramafite-gabbro units of ophiolite are composed of the rocks of peridotite-troctolite assemblage. Only in the north, in the Bikin River area, the ophiolite fragments of peridotite-gabbro-norite assemblage appear among them. As an illustration, the description of two regions of this chain is given below.

Kalinovsky ophiolite nappe. In the southern part of Paleozoic ophiolite zone (Fig. 1), a large (more than 200 km long and about 5 km thick) tectonic nappe recently called as Kalinovsky allochthon occurs [7,8]. It is composed of peridotite-troctolite assemblage of ophiolites. At the bottom, the Kalinovsky nappe is restricted by the Kalinovsky thrust-fault that plunges at low angle (30-50º) to the south or south-east [6]. Along this fault, the ophiolites are overthrusted upon the Jurassic - Lower Cretaceous terrigenous rocks (Udekovskaya suite). Usually, tectonized serpentinites and gabbroids and in some cases metamorphic rocks (crystalline schist, amphibolite, etc.) are situated at the base of the Kalinovsky nappe. At the top, the Kalinovsky nappe is overlapped by the Middle - Late Jurassic olistostrome.

Although the initial interrelations between members of the ophiolitic association were broken by the superposed tectonic processes, the study of some large outcrops made it possible to reconstruct the original sequence [8]. The foot of the ophiolite section is composed of tectonized serpentinized harzburgites, dunites, and lherzolites. Above, the

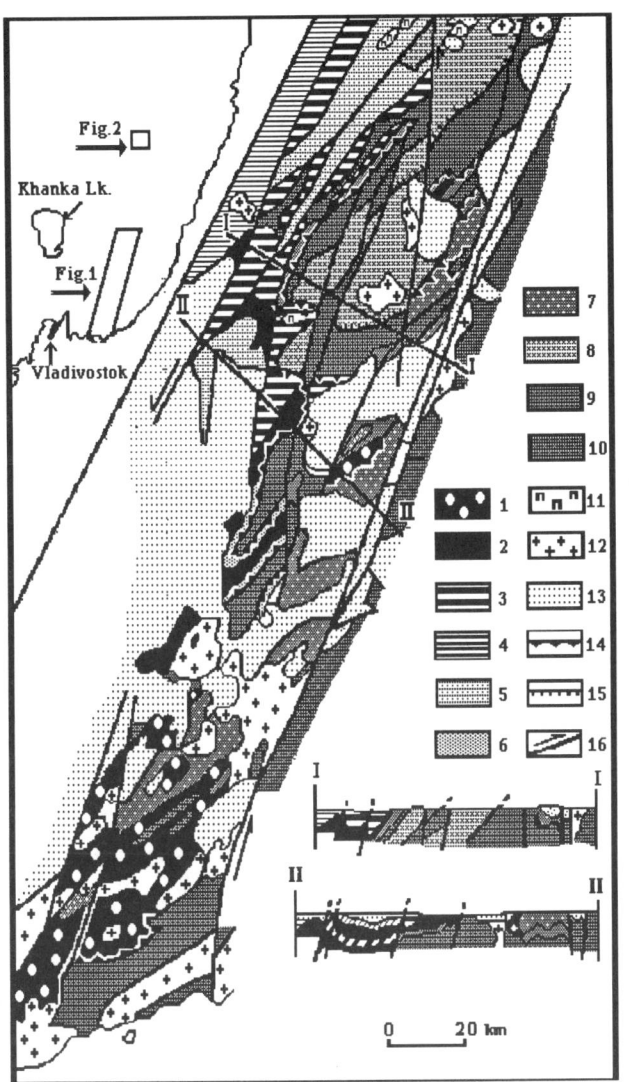

Figure. 1. Geologic scheme of the southern part of the Sikhote-Alin Area [after 6,7,8].

1 - deeply metamorphosed ophiolites of the Sergeevsky nappe (metaultramafites, gabbro-amphibolites, amphibolites, quartzites, and marbles); 2-3 - slightly metamorphosed ophiolites of the Kalinovsky nappe: 2 - ultramafites, gabbroic rocks, plagiogranites (D_3), 3 -diabases, spilites (D_3), cherts (D_3-P_1), limestones (C_1-P_1); 4 - basalts, andesites, dacites, rhyolites, tuffs, terrigenous rocks (P_2); 5 - sandstones, siltstones, siliceous rocks (P_2-J); 6 - sandstones and siltstones; 7 - sandstones, siltstones, in the uppermost section - tuffs of alkaline basalts (P_2-J_3); 8 - siliceous rocks (T_2-J_1); 9 - ophiolite-bearing olistostrome (J_{2-3}); 10 - flysh (K_1); 11 - pyroxenites and monzonites (K_1); 12 - granitoids (K_{1-2}); 13 - rhyolites, basalts, tuffs, terrigenous rocks (K_2-P); 14 - pre-folding overthrust; 15 - syn-folding overthrust; 16 - strike-slip faults.

cumulate complex occurs. It includes plagioclase dunites and troctolites, as well as websterites, wehrlites, clinopyroxenites, and olivine gabbro-norites. Still above, two-pyroxene, clinopyroxene, and amphibole - bearing gabbro are located. The upper part of gabbro unit contains plagiogranite veins. Gabbroides are overlapped by basalts. Boundaries between them are tectonic. In the upper basalts, the relics of pillow with the glassy quenching crusts are preserved. Basalts are overlapped by sedimentary rocks represented by siliceous and calcareous deposits as well as hyaloclastites.

The problem of interrelations between the ultramafic-gabbro base and basalts is not solved yet. In the Kalinovsky nappe area as well as in other ophiolite locations, diabases are common. They are possibly fragments of the sheeted dikes complex. However, it is not completely certain because of their poor exposure.

The age of ophiolites is considered to be Late Devonian. It is based on dating of cherts and limestones with Late Devonian to Early Permian radiolarians and conodonts, which are overlapping the basalts [8].

Bikin ophiolite nappes. In the northern part of the ophiolite zone (the middle Bikin River), relatively large outcrops of ultramafite-gabbro rocks form three isolated massifs - Olonsky, Soldinsky, and Zalominsky. These massifs contain all petrographic varieties of peridotite-gabbro-norite assemblage [14]. Other outcrops of ultramafite-gabbro rocks in this region essentially belong to the monomictic (more rarely polymictic) melange. In rare instances, isolated exposures of those rocks were found as small isometric bodies (probably olistoliths) between volcanogenic, siliceous and terrigenous deposits.

All the massifs are members of the nappes tilted to the south-east at the angle of 30-50°. As an illustration, the scheme of the Olonsky massif, which is the best-studied, is shown on Fig. 2.

The Olonsky massif represents the allochtons plunging to the south-east at the angle of not more than 50°. At the bottom, it is bounded by the tectonic melange, 10-50 m thick. Underlying rocks are basalts and cherts of the Permian age and Early Cretaceous granitoids. The latter intruded along the thrust paleozone. The upper (hanging) contact of the massif is represented by the zone of tectonic clays, 1-10 m thick, dipping to the south-east at the angle of 65°. Overlapping rocks are the same Permian volcanogenic-siliceous deposits. The north part of the massif contacts with the Permian sedimentary rocks along the strike-slip fault trending to the north-west. Its south part is overlapped by Quaternary deposits. The massif is crumpled into isoclinal folds and broken into separate blocks along multidirectional fractures.

Somewhat conventionally, three units of the section may be distinguished - ultramafite, pyroxenite, and gabbro. The ultramafite unit, being the base of the massif, is composed of alternating bands of lherzolites, harzburgites, and websterites, the thickness of which varies from 1 cm to 1m. In the same place, the isolated layers of serpentinized dunites, not more than 2 m thick, occur. The alternating layers are not persistent in strike. They are often rotated relatively each other, so we failed to observe a single section. The thickness of the single fragments, which were not dislocated inside, ranges within 20-150 m whereas their extension is up to 700 m.

In the middle unit of the section, there is an alternation of websterites and orthopyroxenites with interlayers of serpentinized dunites, lherzolites, harzburgites, and, in the upper part, plagioclase pyroxenites. The thickness of this unit reaches 300 m whereas the observed extension is up to 1100 m.

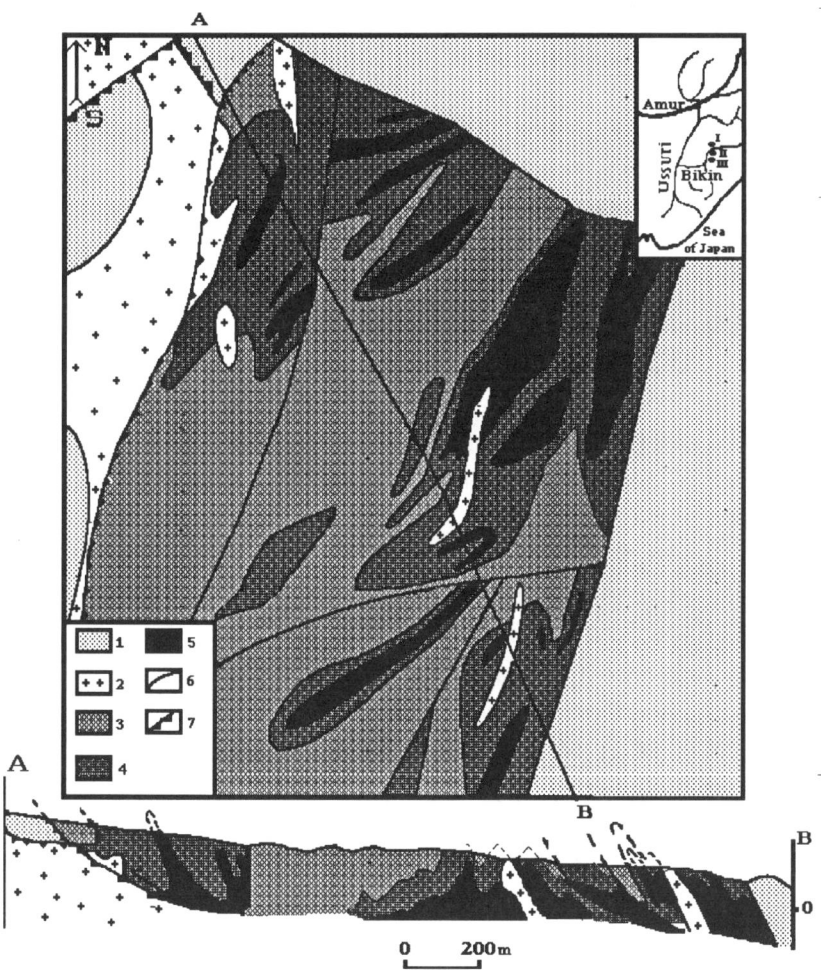

Figure 2. Geological scheme of the Olonsky massif

1 - Late Paleozoic basalts and cherts (Soldinsky complex); 2 - Early Cretaceous granites (Marevsky complex); 3-5 - Olonsky massif: hercynite gabbro-norite (3), pyroxenites, websterites (4), lherzolites, harzburgites, dunites (5); 6,7 - fracture zones: strike-slip faults (6), thrusts (7). Position of massifs is shown in the upper-right corner: I - Olonsky, II - Zalominsky, III - Soldinsky.

The upper unit is composed of bedded hercynite-bearing gabbro-norites. The bedding is resulted from layer-by-layer concentration of plagioclase and pyroxene. The thickness of the plagioclase- and pyroxene-rich layers ranges from 2-3 mm to 1 m. The entire unit reaches 800 m in thickness.

Zalominsky and Soldinsky massifs are similar to the Olonsky one in their structure. The only difference is that ultramafites predominate in Zalominsky massif, as gabbro does in the Soldinsky one. It is not improbable that they all are tectonically separated fragments of the single ultramafite-gabbro sequence.

Basalts and sedimentary rocks framing the massifs are similar to those composing the Kalinovsky nappe in composition and age, but differ from them in the higher extent of dislocation.

PETROGRAPHY, MINERALOGY, AND CHEMICAL COMPOSITION OF THE ROCKS

Peridotite-gabbro-norite assemblage. The assemblage is characterized by banded - lentiform, linear-parallel and gneiss-like (especially in gabbro-norites) structures. The common minerals are monoclinic and rhombic pyroxenes and aluminous green spinel (hercynite). Hercynite was not found only in peridotites. In other rocks, it is presented as either accessory or rock-forming mineral. Peridotites are characterized by high-temperature plastic deformations manifested in the bending cleavage and thin fractures inside pyroxene crystals. Besides, some minerals show the undulatory mosaic extinction. In rare samples, the linear-parallel orientation of minerals was revealed, probably connected with high-temperature recrystallization.

Dunites are usually intensively serpentinized. The least altered samples have the looped texture, on the background of which the relict panidiomorphic texture is seen sometimes. The primary rock contained 94-96% olivine, 3-4% pyroxene, and 1-3% Cr-spinel. The latter is often preserved even in fully serpentinized varieties. As Table 1 shows, dunite minerals are characterized by high magnesium content. Iron number ($f=Fe_{tot}/Fe_{tot}+Mg$) of pyroxenes varies within 6.4-7.0% and olivine - 9.6-10.0%, which are close to those of corresponding minerals from Alpine-type peridotites. The same thing is evidenced by spinel composition, which fall in the field of Alpine-type peridotites by the ratio of chromium and magnesium numbers (Fig. 3).

Spinel lherzolites contain up to 30% of pyroxene, 1-1.5% of Cr-spinel, and 68-75% of serpentinized olivine. Diopside in lherzolites contains more iron ($f=7.4-9.2\%$) and less alumina and chromium, than diopside in dunites. We failed to determine olivine composition, as it is fully serpentinized. Cr-spinel contains more iron and chromium in comparison with spinel of dunites. Evolution of composition towards the increase in magnesium number and decrease in chromium number is well seen in spinel of lherzolite. Initial compositions of spinel are similar to those of plagioclase lherzolites (Fig. 3) from peridotite-troctolite assemblage. However, plagioclase in the rocks considered was not found.

Olivine websterites contain variable amounts of pyroxene and olivine and make up a transitional group of rocks from lherzolites to gabbro-norites. Olivine in websterites has the highest iron number (f=18.2-18.3) and contains lower nickel concentration (Table. 1). Commonly, it is clustered into thin lens-like bands and interbeds 1-2 mm thick. Sometimes, the narrow discontinuous rims of pyroxene and brown-green spinel are observed around olivine.

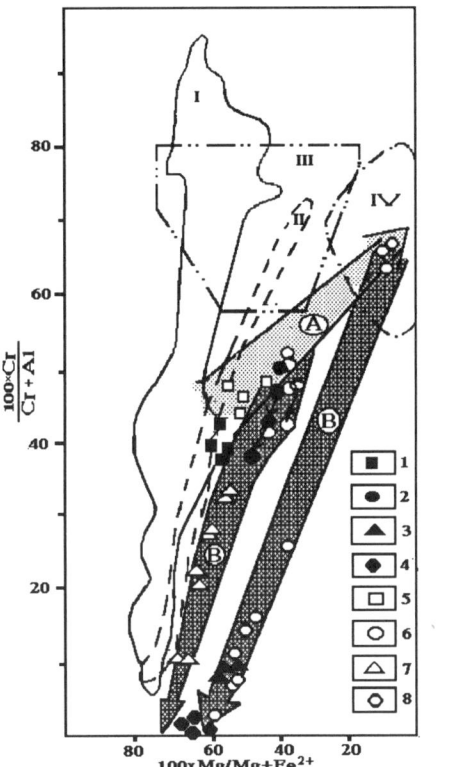

Figure 3. Change of spinel composition in coordinates $X_{Mg\#}$-$X_{Cr\#}$

1-4 - spinels from rocks of the peridotite-gabbro-norite assemblage: dunite (1), lherzolite (2), websterite (3), hercynite gabbro-norite (4); 5-8 - spinels from rocks of peridotite-troctolite assemblage: serpentinized peridotite (5), serpentinized lherzolite (6), plagioclase lherzolite (7), leucocratic troctolite (8). Fields: I-III (after [2]) - Alpine-type peridotites (I), Alpine peridotites (II), layered intrusions (III); IV - ophiolite gabbroides of the outer zone of Tonga arc (from [15]). Thin arrows show the change of zonal spinel composition from core to rim. A - spinel composition change during the olivine-plagioclase fractionation; B - spinel composition change resulted from the reaction between olivine and plagioclase.

Table 1. Mineral composition of ultramafic rocks from the peridotite-gabbro-norite assemblage

	1	2	3	4	5	6	7
SiO_2	40.73	40.80	53.59	52.74	-	-	-
TiO_2	-	-	0.14	0.12	0.25	0.32	0.27
Al_2O_3	0.04	0.01	2.95	2.93	34.93	32.73	33.69
Cr_2O_3	-	-	0.96	0.89	33.05	31.88	32.61
Fe_2O_3	-	-	-	-	1.08	4.28	3.27
FeO	9.43	9.69	2.29	2.13	17.44	18.51	16.54
MnO	0.16	0.15	0.09	0.09	0.27	0.32	0.31
MgO	49.06	48.95	17.08	17.60	13.60	12.23	13.60
CaO	0.02	0.01	23.40	23.13	-	-	-
Na_2O	-	-	0.29	0.16	-	-	-
Total	99.83	100.01	100.78	99.79	100.18	100.28	100.29

	8	9	10	11	12	13	14
SiO_2	-	39.66	54.72	51.96	53.67	53.90	51.31
TiO_2	0.23	-	0.00	0.11	0.07	0.00	0.20
Al_2O_3	31.67	-	2.93	4.53	2.76	4.52	4.39
Cr_2O_3	35.12	-	0.11	0.35	0.24	0.33	0.39
Fe_2O_3	3.86	-	-	-	-	-	-
FeO	16.96	17.19	11.48	4.65	4.19	10.99	4.34
MnO	0.32	0.20	0.17	0.08	0.07	0.17	0.10
MgO	13.34	43.25	31.49	15.56	16.77	29.32	14.93
CaO	-	0.02	0.42	22.96	23.23	0.73	23.19
Na_2O	-	-	0.00	0.38	0.25	0.00	0.35
Total	101.50	100.50	101.32	100.59	101.26	99.96	99.20

	15	16	17	18	19	20	21
SiO_2	-	-	53.61	54.24	53.81	-	-
TiO_2	0.00	0.00	0.00	0.00	0.00	0.07	0.08
Al_2O_3	53.50	55.39	1.97	1.31	1.21	26.03	31.98
Cr_2O_3	7.82	7.09	0.61	0.18	0.13	34.45	29.16
Fe_2O_3	5.01	4.30	-	-	-	7.56	7.69
FeO	19.26	19.14	2.74	2.99	2.49	22.46	20.58
MnO	0.15	0.15	0.05	0.04	0.02	0.39	0.35
MgO	13.59	14.04	16.54	16.62	17.44	8.59	10.67
CaO	-	-	24.13	24.32	24.37	-	-
Na_2O	-	-	0.06	0.18	0.19	-	-
Total	99.34	100.12	99.71	99.88	99.66	99.54	100.51

Note: 1-8 - diopside-bearing dunite, Zalominsky massif [1,2- olivines, Total includes 0.4% NiO; 3,4 - clinopyroxene (3 - core, 4 - rim); 5-8 - spinels (5 - intergranular, 6 - in olivine, 7,8 - intergranular, core (7) and rim (8)]; 9-16 - olivine websterite, Olonsky massif [9 - olivine, Total includes 0.18% NiO, 10-12 - orthopyroxene (10) and clinopyroxene (11 - core, 12 - rim around olivine), 13-14 - large crystals of orthopyroxene (13) and clinopyroxene (14), 15,16 - spinel, core (15) and rim (16)]; 17-23 - apolherzolite serpentinite, Olonsky massif [17-19 - clinopyroxene, core (17) and rim (18) of a large crystal, core (19) of small crystal, 20-21 - spinel, core (20) and rim (21)]. Here and further, the analyses were carried out in the Far East Geological Institute, Far Eastern Branch of the Russian Acad. Sci., on the X-ray microprobe JXA-5A. FeO - common iron, Fe_2O_3 is calculated from the stoichiometric model of spinels. Dash - not analyzed.

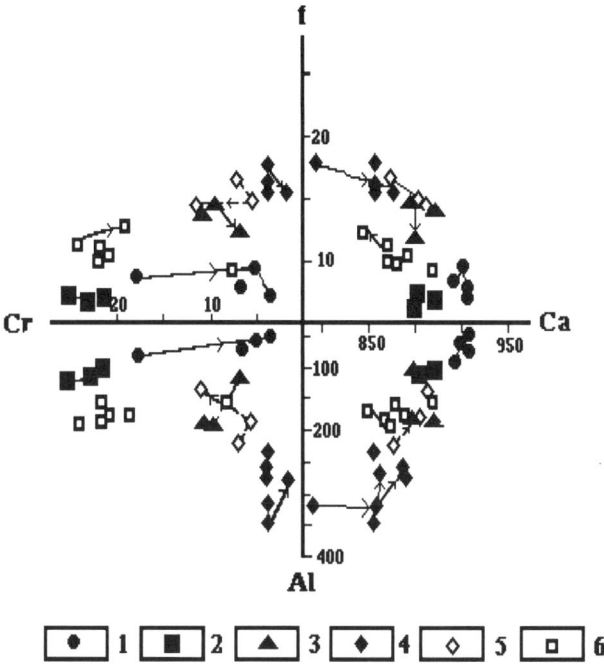

Figure 4. Variations of clinopyroxene composition
1-4 - peridotite-gabbro-norite assemblage: lherzolite (1), dunite (2), websterite (3), hercynite gabbro-norite (4); 5,6 - peridotite-troctolite assemblage: plagioclase lherzolite (5), troctolite (6). Thin arrows show the change of zonal pyroxene composition from core to rim, arrows with dash line - from early to late phases. Cr, Al, Ca - formula coefficient x 1000, $f = (Fe_{tot}/Fe_{tot}+Mg)$, mol. %.

Orthopyroxene (bronzite) in large crystals of groundmass contains more alumina, chromium, and iron than that in small segregations around olivine. Analogous features were revealed in clinopyroxenes also (Table 1). Calcium content is different in ortho- and clinopyroxenes. It is decreasing in orthopyroxene and increasing in clinopyroxene by passing from large crystals of groundmass to small segregations in the olivine rims (Fig. 4). Probably, this is due to the growth of those rims with decreasing temperature.

Spinel in websterites shows high alumina and low chromium contents. It is zonal, with the highest alumina and magnesium contents and the lowest chromium content in the rims.

Hercynite-bearing gabbro-norites are melanocratic and leucocratic. There are all transitions from almost feldspar-free pyroxenites to nearly monomineral plagioclasites, often alternating, among them.

Plagioclase in the samples analyzed (Table 2) is represented by labrador-bytownite containing 69-75% An. Some crystals are zonal. Their zoning is reverse, with calcium increasing from core to margin.

Table 2. Mineral composition of hercynite gabbro-norite

	1	2	3	4	5	6	7	8	9	10	11	12	13
SiO_2	49.52	48.38	52.26	52.38	50.68	50.84	42.43	-	48.89	52.18	50.25	51.15	-
TiO_2	0.02	0.03	0.14	0.11	0.35	0.43	1.75	0.02	-	0.03	0.23	0.29	0.08
Al_2O_3	31.40	32.32	6.33	5.84	7.44	6.02	14.24	62.42	33.23	6.58	7.92	6.59	62.80
Cr_2O_3	-	-	0.11	0.09	0.14	0.15	0.43	1.61	-	0.05	0.12	0.08	0.43
Fe_2O_3	-	-	-	-	-	-	-	2.63	-	-	-	-	2.81
FeO	0.06	0.13	11.68	12.11	5.62	4.62	7.28	17.00	0.09	12.03	5.01	4.66	15.58
MnO	0.01	0.03	0.21	0.22	0.13	0.12	0.09	0.15	0.00	0.22	0.14	0.16	0.18
MgO	0.00	0.00	27.34	28.44	14.62	14.29	16.15	16.08	0.01	28.63	13.17	14.32	16.89
CaO	14.45	15.11	1.01	0.73	20.21	22.49	13.57	-	15.66	0.80	21.62	21.97	-
Na_2O	3.53	3.10	0.08	0.08	0.83	0.46	3.02	-	0.06	0.01	0.01	0.02	-
K_2O	0.02	0.02	0.01	0.00	0.00	0.00	0.19	-	0.06	0.01	0.01	0.02	-
NiO	-	-	0.06	0.04	0.04	0.04	0.05	0.19	-	0.04	0.04	0.05	-
Total	99.02	99.12	99.23	100.04	100.06	99.46	99.21	100.09	100.76	100.63	99.36	100.12	98.78
f (An)	(69)	(73)	19.3	19.3	17.7	15.3	20.2	40.3	(75)	19.1	17.6	15.4	37.4

Note: 1,2,9 - plagioclases (1 - core, 2 - rim of one crystal); 3,4,10 - orthopyroxenes (3 - core, 4 - rim of one crystal); 5,6,11,12 - clinopyroxenes (5,11 - core, 6,12 - rim), 7 - amphibole; 8,13 - spinels. 1-8 - minerals of sample B-530/4; 9-13 - minerals of sample B-530/6.

Orthopyroxene is represented by hypersthene with relatively stable iron number (f=19.2+/-0.1), and variable concentrations of some other elements. Zoning is well manifested in calcium distribution. CaO decreases from the core to rim that indicates the crystal growth with lowering temperatures. At the same time, chromium and alumina concentrations decrease. In iron number, hypersthene of gabbro-norite is similar to hypersthene of leucocratic troctolites from peridotite-troctolite assemblage, but differs from that in higher alumina and calcium and lower chromium contents.

Clinopyroxene, like hypersthene, possesses a strongly pronounced zoning connected with the temperature lowering during the crystal formation. In clinopyroxene, iron, alumina, and chromium decrease from core to margin . At the same time, calcium concentration sharply increases (Fig. 4). According to chemical parameters, pyroxenes of gabbro-norites are similar to pyroxenes of basic granulites from Zverevskaya series of the Stanovoy Ridge [12]. High alumina content characterizes pyroxenes and amphibole, (Table 2).

Spinel is a typical hercynite and has high alumina and low chromium concentrations. In composition, it is close to spinel belonging the exsolution textures in leucocratic troctolites from the peridotite-troctolite assemblage. Both of them are situated near each other in Figure 3.

Peridotite-troctolite assemblage. The rocks of peridotite-troctolite assemblage contain usually the minerals of several generations, among which the primary (magmatic) and secondary (metamorphic) parageneses were distinguished.

Serpentinites formed over harzburgites and, probably, over dunites partly. Primary minerals in serpentinites are rarely preserved, although olivine, pyroxene, and spinel relics are observed in some samples. However, it is not impossible, that in some cases, olivine is a secondary mineral formed during serpentine dehydration. It is indicated, in particular, by both the crystal segregation (isometric crystals forming glomeroporphyritic whisk-shaped aggregates in serpentine) and sharp variations of composition from crystal to crystal (Table. 3).

Table 3 Mineral composition of serpentinites of the peridotite-troctolite assemblage (oxides in mass. %)

	1	2	3	4	5	6	7	8	9	10	11
SiO_2	40.89	41.06	40.68	-	-	-	-	-	-	-	-
TiO_2	-	-	-	0.07	0.13	0.16	0.19	0.33	0.33	0.41	0.30
Al_2O_3	0.25	0.00	0.05	30.18	26.32	27.39	24.31	24.85	27.80	22.40	28.74
Cr_2O_3	-	-	-	34.82	35.61	34.86	34.77	33.24	30.41	35.99	29.82
Fe_2O_3	-	-	-	4.13	8.44	7.26	10.35	9.91	9.29	10.12	9.59
FeO	7.95	6.93	8.61	18.07	17.62	19.29	20.92	25.09	24.40	22.50	21.81
MnO	0.12	0.50	0.19	0.36	0.43	0.39	0.48	0.33	0.29	1.22	1.00
MgO	50.85	51.03	49.71	11.39	12.00	11.04	9.64	7.23	7.90	8.07	9.20
CaO	0.02	0.02	0.02	-	-	-	-	-	-	-	-
Total	100.08	99.54	99.26	100.02	100.55	100.39	100.66	100.98	100.42	100.62	100.46
f	8.0	7.1	8.8	52.9	54.1	56.7	63.8	72.5	69.9	68.7	65.0

Note: 1-7 - minerals from serpentinized harzburgite (sample B-532/1): olivine (1-3) and spinels (4,5 - cores, 6 - rim of large crystals, 7 - core of small crystal); 8-11 - spinels from serpentinized plagioclase peridotite (8,10 - cores, 9,11 - rims).

Iron number of olivine varies within 7-9%, conforming to that of primary olivines from Alpine-type peridotites. However, the latter are not characterized by such a sharp variation of iron number and high admixture content.

In the same sample, the relics of Cr-spinel have been preserved. In composition (Table 3), they are typical chromopicotite and similar to spinel of Alpine-type peridotites (Fig. 3).

Plagioclase peridotites contain 75-80% of serpentinized olivine (f=12-12.3%), 5-7% of pyroxenes, 13-15% of basic plagioclase, replaced by grossular and hydrogrossular saussurite, as well as 1.5-2% of aluminous spinel variable in composition and replaced by magnetite and corundum. Mineral composition is given in Table 4. In tectonic zones, plagioclase peridotites were turned into serpentine-carbonate rock. However, the spinel relics with primary magmatic core and metamorphic rims are preserved in them. The original composition of spinels in core corresponds to that of classic plagioclase peridotites. They both fall on the trend of olivine-plagioclase fractionation on the diagram (Fig. 3). The spinel cores are surrounded by metamorphic rims, in which magnesium content sharply increases and chromium content decreases. It is due to the redistribution of iron and chromium between spinel and enclosing mass during the late-magmatic transformations.

In silicate minerals, the reactionary margins formed at the boundary of plagioclase and olivine at this stage. These margins usually consist of ortho- and clinopyroxenes with the admixture of high-alumina hornblende and brown-green spinel variable in composition.

Troctolites contain 50-70% of basic plagioclase (76-80% An), 20-25% of olivine (f=19.7-20%), 5-10% of pyroxenes, 5-7% of amphibole, and up to 1% of spinel. Troctolites are the most complicated rocks containing several mineral parageneses. The first, high-temperature magmatic paragenesis includes most large crystals of plagioclase, all olivine, and individual crystals of spinel, preserved in plagioclase. This spinel, containing the highest iron and chromium amount (Table 5), falls on the trend of olivine-plagioclase fractionation on the diagram (Fig. 3). It is probable that a part of pyroxenes also formed at the first stage of the rocks' generation. However, it is still impossible to identify them.

The second mineral paragenesis resulted from the reaction between olivine and plagioclase. Commonly, olivine is surrounded by rims of ortho- and clinopyroxene, high-alumina amphibole, and aluminous spinel. All these minerals form the second paragenesis. Besides, the areas of "graphic" intergrowth occur in the rocks. They consist of green spinel, olivine, plagioclase, pyroxenes, and hornblende. In these areas, plagioclase is more basic (84% An), olivine and orthopyroxene have the highest iron number, and clinopyroxene contains the most magnesium amount. Spinel in graphic aggregates is the most aluminous. All minerals inside such areas belong to the second paragenesis. Finally, the third paragenesis is formed by low-temperature minerals - serpentine and magnetite on olivine, sericite on plagioclase, corundum on hercynite, etc.

In addition to the described rocks, in the same assemblage, leucocratic olivine and clinopyroxene gabbro are presented. The former show the same features as leucocratic troctolite, whereas the latter are strongly altered, with only relics of the primary minerals preserved.

In petrochemical characteristics (Table 6), both assemblages are of a single series. It shows tholeiitic trend of differentiation, high magnesium number, and enrichment in

Table 4 Mineral composition of plagioclase lherzolite of the peridotite-troctolite assemblage

	1	2	3	4	5	6	7	8	9	10	11	12
SiO_2	40.04	40.44	55.73	55.80	53.01	52.10	53.05	42.22	38.99	-	-	-
TiO_2	-	-	0.04	0.04	0.12	0.59	0.36	1.68	-	0.73	0.13	0.00
Al_2O_3	0.06	0.03	3.51	3.10	3.58	4.49	4.19	14.96	23.38	36.74	43.50	55.35
Cr_2O_3	0.04	0.05	0.30	0.28	0.28	0.88	0.70	1.33	-	26.99	22.49	10.46
Fe_2O_3	-	-	-	-	-	-	-	-	-	3.43	2.94	3.12
FeO	11.88	11.52	7.86	7.47	2.92	3.56	4.04	6.88	0.15	18.89	16.84	13.91
MnO	0.20	0.17	0.21	0.21	0.10	0.12	0.10	0.08	0.06	0.55	0.27	0.19
MgO	47.49	47.50	32.16	32.72	16.64	15.83	15.89	17.37	0.00	12.39	14.43	17.53
CaO	-	-	0.66	0.61	23.68	22.32	21.85	10.09	37.03	-	-	-
Na_2O	-	-	0.01	0.01	0.55	0.49	0.63	3.51	-	-	-	-
K_2O	-	-	0.00	0.00	0.00	0.00	0.00	0.19	-	-	-	-
NiO	0.24	0.25	0.04	0.04	0.02	0.02	0.04	0.06	-	-	0.19	0.30
Total	100.31	99.96	100.52	100.29	100.90	100.39	100.84	98.37	99.62	99.72	100.59	100.56

Note: 1-2 - olivines; 3-4 - orthopyroxenes; 5-7 - clinopyroxenes (5,6 - core, 7 - rim); 8 - amphibole; 9 - grossular on plagioclase; 10-12 - spinels (10 - in garnet, 11 - intergranular among olivine, 12 - intergranular among pyroxenes).

Table 5. Mineral composition of leucocratic troctolite of peridotite-troctolite assemblage

	1	2	3	4	5	6	7	8	9	10	11	12
SiO_2	39.04	39.26	48.59	48.49	47.62	-	53.49	53.51	50.14	41.79	-	-
TiO_2	-	-	-	-	-	5.93	0.06	0.01	0.14	1.02	0.13	0.00
Al_2O_3	0.03	0.03	33.01	33.18	33.40	6.75	3.78	3.69	5.11	15.20	47.59	62.48
Cr_2O_3	0.00	0.04	-	-	-	17.35	0.27	0.23	0.27	0.41	13.65	1.29
Fe_2O_3	-	-	-	-	-	34.87	-	-	-	-	5.37	2.36
FeO	18.30	18.50	0.08	0.08	0.06	35.55	11.77	12.10	5.45	7.94	22.75	16.98
MnO	0.36	0.35	0.01	0.02	0.01	0.63	0.31	0.30	0.16	0.11	0.34	0.18
MgO	41.86	41.96	0.01	0.00	0.01	1.75	29.61	30.41	15.73	16.40	11.12	16.03
CaO	-	-	16.52	16.35	15.76	-	0.60	0.62	22.13	11.62	-	-
Na_2O	-	-	2.61	2.21	2.71	-	0.06	0.04	0.72	3.42	-	-
K_2O	-	-	0.04	0.03	0.04	-	0.01	0.02	0.02	0.35	-	-
NiO	0.12	0.12	-	-	-	-	0.03	0.03	0.03	0.04	-	-
Total	99.71	100.26	100.87	100.36	99.61	102.82	99.99	100.96	99.90	98.30	100.95	99.32

Note: 1,2 - olivines. 3-5 - plagioclases; 6 - spinel in plagioclase; 7,8 - orthopyroxenes; 9 - clinopyroxene, 10 - amphibole, 11,12 - spinels, 7-12 - minerals of reactionary zones at the olivine-plagioclase boundary.

Table 6. Chemical composition of the rocks of the peridotite - troctolite and pyroxenite-gabbro-norite assemblages (oxides in mass. %, trace elements in ppm)

	1	2	3	4	5	6	7	8	9	10	11	12	13
SiO_2	40.25	38.45	37.50	44.00	43.75	48.10	36.26	43.35	48.85	49.25	42.05	44.80	47.95
TiO_2	0.03	0.17	0.10	0.14	0.20	0.24	0.06	0.10	0.20	0.15	0.09	0.09	0.29
Al_2O_3	0.64	1.04	7.03	21.61	26.22	19.76	1.02	0.95	4.51	4.30	21.55	20.95	15.78
Fe_2O_3	6.33	8.74	6.81	4.92	2.25	2.20	7.24	6.41	5.17	2.37	1.83	1.99	1.90
FeO	2.52	2.60	2.89	2.48	2.00	2.40	2.13	1.60	4.34	5.70	2.55	3.74	5.61
MnO	0.16	0.14	0.13	0.11	0.07	0.09	0.24	0.13	0.15	0.15	0.11	0.07	0.12
MgO	36.53	35.06	31.70	11.33	6.50	8.57	38.98	31.22	20.86	22.93	9.77	12.96	15.02
CaO	0.28	1.17	3.14	9.34	13.25	13.07	1.11	6.52	11.55	12.25	15.86	11.43	11.00
Na_2O	0.10	0.20	0.24	1.80	1.15	1.67	0.08	0.17	0.35	0.25	0.65	0.98	1.32
K_2O	0.05	0.02	0.08	1.17	0.20	0.91	0.10	0.04	0.07	0.05	0.18	0.15	0.12
P_2O_5	0.14	0.02	0.04	0.02	0.04	0.01	0.05	0.04	0.03	0.04	0.04	0.03	0.03
H_2O^-	0.50	0.40	0.33	0.28	0.21	0.28	0.42	0.40	0.28	0.22	0.29	0.28	0.08
LOI	12.06	11.77	9.81	3.49	3.80	2.94	11.58	8.99	3.17	1.21	4.68	2.06	0.80
Total	99.59	99.78	99.80	99.69	100.1	100.2	99.27	99.92	99.53	99.50	99.65	99.53	100.0
f(%)	11.2	14.3	18.8	25.5	25.8	22.3	11.1	11.7	19.5	16.1	19.4	19.3	21.4
Ni	1100	640	650	190	120	64	1815	1000	460	410	190	190	310
Co	88	170	120	45	20	26	63	130	110	93	72	71	57
Cr	1400	480	970	560	170	120	1170	1100	930	1900	580	200	1000
V	31	34	38	36	20	63	14	64	130	100	60	49	110
Cu	11	34	45	45	31	92	20	91	120	70	14	17	90
Zn	78	54	46	46	30	33	47	41	54	30	36	52	42

Note: 1-6 - peridotite-troctolite assemblage: 1 - serpentinized harzburgite, 2 - serpentinized lherzolite, 3 - serpentinized plagioclase lherzolite, 4 - leucocratic troctolite, 5 - leucocratic olivine gabbro, 6 - leucocratic gabbro. 7-13 - peridotite-gabbro-norite assemblage: 7 - diopside-bearing dunite, 8 - lherzolite, 9,10 - olivine websterite, 11-13 - hercynite gabbro-norites. Samples 1-6,12,13 - Soldinsky massif, 7 - Zalominsky massif, 8-11 - Olonsky massif. Analyses were made in Far East Geological Institute, analysts S.P. Batalova (major-elements, wet chemical analysis), and T.V. Lankova and T.K. Babova (trace elements, spectrophotometry).

nickel and chromium at low titanium and alkali concentrations. On AFM diagram (Fig. 5), the rocks of both associations occupy the fields of cumulates and upper gabbro of classic ophiolites of the world. The Table and Figure show that the rocks of two associations, sharply different in mineral composition, are very close in chemical composition of the entire rock.

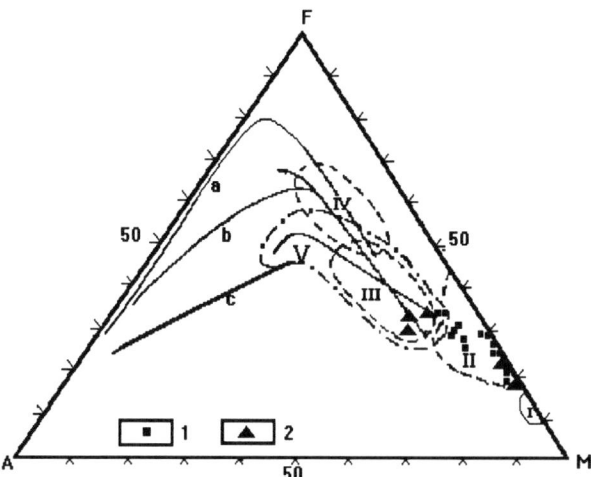

Figure 5. Crystalline rocks of ophiolite assemblages on diagram AFM
1 - peridotite-gabbro-norite assemblage; 2 - peridotite-troctolite assemblage. Fields: (from [1,3]) - I - restite, II - cumulates and lower gabbro, III - upper gabbro and rocks of dike complex, IV - oceanic tholeiites, V - rocks of marianite-boninite and island-arc series. Differentiation trends: a - Skaergaard, b - Tingmuli, c - calcareous-alkaline volcanites of Cascade Mountains (from [1]). Arrows show trends of differentiation of oceanic tholeiites and island-arc dolerites (from [3]).

DISCUSSION

The data given above suggest, that both associations of crystalline rocks are generated from one-type magma, but under the different P-T conditions. The reconstruction of P-T conditions using the two-pyroxene thermobarometers [4,9], shows that rocks of the peridotite-gabbro-norite assemblage formed at higher temperatures and pressures than those of the peridotite- troctolite one (Fig. 6). Thus, for the cores of zonal pyroxenes from hercynite gabbro-norite, the crystallization temperature was about 1200 °C, whereas the most high-temperature pyroxenes from plagioclase lherzolites showed temperature 100° less. Regular change in the composition of zonal pyroxenes from gabbro-norites is well explained by the temperature drop during the rocks formation. Then, the temperature change was not less that 100 °C. Fig. 6 shows that rims of the zonal pyroxenes from gabbro-norites formed at temperature 1100+/-50 °C, that is in the same interval as most

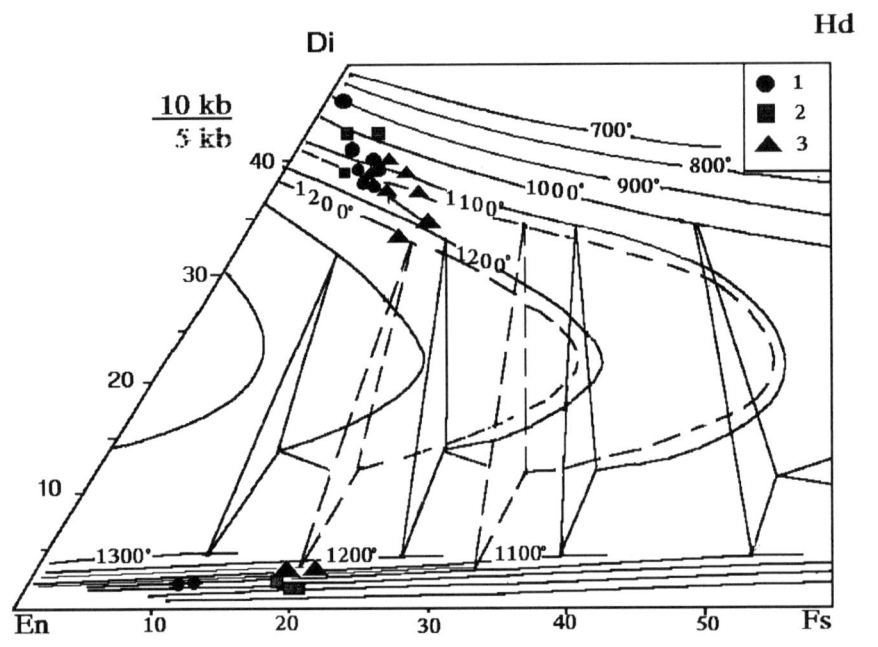

Figure 6. Pyroxenes of crystalline rocks of ophiolite assemblages on diagram by D. Lindsley (from [9]) 1,2 - peridotite-troctolite assemblage: plagioclase lherzolite (1), troctolite (2); 3 - hercynite gabbro-norites. Solid line shows isotherms at P = 10 kbar, dash line - at P = 5 kbar.

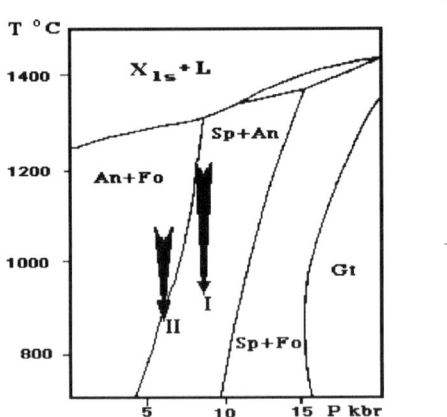

Figure 7. Phase diagram for system $CaO-MgO-Al_2O_3-SiO_2$ (CMAS) by T. Gasparik [4]
An - anorthite, Fo - forsterite, Cpx - clinopyroxene, Opx - orthopyroxene, Sp - spinel, Gt - garnet. Arrows show crystallization trends of rocks of peridotite-gabbro-norite (I) and peridotite-troctolite (II) assemblages.

high-temperature pyroxenes from plagioclase lherzolites and troctolites. At the same time, the most low-temperature pyroxenes of plagioclase lherzolites and troctolites were generated at 1000-900 °C. Thus, the intervals of pyroxene crystallization in both rock's groups are approximately the same (~200 °C), while the initial and final temperatures of pyroxene formation in hercynite gabbro-norites are higher than those in troctolites and plagioclase lherzolites. This fact may be explained if peridotite-gabbro-norite assemblage formed at higher pressures than peridotite-troctolite one.

Accepting this assumption, one may estimate P-T conditions of the rock formation with the help of T. Gaspariks two-pyroxene thermobarometer [4]. In this case, hercynite gabbro-norites developed at 1200-1000 °C and 8-10 kbar, while troctolites and plagioclase lherzolites formed at 1100-900 °C and 5-7 kbar (Fig. 7). It should be noted, that pressures obtained are somewhat excessive, as the thermobarometer was made for the magnesian iron-less systems. In our case, pyroxenes contain large amount of iron, the presence of which, as known, shifts the reaction to the field of lower pressures. However, if the absolute values may be excessive, then the pressure variation is not changing. This pressure variation explains why the rocks similar in bulk chemical composition so differ in mineralogy. The formation of plagioclase peridotites and troctolites (as probably, the whole assemblage) occurred mostly within the field of the olivine-plagioclase stability and along the border of the orthopyroxene-clinopyroxene-spinel stability under the constant pressure. Finally, at temperature about 900°C, the crystallization trend crossed the invariant curve. Then, the olivine-plagioclase association became unstable. As a result, reactionary rims consisting of two pyroxenes and hercynite were originated between olivine and plagioclase. Since this episode was relatively short in time, no complete recrystallization occured. We can see only the regular change in composition of some minerals, which indicates the reaction trend.

At the same time, the peridotite-gabbro-norite association originally formed at the higher pressure when olivine and plagioclase could not exist together. In this case, the ortopyroxene-clinopyroxene-spinel paragenesis is stable. As a result, the two-pyroxene - spinel and two-pyroxene - plagioclase rocks replaced the olivine-plagioclase ones. The former were crystallized at both the higher pressure and temperature.

The banded - lentiform and gneiss-like textures of rocks from the peridotite-gabbro-norite assemblage suggest that they formed under the lateral stress. Such conditions may appear in the zones of local spreading associated with strike-slip faults. In these zones, the regime of extension, given rise to the magmatic chambers in the crust, may be quickly changed for the regime of local compression. The latter may be resulted in crystallization of high-pressure mineral parageneses. We may predict a discovery of the high-pressure peridotite-gabbro-norite assemblages in transform-fault zones located in modern oceans and back-arc basins (for instance, Parece-Vela Basin, Lau Basin and Cayman Trough).

Acknowledgments: The author is deeply thankful to S.A. Shcheka, whose critical remarks and advises improved the manuscript.

References:

1. R.G. Coleman. *Ophiolites.* Mir Publishing House, Moscow (1979) *(in Russian).*
2. H.J.B. Dick and T. Bullen. Chromian spinel as a petrogenetic indicator in abyssal and Alpine-type peridotites and spatially associated lavas, *Contrib. Miner. and Petrol.* **86**, 54-76 (1984)
3. N.L. Dobretsov, E.G. Konnikov, V.P. Medvedev and E.V. Sklyarov. Ophiolites and olistostrome of the East Sayan area. In: *Riphean-Lower Paleozoic ophiolites of the North Eurasia* N.L. Dobretsov (Ed.). pp.34-57 Nauka Publishing House, Siberian Branch, Novosibirsk, (1985) *(in Russian).*
4. T. Gasparik. Two-pyroxene thermobarometry with new experimental data in the system CaO-MgO-Al_2O_3-SiO_2, *Contrib. Miner. and Petrol.* **87**, 87-97 (1984)
5. I.I. Bersenev (Ed.). *Geology of the USSR. V.XXXII. Primorye region.* Nedra Publishing House, Moscow (1969). *(in Russian).*
6. V.V. Golozubov and N.G. Melnikov. *Tectonics of the geosynclinal complexes of the South Sikhote-Alin Area.* Far East Scientific Centre Publishing House, Acad. Sci. of the USSR. Vladivostok, (1986).*(in Russian).*
7. A.I. Khanchuk,I.V. Panchenko and I.V. Kemkin. *Geodynamic evolution of the Paleozoic and Mesozoic Sikhote-Alin and Sakhalin.*Pabl. Far East Branch Acad. Sci.USSR, Vladivostok, (1988). *(in Russian).*
8. A.I. Khanchuk, I.V. Kemkin and I.V. Panchenko. Geodynamic evolution of the Sikhote-Alin and Sakhalin in Paleozoic and Mesozoic .In: *Pacific Margin of Asia. Geology.*A.D. Shcheglov and V.I.Shuldiner (Eds). pp. 218-254, Chap. 13, Nauka Publishing House, Moscow (1989). *(in Russian).*
9. D.H. Lindsley. Pyroxene thermometer. *Amer. Miner.* **68**, 477-493 (1983).
10. S.A. Scheka (Ed.). *Magmatic rocks of the Far East.* Far East Scientific Centre Publishing House, Acad. Sci. of the USSR. Vladivostok, (1973). *(in Russian).*
11. A.O. Mazarovich *Tectonics development of the South Primorye in Paleozoic and Early Mesozoic.* Nauka Publishing House, Moscow (1985). *(in Russian).*
12. I.V. Panchenko. *Geology and evolution of metamorphism of Lower-Pre-Cambrian complexes of the Stanovoy Ridge.* Far East Scientific Centre Publishing House, Acad. Sci. of the USSR, Vladivostok (1985). *(in Russian).*
13. A.D. Shcheglov (Ed.). *Volcanic belts of the Eastern Asia.* Nauka Publishing House, Moscow (1984). *(in Russian).*
14. S.V. Vysotskiy and V.N. Okovity. Ophiolites of the North Primorye: petrology of ultramafic-gabbro association. *Tikhookeanskaya Geologiya (Pacific geology).***5**, 76-87 (1990). *(in Russian).*
15. S.V. Vysotskiy. *Ophiolite association of the Pacific island-arc systems.*Far East Scientific Centre Publishing House, Acad. Sci. of the USSR, Vladivostok (1989). *(in Russian).*
16. S.S. Zimin. *Parageneses of ophiolites and upper mantle.* Nauka Publishing House, Moscow (1973). *(in Russian).*

Geology and petrology of the Shimokawa ophiolite (Hokkaido, Japan): ophiolite possibly generated near R-T-T triple junction

S. MIYASHITA[1] and A. YOSHIDA[2]

[1]Department of Geology and Mineralogy, Faculty of Science, Niigata, University, Ikarashi, Niigata, 950-21 Japan
[2]Nagareyama High School, Nagareyama, Japan

Abstract The Shimokawa ophiolite, exposed in the central part of the northern Hidaka zone (Cretaceous accretionary complex), Hokkaido, comprises serpentinite, cumulate gabbros, composite main body of sediments and basaltic rocks, and breccia in ascending order. The main body is further subdivided into three parts from lower to upper: 1) alternation of gabbroic dolerite sheets and clastic sediments (black shale and fine-grained sandstone), 2) complicated association of massive and pillow lavas, dolerites, black shale and fine-grained sandstone, and 3) thick pillow lava. Many lines of field evidence reveal that the basaltic rocks were extruded or intruded into thickly accumulating sediments, though melange-like occurrences are locally observed.

Major and minor element compositions of the basaltic rocks of the main body show N-MORB features. Mineral compositions of clinopyroxene and Cr-spinel are also comparable with those of MORBs. Basalts are classified into 5 groups based on the phenocryst assemblages, as follows: Ol-Pl(-Cr-Spinel), Ol-Pl-Cpx, Pl-Cpx, Pl and aphyric basalts. General crystallization order is olivine (+ Cr-spinel) - plagioclase - clinopyroxene, which is in good agreement with the low pressure crystallization order in MORBs. However, the occurrence of resorbed clinopyroxenes with high Mg-values in some basalts suggests that the Shimokawa complex underwent polybaric fractional crystallization.

The local appearance of breccias consisting mainly of the same constituents as the main body suggests that the collapse of oceanic crust happened immediately after the formation of the ophiolite, possibly at transform faults or normal faults. However, the appearance of exotic limestone pebbles suggests that the sedimentation was also occurring. The provenance of the exotic blocks could have been the trench slope of the continent where large limestone blocks might have been exposed, because the western side of the Shimokawa complex is composed of accretionary complexes which include many limestone blocks.

On the basis of above evidence, it is concluded that the Shimokawa ophiolite was probably generated at or near a RTT (ridge-trench-trench) triple junction. Subsequently, the ophiolite was accreted and melange-like lithofacies were produced in places. Although the age of the Shimokawa ophiolite remains unsolved, the ophiolite might have been generated at the Farralon-Izanagi Ridge or Kula-Pacific Ridge.

key words: ophiolite, ridge-trench-trench triple junction, Hidaka zone, Hokkaido

INTRODUCTION

The Shimokawa ophiolite, situated in the central part of the northern Hidaka zone (Fig. 1), Hokkaido, is composed of ophiolitic but peculiar assemblage of lithofacies, i.e. serpentinites, anorthositic gabbro, thick alternation of dolerite sheets with terrigenous sediments, pillow and massive basalts interlayered with black shale, and breccia in ascending order. The petrogenesis of the Shimokawa ophiolite has long been controversial. Kosaka (1975) suggested first that the metabasaltic rocks have MORB features in respect to bulk chemistry, though many of the analyzed rocks were considerably altered. Bamba (1985) and Koshimizu *et al.* (1986) believed in a forearc origin of the complex based on the contemporaneous occurrence of terrigenous clastics and the behavior of REE contents of basaltic rocks. On the other hand, Miyake *et al.* (1981)

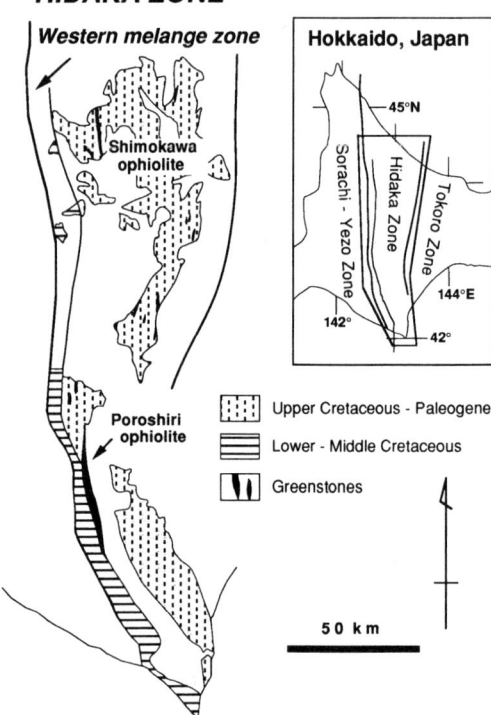

Fig. 1. Tectonic subdivisions of Hokkaido and location of the Shimokawa ophiolite. The Shimokawa ophiolite is located in the northern part of the Hidaka zone.

and Miyake (1988) claimed that the complex is a tectonically mixed melange. Mariko (1984), however, emphasized that the complex has an ophiolitic succession and proposed a Guaymas basin model. Similarly, Miyashita and Yoshida (1988), and Miyashita and Watanabe (1988) claimed coherent relations between the basalts and clastic sediments, and proposed that the complex was generated at an ocean ridge located near a trench.

Basaltic rocks comparable to those of the Shimokawa ophiolite were recently reported from many localities in the northern Hidaka zone (Miyashita and Katsushima, 1986; Miyashita, 1987; Kiminami et al., 1990), and also from the Shimanto zone distributed in southwest Japan (Kiminami and Miyashita, 1992; Kiminami et al., 1992). Kiminami and Miyashita (1989) believed that these basaltic rocks are in situ with surrounding clastic sediments due to the collision of the Kula-Pacific Ridge with the trench. The Shimokawa ophiolite is the largest complex among such in situ basaltic bodies.

In this paper, mode of occurrence, petrographical features, major and minor element chemistry of the Shimokawa ophiolite are described. Based on these data, the origin of the Shimokawa ophiolite is discussed.

GEOLOGIC SETTING

Though the Hidaka zone, consisting of tubiditic clastic sediments and melanges, was regarded as an accretionary complex formed during Early to Late Cretaceous (e.g., Kiminami et al., 1986), recent studies revealed that the ages of this zone, except for the western zone (the Idonnappu zone), range from latest Cretaceous to Paleogene (Kiminami et al., 1990; Nanayama, 1992; Kiminami et al., in press). Many greenstone bodies with MORB features are exposed in the northern Hidaka zone (Miyashita and Katsushima, 1986; Miyashita, 1987). The Shimokawa ophiolite, the largest body among them, extends in a N-S direction over 20 km with a width of 1

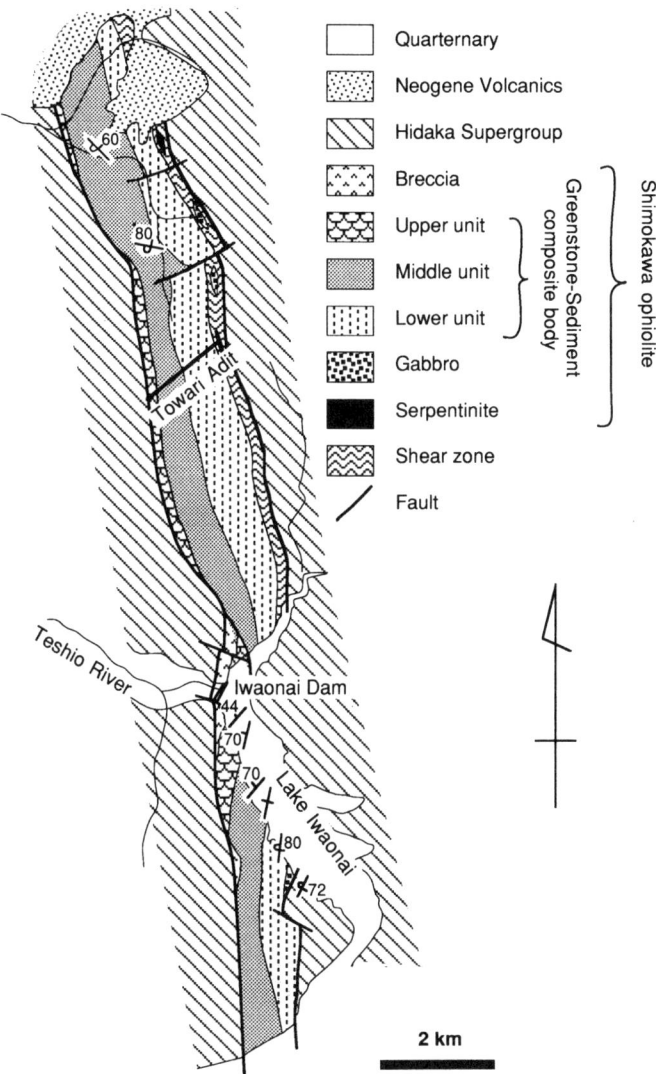

Fig. 2. Simplified geologic map of the Shimokawa ophiolite after Miyashita and Watanabe (1988).

to 2 km in the central part of the northern Hidaka zone (Fig. 2).

The Shimokawa complex consists of two segments separated by the Iwaonai Lake. The north segment shows eastward reversed dips (Mariko et al., 1982), whereas the south segment exhibits westward normal dips. The east side of the complex, i.e. the lowest part of the ophiolite, is bounded by a wide shear zone to the Hidaka Supergroup which is composed of tubiditic sediments. The western side of the complex is also faulted aggainst the Hidaka Supergroup. The internal structures represented by bedded sediments and pillow lavas are approximately parallel with the trend of distribution of the complex, though the strikes of strata are locally bent to WNW-ESE in the northern part of the north segment, and to NE-SW in the northern part of the south segment (Fig. 2). Although the primary structures of the ophiolite are generally well preserved, e.g. pillow form and intrusive boundary of dolerite, faults and sheared structures are locally but ubiquitously observed. Slaty cleavages trending N-S with steep dips are locally developed in black shales. Faults roughly concordant to the cleavage are also observed. Other faults trending NE-SW to ENE-WSW and dipping northward about 40° to 70° are locally present. This fault produced fault breccias a few tens of cm wide, that are calcified and/or silicified.

The complex consists of four lithologic units: serpentinite, metagabbroic cumulate including anorthosite, composite basaltic rocks and sediment (main body), and breccia in ascending order. The serpentinite and metacumulate occur as small lenticular bodies along the eastern marginal shear zone.

The composite basalts-sediments constitute the main portion of the complex, and are subdivided into three parts (Fig. 2). The lower part is characterized by dolerite sheets intruded into alternating black shale and fine-grained sandstone. The largest dolerite sheet attains a thickness of about 200 m in which dolerites show a holocrystalline gabbroic texture, and are called gabbroic dolerite. The middle part is defined by the appearance of pillow lava and is composed of a complicated association of dolerite sheets, massive and pillow basalts, and black shale with intercalation of fine-grained sandstone. The upper part consists mainly of pillow lavas and subordinate massive lavas with a thin intercalation of black and gray shale. Feeder dikes ranging from a few tens of cm to a few m thick are intruded into the main body in places.

Parallel and cross laminae are commonly observed and graded bedding is locally developed in the clastic sediments. Rare coarse-grained sandstones with quartz-feldspathic wacke appear in the alternation of the lower part. The sandstones are composed mainly of quartz, plagioclase and fragments of shale with accessory microcline, biotite, muscovite, epidote, zircon, tourmaline, sphene, opaque minerals and chert fragments. Shales are dark gray to black and exhibit slaty cleavages.

Breccias having a maximum thickness of 200 m overlie the pillow lavas of the main body in the southern segment. The breccias consist mainly of very angular fragments of basalts, dolerites, black shales, and fine-grained sandstones that are the same composition as the main body of the Shimokawa ophiolite. Limestone boulders also appear in the breccias. Rare similar breccias are intercalated also within the middle to upper part of the main body. Sorting is generally very poor and the size of blocks ranges from millimeters to one meter. The breccias are heterogeneous from place to place and intercalated rarely with a thin laminated sandstone layer.

The coherent relation between basalts and terrigenous sediments is confirmed from the following facts. Intrusive boundaries of dolerite are observed at several localities where chilled glassy margins of dolerite exist. Very irregular shaped xenoliths of black shales are occasionally found near the margin of the dolerite sheet. The intruded sediments are not metamorphosed to hornfels but are silicified along the contact, suggesting that a significant chemical interaction occurred due to an expulsion of a large amount of hydrothermal solution from the unlithified sediments by contact with basaltic magma. Furthermore, interpillow black shale indicates that the magma extruded upon unsolidified sediments and rolled up them into interpillow space. The interpillowed sediments are also silicified in places.

On the other hand, melange-like occurrences where basaltic blocks are enclosed in sediment are found in the middle part of the main body along the southern shore of the Iwaonai Lake as emphasized by Miyake (1988). Such melange-like occurrences may be explained by later tectonic movement during the accretionary process or collapse of the Shimokawa ophiolite, as discussed below. Despite the appearance of such melange-like lithofacies, the general stratigraphic succession can be traced along the whole complex as shown in Fig. 2. Detailed correlation of columnar sections is given in Miyashita and Watanabe (1988: Fig. 3).

Andesite to basaltic andesite dikes less than a few m thick are exposed at four localities in the main body. These dikes are free from metamorphism and deformation though the host rocks underwent metamorphism (Mariko, 1984; Yoshida, 1986). Therefore, they are ascribed to a later event after the tectonic emplacement of the ophiolite. Also, metabasalt dikes about one meter wide intruded into metagabbroic cumulates. These basalt dikes suffered from metamorphism and deformation along with the host cumulates, suggesting that the basalt dikes intruded before the tectonic emplacement of this ophiolite. As described later, the basalt dikes show different chemical signatures than the basaltic rocks in the main body of the Shimokawa ophiolite.

PETROGRAPHY OF IGNEOUS ROCKS
Serpentinites
Small bodies of serpentinite less than ten m thick occur along the eastern margin of the complex. These rocks are composed of serpentine minerals such as lizardite and chrysotile, magnetite, calcite and chlorite. Cr-spinel (Cr/(Cr+Al): 0.52-0.56) is the only primary mineral preserved. Rare pseudomorphs of orthopyroxene are observed, so the protoliths of the serpentinites are assumed to be harzburgite to dunite.

Metagabbros
Metagabbro masses appear along the eastern margin of the complex, associated with the serpentinites. The largest mass of metagabbro, about 100 m thick, is exposed along the southern shore of the Iwaonai Lake. This mass comprises anorthosite, anorthositic gabbro, gabbro and clinopyroxenite; anorthositic gabbro is predominant. Anorthosite and pyroxenite occur as a thin layer, a few tens to several cm thick, within the anorthositic gabbro.

Primary minerals observed are plagioclase (An88.2-82.1) and clinopyroxene (Mg value(Mg/ (Mg+Fe)): 0.88-0.87). Plagioclase is euhedral and shows weak normal zoning at the margin. Clinopyroxene is ophitic to poikilitic in anorthositic gabbro and granular in the pyroxenite layer. Metamorphic minerals of hornblende, actinolite, sodic plagioclase and chlorite replace the primary minerals in variable amounts. Abundant clinozoisite, chlorite and hydogrossular occur in anorthositic rocks.

Gabbroic dolerite
Gabbroic dolerites consisting of coarse-grained euhedral plagioclase and ophitic to poikilitic clinopyroxene appear in a thick sheet within the lower part of the main body. Pseudomorphs of olivine and orthopyroxene replaced by chlorite, serpentine minerals and quartz, are common. A small amount of brown hornblende occurs in association with pyroxenes. Considerable amounts of ilmenite appear in evolved gabbroic dolerites. Rare porphyritic plagioclases attaining 1 cm in diameter contain rounded tiny inclusions of Cr-spinel.

Plagioclases and clinopyroxenes of gabbroic dolerite typically show conspicuous normal zoning, suggesting that intensive fractional crystallization occurred on the small scale. The analyses of gabbroic dolerites from IW12 to IW34 are arranged from the lower to upper horizon of a thick gaabbroic dolerite sheet (Table 1). Upward increase in FeO^*/MgO in the gabbroic dolerite sheet reveals that an intensive fractionation occurred on the large scale.

Dolerite and massive basalt
Dolerites and massive basalts are characterized by the coexistence of coarse-grained dolerite with hyalopilitic textures even in thin section. The coarse-grained part with typically poikilitic texture is composed mainly of plagioclase and clinopyroxene in a groundmass of dendritic clinopyroxene, plagioclase and opaque minerals. A considerable amount of ilmenite or olivine (now replaced by secondary minerals) is sometimes observed. Porphyritic plagioclases attaining 1 cm in diameter are common. These porphyritic grains contain a large amount of dusty material which is considered to have been glass originally. Such big plagioclases show complex zoning. Rare Cr-spinel occurs in the porphyritic plagioclase and also in the groundmass.

Pillow basalt
Pillow basalts appear in the middle part and more predominantly in the upper part of the main body. Some pillows are greater than 1 meter in diameter. Close packed pillow structure is usual but inter-pillow black shale is locally present. Vesicularity is extremely low in most basalts.

Some relatively evolved basalts are moderately vesiculated but the vesicles do not exceed 10 modal %.

On the basis of the phenocryst assemblages, five groups are distinguished, as follows: 1) plagioclase-olivine, 2) plagioclase-clinopyroxene-olivine, 3) plagioclase-clinopyroxene, 4) plagioclase, and 5) aphyric type. The first group is predominant. A small amount of Cr-spinel occurs in the first two groups. Magnetite and ilmenite microphenocrysts appear in the plagioclase-clinopyroxene and plagioclase-olivine basalts. The amount and size of phenocrysts ranges from a few to 20 modal % and 0.5 to 2 mm in diameter, respectively. Plagioclases of the plagioclase phyric basalts are sometimes very coarse attaining about 6 mm in diameter. Such large plagioclases have resorbed cores.

It is noted that some basalts contain resorbed xenocrysts of clinopyroxene with rounded to irregular shapes (Miyashita, 1992). As mentioned below, the Pl-Ol basalts with clinopyroxene xenocrysts have the lowest FeO^*/MgO ratios, whereas the Pl-Cpx-Ol basalts without xenocrysts have higher FeO^*/MgO ratios than Ol-Pl basalts.

Metabasalt dike

Several metabasalt dikes less than a few m wide are intruded into anorthositic gabbros. These dikes are composed of aphyric basalt with intersertal to subophitic texture. Though the primary texture is preserved, most plagioclases are altered to albite, and clinopyroxenes to actinolite and epidote. Abundant chlorites appear in the matrix. Accordingly the dikes underwent greenschist facies metamorphism. Furthermore, both the dikes and the host gabbros are partly mylonitized. The basalt dikes are not distinguished petrogragically from the basaltic rocks of the main body, but they probably are derived from a different source than the main body, because of different geochemical signatures.

Later andesite intrusions

Basaltic andesite dikes exhibit intersertal textures and are composed of olivine, hypersthene and augite phenocrysts, and groundmass with plagioclase laths. Hypersthene and augite occur also in the groundmass. Olivine phenocrysts replaced by altered material are observed in less-evolved rocks in which microphenocrysts of Cr-spinel (Cr/Cr+Al: 0.71-0.73) appear. These rocks are moderately to highly vesiculated and vesicles are filled by calcite and clay minerals. Except for olivines, other minerals are very fresh in contrast to the host rocks of the Shimokawa ophiolite. Furthermore, these dikes are free from deformation. Therefore, the basaltic andesites intruded after the tectonic emplacement of the Shimokawa complex.

MAJOR AND MINOR ELEMENT CHEMISTY

Analytical method and secondary compositional change

Sixty-nine samples of the Shimokawa ophiolite (12: gabbroic dolerite, 13: dolerite, 5: massive lava, 29: pillow lava, 3: anorthositic gabbro, 1: serpentinite, 2: metabasalt dike, 4: basaltic andesite) have been analyzed by the XRF method using IKF-3064 (Rigaku Denki). Analytical method of major elements was followed after Nakagawa and Komatsu (1983), and minor elements after Tamura *et al.* (1989). Geo-standard sample of JB-1 was analyzed each six analyses to check the accuracy of analyses. The results are listed in Table 1.

The Shimokawa ophiolite underwent ocean floor metamorphism (Mariko, 1984; Yoshida, 1986) and several elements may be considerably modified from the primary composition. The secondary change of the bulk composition is not the main purpose of this paper, so it is described briefly here. Although several different processes are known for secondary alteration, the degree of modification of the elements usually increases with increasing H_2O (e.g., Hart, 1970; Mottle & Holland, 1978; Miyashiro *et al.*, 1971). K_2O, Rb and Ba contents of basaltic rocks increase with increasing H_2O. On the contrary, CaO contents decrease with increasing H_2O. Therefore, the high concentration of K_2O, Rb and Ba, and low contents of CaO in the rocks rich in H_2O are ascribed to secondary modification. Except for these elements, systematic correlation with the H_2O contents is not observed. Most analyses of the Shimokawa basaltic

Table 1. Bulk compositions of basic rocks of the Shimokawa ophiolite and later intrusives. Total Fe as FeO where the values of Fe_2O_3 are not shown.

	Gabbroic dolerites												Massive basalts		
Sample No.	IWL10	IW12	IW14	IW15	IW18	IW19	IW20	IW24	IW25	IW27	IW28	IW34	IW40	IW42	IW48 V
SiO2 wt%	48.90	48.08	48.81	48.96	49.01	49.19	48.21	50.96	50.04	47.50	49.77	50.38	50.16	49.98	50.53
TiO2	1.07	1.18	1.33	1.39	1.23	1.17	1.13	1.55	1.73	2.23	1.30	1.37	1.34	1.85	1.65
Al2O3	17.08	17.32	16.16	15.84	17.34	20.67	17.70	14.56	16.61	13.15	15.95	15.88	15.34	15.05	14.70
Fe2O3	1.62	2.53	2.58		2.26			1.84	2.44	2.60		2.43	2.39	2.20	2.75
FeO	6.58	6.27	6.32	8.81	5.59	6.83	8.25	8.40	7.10	10.32	9.00	6.36	6.16	7.68	7.19
MnO	0.13	0.12	0.15	0.16	0.14	0.11	0.12	0.18	0.17	0.22	0.14	0.16	0.16	0.18	0.19
MgO	8.09	7.31	9.35	9.08	6.66	6.48	7.71	6.54	6.47	6.56	7.87	7.45	7.43	6.49	7.18
CaO	12.31	11.62	10.63	11.21	11.03	10.66	10.81	10.29	11.22	9.69	11.38	11.63	11.93	8.82	9.50
Na2O	1.51	1.91	1.93	2.15	2.26	2.93	2.10	2.62	2.15	1.81	2.28	2.06	1.71	3.37	4.02
K2O	0.00	0.01	0.13	0.18	0.17	0.26	0.13	0.02	0.00	0.00	0.10	0.00	0.00	0.28	0.23
P2O5	0.07	0.09	0.11	0.15	0.11	0.15	0.12	0.12	0.16	0.20	0.13	0.10	0.10	0.20	0.15
LOI	2.21	2.26	3.00	2.22	2.78	2.16	3.97	2.13	2.10	4.73	2.36	2.30	2.57	2.68	2.83
Total	99.57	98.70	100.50	100.15	98.58	100.61	100.25	99.21	100.19	99.01	100.28	100.12	99.29	98.78	100.92
FeO*/MgO	0.994	1.170	0.924	0.970	1.145	1.054	1.070	1.538	1.437	1.930	1.144	1.147	1.119	1.488	1.346
Sr ppm			160		161			148				135		240	216
Rb			2.5		9.0			2.6				3.5		7.6	8.1
Ni			235		98			78				91		95	82
Zr			94		101			134				96		168	137
Y			34		33			44				36		50	46
Cu			57		40			49				64		44	80
Zn			77		66			89				74		104	93
Nb			0.6		2.9			1.2				1.5		1.6	1.4
Ba			22		29			19				19		68	68
Cr			388		310			310				387		211	260

	Massive basalts and dolerites												IW44	IW58
Sample No.	IW50 V	IW51 (Pl-C)	IW56	IW96	IW97	88-119	88-138	88-139 V	88-140	88-152	88-153	88-153C	88-163	Pl-C	Pl
SiO2 wt%	50.32	50.48	49.61	49.66	49.57	50.20	51.48	50.01	48.15	49.83	49.38	48.96	49.05	48.68	50.95
TiO2	1.66	1.68	1.71	1.49	1.28	1.53	1.99	1.55	2.03	1.16	1.02	1.74	2.38	1.86	1.68
Al2O3	14.89	15.02	14.93	17.24	17.03	16.50	14.72	14.12	16.28	16.64	17.67	16.21	15.01	14.93	15.73
Fe2O3	2.96				2.23									1.79	2.32
FeO	7.14	8.62	7.86	8.35	5.98	8.75	9.50	9.58	10.52	8.38	7.17	9.31	10.88	8.40	6.24
MnO	0.18	0.14	0.12	0.13	0.15	0.13	0.17	0.15	0.14	0.11	0.12	0.16	0.15	0.18	0.17
MgO	7.27	7.23	6.90	6.51	7.42	6.87	6.57	7.02	5.96	7.19	7.10	6.81	5.95	6.90	6.72
CaO	9.13	11.19	11.00	11.26	12.40	10.66	7.18	9.43	10.52	12.36	12.72	11.22	9.78	9.24	12.05
Na2O	3.99	2.64	3.03	2.76	2.27	2.68	4.79	4.35	3.14	2.51	2.33	2.77	2.32	2.16	2.25
K2O	0.29	0.08	0.21	0.13	0.00	0.21	0.14	0.14	0.20	0.13	0.04	0.11	0.19	0.21	0.11
P2O5	0.16	0.19	0.17	0.19	0.10	0.17	0.31	0.19	0.33	0.12	0.12	0.23	0.28	0.21	0.16
LOI	2.93	1.94	3.39	2.06	2.36	1.81	2.21	3.30	1.94	1.54	1.58	1.94	2.55	3.60	1.75
Total	100.92	100.16	99.79	100.77	100.79	99.51	99.06	99.84	99.21	99.97	99.25	99.46	98.54	98.16	100.13
FeO*/MgO	1.349	1.192	1.139	1.283	1.076	1.274	1.446	1.365	1.765	1.166	1.010	1.367	1.829	1.451	1.239
Sr ppm	237		396	200	156	214	54	128	230	140	129	175	205		183
Rb	7.1		5.6	1.5	2.3	2.3	2.4	2.7	4.9	2.5	2.4	2.6	2.7		2.2
Ni	85		84	87	94	74	61	73	56	85	114	93	48		80
Zr	136		138	139	95	118	201	122	211	89	80	150	212		132
Y	46		48	43	37	38	51	38	56	31	27	47	61		46
Cu	30		55	63	52	43	11	57	82	63	62	34	43		61
Zn	76		69	79	68	76	84	97	92	66	60	79	96		75
Nb	3.4		0.6	2.1	0.7	6.0	9.3	5.7	9.1	3.7	5.1	6.1	6.5		0.9
Ba	63		37	16	15	31	14	23	52	23	12	22	37		19
Cr	294		295	229	361	328	185	262	170	303	381	352	232		283

	Pillow basalts												
Sample No.	IW69 Ol-Pl	IW85 Ol-Pl	IW86B Ol-Pl	IW91 Pl	IW100 Ol-Pl-C	IW101 Ol-Pl	IW105A Aph	IW105B Aph	87IW27 Ol-Pl	87IW30 Ol-Pl	87IW32 Pl	88IW1 Ol-Pl	88-101 Pl-C	88-112A Ol-Pl-C	88-114A Pl-C
SiO2 wt%	47.83	47.71	48.71	49.17	49.75	49.66	49.86	49.96	50.79	49.98	50.94	49.42	50.32	49.42	50.25
TiO2	0.91	1.23	1.05	1.18	1.66	0.90	1.29	1.36	1.36	1.33	1.48	1.07	1.29	1.32	1.42
Al2O3	16.76	18.28	16.49	16.89	15.22	16.42	16.42	15.32	15.39	15.55	16.97	17.32	15.45	16.51	15.19
Fe2O3		3.24		1.71			2.59	1.10							
FeO	7.48	5.98	7.66	6.76	8.87	6.94	6.24	7.88	9.00	8.24	8.62	7.89	8.83	8.69	9.59
MnO	0.10	0.13	0.11	0.14	0.14	0.13	0.15	0.16	0.15	0.14	0.11	0.11	0.14	0.16	0.15
MgO	7.83	7.23	7.47	7.42	6.59	7.91	7.22	6.97	7.20	7.49	5.22	7.18	7.10	7.00	7.12
CaO	8.82	11.99	11.67	12.79	11.08	11.69	12.56	9.58	10.40	11.40	11.39	12.45	11.27	11.50	12.22
Na2O	3.51	2.00	3.22	1.80	3.62	3.12	1.90	2.85	3.41	2.29	2.61	2.55	2.55	2.31	2.28
K2O	0.54	0.19	0.14	0.04	0.17	0.15	0.00	0.64	0.17	0.13	0.10	0.14	0.10	0.09	0.08
P2O5	0.08	0.09	0.11	0.08	0.22	0.10	0.09	0.14	0.17	0.14	0.19	0.11	0.16	0.17	0.15
LOI	4.99	2.83	2.57	1.72	1.61	2.16	2.13	3.17	2.00	1.71	2.04	1.67	2.48	1.94	1.09
Total	99.67	100.90	100.05	99.70	99.91	99.94	100.45	99.13	100.04	98.40	99.67	99.91	99.69	99.61	99.54
FeO*/MgO	0.955	1.230	1.025	1.118	1.346	0.877	1.187	1.273	1.250	1.100	1.651	1.099	1.244	1.241	1.347
Sr ppm	281	183	198	143	172	280	163			156	162	153	142	125	
Rb	12.0	2.2	2.7	3.0	4.0	3.6	2.8			3.5	3.6	2.5	3.4	3.1	
Ni	127	80	130	113	97	116	85			62	108	83	87	61	
Zr	62	132	89	86	155	70	92			149	81	91	120	105	
Y	27	46	32	34	49	26	33			44	28	40	42	37	
Cu	62	72	61	67	56	55	69			83	54	50	54	63	
Zn	65	68	68	70	82	58	73			64	69	80	75	84	
Nb	1.8	2.4	2.3	0.9	2.2	1.9				1.7	3.7	5.4	6.1	4.6	
Ba	146	25	21	19	26	41	15			15	14	16	21	2	
Cr	412	368	335	379	232	418	367			203	373	355	534	182	

Table 1. (continued)

	Pillow basalts												
Sample No.	88-114B	88-115A	88-118A	88-122	88-123	88-132A	88-132B	88-134	88-141A	88-141B	88-155A	88-155B	88-156	88-157B
	Pl-C	Aph	Pl-C	Aph	Ol-Pl	Pl-C	Pl-C	Pl-C	Ol-Pl	Ol-Pl	Ol-Pl	Ol-Pl	Ol-Pl	Pl-C
SiO2 wt%	50.32	51.35	50.95	51.66	48.95	49.96	49.06	49.40	52.24	51.75	49.68	48.34	48.83	49.46
TiO2	1.35	1.54	1.76	1.28	1.33	1.53	1.52	1.83	1.18	1.27	1.28	1.30	1.45	1.44
Al2O3	15.26	13.68	15.11	15.25	16.26	15.91	16.09	15.07	15.65	14.31	16.99	17.59	16.86	17.07
Fe2O3														
FeO	9.43	9.03	9.06	8.72	8.57	8.78	9.14	9.89	6.98	7.63	8.07	8.44	8.84	8.55
MnO	0.15	0.16	0.17	0.13	0.15	0.14	0.15	0.17	0.14	0.15	0.13	0.13	0.13	0.12
MgO	7.25	7.18	5.82	7.28	7.15	7.06	7.40	7.37	7.23	7.80	6.55	7.01	5.86	5.69
CaO	12.31	9.23	10.08	10.81	10.97	12.31	12.15	9.07	11.04	11.94	12.04	10.99	11.64	11.86
Na2O	2.28	3.71	3.25	2.61	3.25	2.65	2.55	3.66	3.53	3.06	3.01	2.77	2.32	2.52
K2O	0.07	0.09	0.12	0.13	0.13	0.12	0.11	0.25	0.20	0.13	0.17	0.23	0.07	0.11
P2O5	0.15	0.16	0.21	0.15	0.17	0.20	0.19	0.23	0.13	0.13	0.15	0.15	0.18	0.18
LOI	1.15	3.02	2.40	1.64	2.02	1.21	1.29	2.48	1.66	1.63	1.83	2.51	2.39	2.84
Total	99.72	99.15	98.93	99.66	98.95	99.87	99.65	99.42	99.98	99.82	99.90	99.46	98.57	99.84
FeO*/MgO	1.301	1.258	1.557	1.198	1.199	1.244	1.235	1.342	0.965	0.978	1.232	1.204	1.509	1.503
Sr ppm	122	188	223	157	191	138	138	172	280	213	167	194	161	187
Rb	1.9	1.3	4.4	3.6	3.6	2.6	1.8	5.5	3.4	2.3	3.4	6.0	2.4	2.2
Ni	64	59	70	55	83	102	103	88	112	110	110	127	81	76
Zr	102	123	147	138	123	142	139	171	92	103	108	109	134	135
Y	35.5	40	52	38.8	39.5	42.1	41.8	48.9	29.5	31.7	34.5	34.5	41	38.9
Cu	63	70	54	52	51	54	54	28	29	44	54	55	54	52
Zn	84	84	84	77	80	85	88	91	64	68	74	75	79	76
Nb	4.4	5.7	5.8	5.9	7.2	6.5	6.4	7.4	3.9	4.4	4.4	4.9	4.5	5.5
Ba	13	35	29	17	22	16	23	32	44	30	22	26	11	19
Cr	244	310	391	298	335	328	304	274	319	323	317	328	330	262

	Later andesitic dikes				Metabasalt dikes		Gabbroic cumulates			Serpentinite
Sample No.	IW22	IW23	88-150	88-151	IWL1	IWL6	IWL2	IWL3	IWL5	IWL7
SiO2 wt%	53.59	55.79	54.90	54.19	47.63	46.52	40.54	45.58	42.94	33.80
TiO2	0.72	0.73	0.74	0.73	1.05	0.98	0.15	0.17	0.14	0.04
Al2O3	15.56	16.33	16.30	16.12	15.97	14.79	24.99	24.06	20.57	2.09
Fe2O3	4.56	4.39			3.56	2.87	3.22	1.08	2.11	5.30
FeO	3.72	1.87	6.80	6.43	5.50	6.93	1.75	2.47	3.22	2.57
MnO	0.13	0.12	0.11	0.10	0.15	0.15	0.10	0.06	0.09	0.12
MgO	7.85	3.67	4.34	4.38	9.98	10.78	8.59	8.99	14.99	34.69
CaO	10.11	9.03	7.62	7.30	10.35	9.30	15.52	10.59	11.23	5.40
Na2O	1.62	1.97	2.43	2.40	2.26	2.59	0.95	2.20	0.72	0.06
K2O	0.32	0.87	1.18	1.22	0.01	0.14	0.05	0.19	0.06	0.07
P2O5	0.06	0.08	0.11	0.10	0.02	0.03	0.00	0.00	0.00	0.05
LOI	1.85	3.31	4.30	6.27	2.46	3.36	5.05	4.94	4.81	16.09
Total	100.09	98.16	98.83	99.24	98.94	98.44	100.91	100.33	100.88	100.28
FeO*/MgO	0.997	1.590	1.570	1.470	0.872	0.882	0.541	0.383	0.341	0.210
Sr ppm	170	205	205	201	156	168	251	219	103	36
Rb	13	33	33	35	9.6	4.6	2.6	8.8	4.9	0
Ni	81	37	48	41	89	99	412	263	645	2392
Zr	74	102	98	97	57	52	16	14	11	8
Y	23	23	21	21	31	25	7	5	7	5
Cu	34	18	15	13	20	55	7	102	60	36
Zn	83	81	74	74	59	67	32	33	38	36
Nb	2.8	2.6	3.8	3.3	0.2	0.9	0	0	3.2	3
Ba	159	291	302	309	59	26	0	48	0	8
Cr	265	139	154	133	516	351	462	355	486	2586

rocks contain less than 3 wt% of H_2O and the secondary modification seems to be less serious for these analyses. The analysis of a highly altered rock rich in H_2O (IW69) is omitted from the following discussion.

Major element chemistry

The analyses of basaltic rocks (pillow lavas, feeder dikes, massive lavas, dolerite and gabbroic dolerite) reveal MORB features in respect to very low K_2O contents of less than 0.3 wt%. SiO_2 contents of these basaltic rocks show a narrow range about from 48 to 53 wt% (anhydrous basis) over a wide range of FeO*/MgO from 0.88 to 1.93 (Fig. 3b). On the other hand, TiO_2 contents increase systematically from 0.91 to 2.38 wt% with increasing FeO*/MgO. It is apparent that they follow the trend of MORB and clearly are distinctive from those of IAT and OIT as shown in Fig. 3a. The analyses of andesite dikes show a different trend from that of the Shimokawa basaltics and plot in IAT field (Fig. 3a). The K_2O contents of andesites are considerably higher than those of the Shimokawa basaltics and attain 1.25 wt%, whereas metabasalt dikes are characterized by primitive features as indicated by low FeO*/MgO ratios (0.87 and

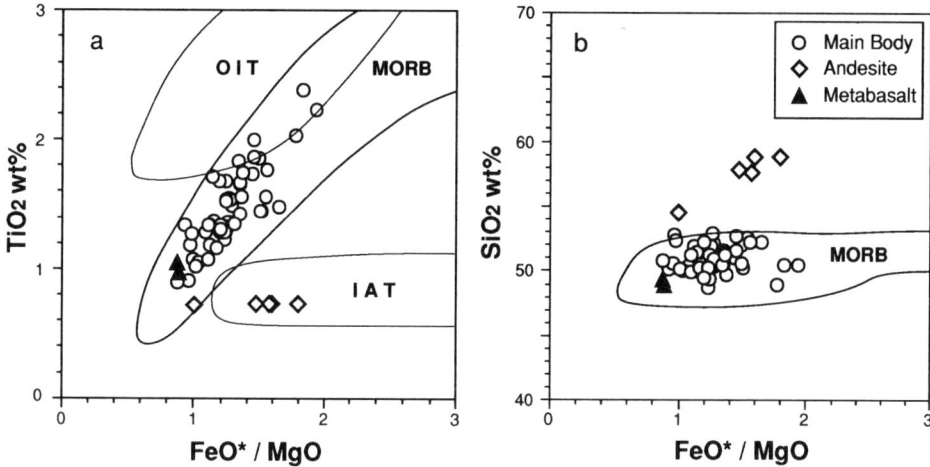

Fig. 3. TiO$_2$ and SiO$_2$ versus FeO*/MgO plots of the igneous rocks of the Shimokawa complex. Fields of MORB, IAT(island arc tholeiite) and OIT(oceanic island tholeiite) are after Basaltic Volcanism Study Project (1981).

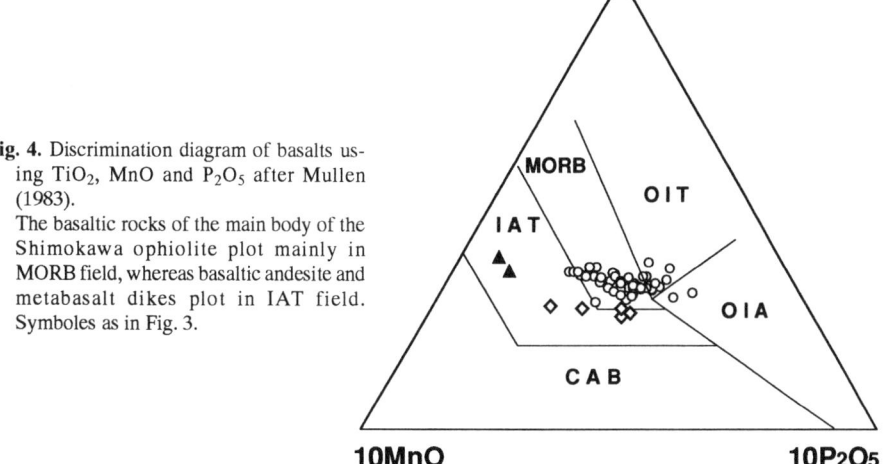

Fig. 4. Discrimination diagram of basalts using TiO$_2$, MnO and P$_2$O$_5$ after Mullen (1983).
The basaltic rocks of the main body of the Shimokawa ophiolite plot mainly in MORB field, whereas basaltic andesite and metabasalt dikes plot in IAT field. Symboles as in Fig. 3.

0.88) and high MgO contents (9.98 and 10.78wt%). As compared with less evolved basaltic rocks of the main body which are lower in FeO*/MgO than 1.0, the metabasalt dikes tend to be lower in SiO$_2$, Al$_2$O$_3$ and CaO contents and higher in MgO and FeO* (Table 1). The metabasalts show also very low contents of K$_2$O. Although the metabasalts plot in a MORB field on Fig. 3, they have different chemical signatures than the main body of the ophiolite with respect to minor element chemistries.

It is noted that dolerites and gabbroic dolerites exhibit a wider compositional range than the basalts, suggesting that intensive fractional crystallization occurred in these dolerites *in situ*. This is indicated from the upward increase in FeO*/MgO ratios in a thick gabbroic dolerite sheet. On the other hand, basalts of each group have narrow compositional ranges, and Pl-Cpx basalts are most evolved among them. Pl-Ol basalts have lower FeO*/MgO ratios than those of Pl-Cpx basalts. Pl-Cpx-Ol basalts are intermediate in FeO*/MgO, falling between Pl-Ol and Pl-

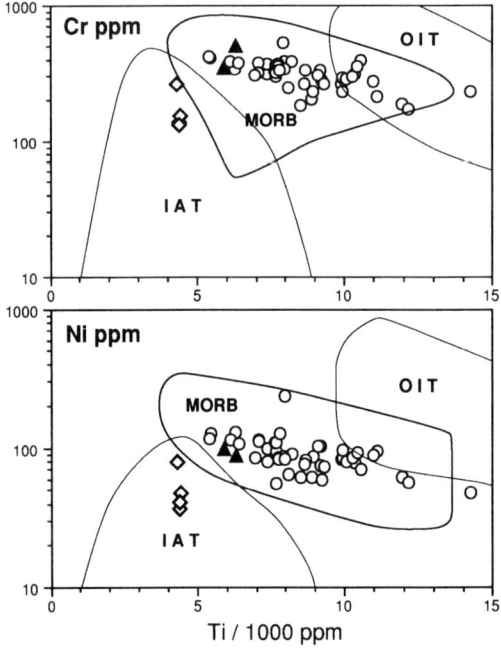

Fig. 5. Discrimination diagram of basalts using Cr, Ni and Ti after Ishizuka (1981).
Symboles as in Fig. 3.

Cpx basalts.

In a discrimination diagram using TiO_2, MnO and P_2O_5, most analyses of the Shimokawa basalts plot in a MORB field whereas basaltic andesites plot in an island arc tholeiite field (Fig. 4). Metabasalt dikes differ from those of the main body and andesites, falling in the IAT field due to very low contents in P_2O_5.

Minor element chemistry

Rb and Ba contents are lower than about 5 and 30 ppm respectively, for most analyses of the main body of the Shimokawa ophiolite. On the other hand, considerable variation in Zr, Y, Cr, and Ni contents are arisen from fractional crystallization. Zr, Y and Nb contents increase systematically with increasing FeO^*/MgO, from 60 to 210, 27 to 61 and 0.6 to 9.3 ppm. Cr and Ni contents ranging from 530 to 170 and 235 to 48 ppm, respectively, tend to decrease with increasing FeO^*/MgO. Sr contents ranging from 129 to 280 ppm, except for two analyses, show no systematic change with FeO^*/MgO.

It is well established that the abundances of certain minor elements of basalts give an important constraint for the tectonic setting in which the basalts were generated. Fig. 5 shows Cr and Ni versus Ti diagrams in which MORB, IAT and OIT plot in distinct fields (Ishizuka, 1981). It is obvious that the Shimokawa basalts plot in the MORB field and are different from IAT and OIT. In a triangle diagram using Ti, Zr and Y (Pearce and Cann, 1973), the Shimokawa rocks plot in the MORB field and are distinct from those of OIT and calc-alkali basalt (Fig. 6a), though the fields of MORB and IAT partly overlap in this diagram. In a second step of discrimination of MORB and IAT using Sr, Zr and Y (Pearce and Cann, 1973), most analyses of the Shimokawa rocks plot in MORB field and are distinct from IAT and calc-alkali basalt (Fig. 6b).

On the other hand, basaltic andesites plot in different fields than those of the Shimokawa ophiolite in every diagram. They plot in IAT field in Cr and Ni versus Ti diagrams, and in calc-alkali basalt field of Fig. 6a and Fig. 6b, which agrees with the characteristics of their major element chemistries. Metabasalt dikes plot in the overlapped field of MORB and IAT in Fig. 6a, and near the boundary between MORB and IAT in Fig. 6b. Taking into consideration the discrimination diagram using P_2O_5, TiO_2 and MnO (Fig. 4), the metabasalt dikes seem to be primi-

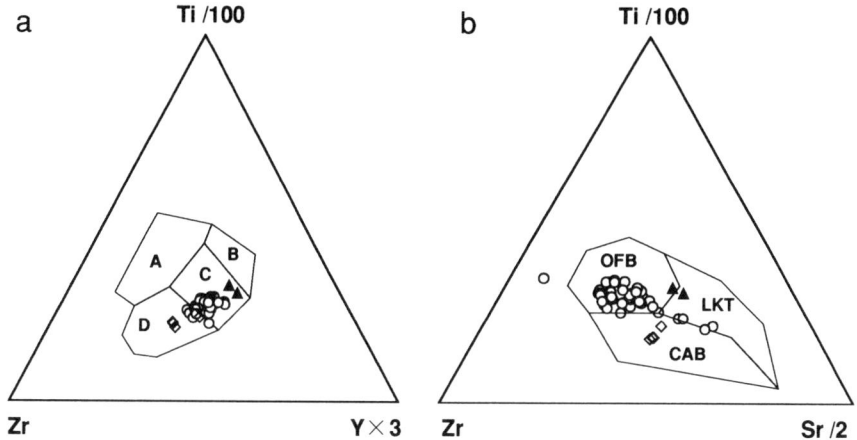

Fig. 6. Discrimination diagrams of basalts after Pearce and Cann (1973).
a: first step of discrimination by Ti, Zr and Y, A: within plate basalt, B: low K tholeiite, C: ocean floor basalt (OFB) and low K tholeiite (LKT), D: calc-alkali basalt (CAB). b: second step of discrimination by Sr, Zr and Y. Symboles as in Fig. 3.

tive volcanic arc basalt.

Three different types of MORBs are distinguished :N-MORB, E-MORB and T-MORB. N-MORB is characterized by depleted LIL elements and Nb, whereas E-MORB is characterized by high concentrations of light REE, Nb and LIL elements such as Ba, K and Rb (e.g., Sun *et al.*, 1979; Basaltic Volcanism Study Project, 1981).

Fig. 7 is a spider diagram of the basaltic rocks of the Shimokawa ophiolite, basaltic andesite, E-MORB, WPB and IAT normalized by average N-MORB composition (Pearce, 1982). E-MORB, WPB and IAT exhibit much higher concentrations of Rb, Ba and K than N-MORB. VAB is generally distinguished from E-MORB and OIT by very low contents of Nb and HFS elements such as P, Zr,Ti and Y.

In Fig. 7a, the Shimokawa basalts are classified into 7 groups based on the phenocryst assemblage and FeO^*/MgO values, and average values for each are shown. Pl-Ol basalts are divided into two groups, Pl-Ol 1 with low FeO^*/MgO and clinopyroxene xenocrysts, and Pl-Ol 2 with high FeO^*/MgO. Dolerites show a wide variation in FeO^*/MgO so they are divided into three groups, dolerite 1 being lower than 1.20, dolerite 3 higher than 1.45 and dolerite 2 being intermediate. All these basaltic rocks have flat patterns similar to those of N-MORB and different from those of VAB, E-MORB and WPB. However, except for the Pl-Ol-Cpx 1 basalt, the other six groups show approximately parallel flat patterns, consistent with the interpretation that these rocks fractionated from the same magma. The Pl-Ol basalt 1 is most depleted in HFS elements but slightly enriched in LIL elements. Therefore, the Pl-Ol basalts may be derived from slightly different sources or degrees of partial melting.

It is obvious that the basaltic andesites demonstrate a pattern similar to that of VAB and are clearly distinguished from the basaltic rocks of the Shimokawa ophiolite (Fig. 7b). In a discrimination diagram using Nb, Y and Zr after Meschede (1986), basaltic rocks of the Shimokawa ophiolite plot in N-MORB and VAB field whereas basaltic andesite dikes plot in WPT and VAB field (Fig. 8). On the other hand, metabasalt dikes plot in N-MORB and VAB fields though they occupy different fields from those of the main body. In the spider diagram of Fig. 7, the metabasalt dikes reveal a pattern different from flat patterns of the main body and they are characterized by very low contents of Nb, P and Zr and by enrichment in Rb and Ba.

Thus, we conclude that the main body of the Shimokawa ophiolite has N-MORB features

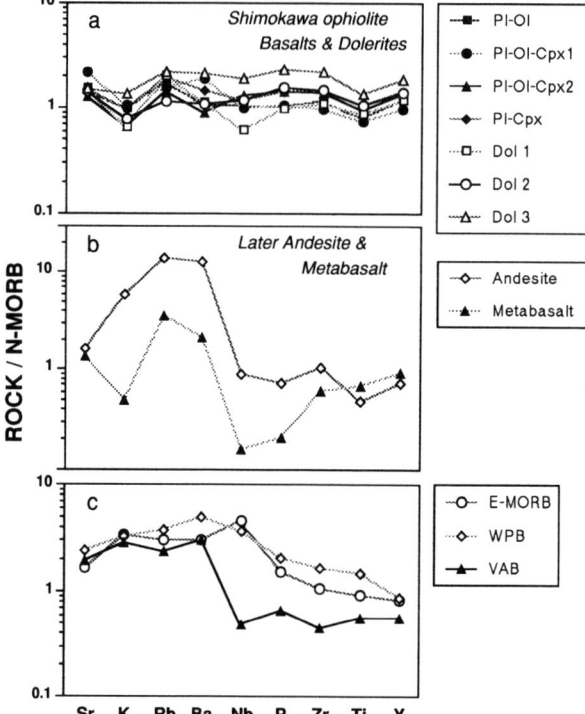

Fig. 7. Spider diagram of igneous rocks of the Shimokawa complex normalized by averaged N-MORB after Pearce (1982).
a: Basaltic rocks of the main body of the ophiolite are classified into 7 groups based on the phenocryst assemblages of basalts and FeO*/MgO values of dolerites.
b: Averages of later basaltic andesites (4 analyses) and metabasalts (2 analyses).

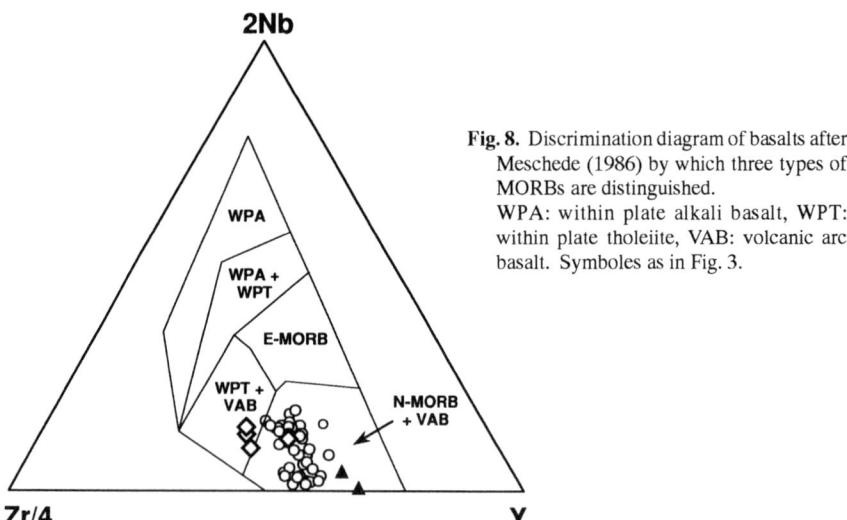

Fig. 8. Discrimination diagram of basalts after Meschede (1986) by which three types of MORBs are distinguished.
WPA: within plate alkali basalt, WPT: within plate tholeiite, VAB: volcanic arc basalt. Symboles as in Fig. 3.

and is clearly distinct from IAT and OIT. On the other hand, metabasalt dikes and basaltic andesite dikes show an affinity to arc volcanism. However, the metabasalt dikes are considerably recrystallized, and may have suffered secondary compositional modification.

MINERALOGY

Details of mineral chemistry of the Shimokawa ophiolite will be described in a separate paper, so they are mentioned here only briefly. Minerals were analysed by the electron microprobe analyser (JXA8600) of Niigata University under the conditons of probe currents: 1.3×10^{-8}A and accelerating voltages :15KV.

Plagioclase

Plagioclase is the most predominant minerals throughout the Shimokawa ophiolite. Selected analyses of plagioclases are presented in Table 2.

Plagioclase occurs in most basalts as coarse-grained euhedral phenocrysts attaining a maximum 1 cm diameter. Very coarse grained plagioclases appear in some dolerites. Most of these coarse crystals contain a dust inclusion zone. The cores of basalt plagioclases are usually altered to secondary minerals, whereas those of dolerites and gabbroic dolerites are commonly fresh. These coarse plagioclases exhibit normal zoning with broad cores being of comparatively uniform composition. Oscillatory zoning and irregular zoning are also found in places, suggesting complex crystallization history of plagioclases (Miyashita, 1992).

An contents of plagioclase cores are different from sample to sample and range about 84 to 78 in olivine dolerite, 83.5 to 81.5 in gabbroic dolerite, 83.5 to 72.5 in Pl-Ol basalt, 80.5 in Pl-

Table 2. Representative analyses of plagioclases.
* means total Fe as Fe_2O_3.

	Basalt	Dolerite		Gabbroic dolerite		Gabbro
Sample No	88IW1	88-153		IW27		IWL5
	core	core	rim	core	rim	core
SiO2 wt%	48.13	47.21	57.08	54.29	61.02	45.14
TiO2	0.02	0.02	0.06	0.01	0.04	0.01
Al2O3	32.12	33.23	26.55	28.41	24.82	33.92
Fe2O3*	0.40	0.49	1.10	0.77	0.59	0.36
MnO	0.03	0.00	0.00	0.00	0.00	0.03
MgO	0.22	0.23	0.12	0.16	0.03	0.04
CaO	16.06	16.30	9.75	11.37	6.69	17.87
Na2O	2.17	1.72	5.87	5.23	7.86	1.37
K2O	0.00	0.00	0.04	0.00	0.11	0.01
Total	99.15	99.20	100.53	100.24	101.05	98.75
			O = 8			
Si	2.223	2.180	2.554	2.451	2.689	2.109
Al	1.748	1.809	1.400	1.512	1.289	1.868
Ti	0.001	0.001	0.002	0.000	0.001	0.000
Cr	0.014	0.017	0.037	0.026	0.020	0.012
Fe	0.001	0.000	0.000	0.000	0.000	0.001
Mn	0.015	0.016	0.008	0.011	0.002	0.003
Mg	0.795	0.807	0.467	0.550	0.316	0.894
Ca	0.194	0.154	0.509	0.458	0.671	0.124
Na	0.000	0.000	0.002	0.000	0.006	0.001
Total	4.991	4.984	4.979	5.008	4.994	5.012
An%	80.4	84.0	47.9	54.6	32.0	87.8

Table 3. Representative analyses of clinopyroxene.
* means total Fe as FeO. X: xenocryst, GM: groundmass, Ph: phenocryst.

	Main Basalt				Body Dolerite		Gabbroic dolerite		Gabbro	Andesite (Later)
Sample No	87-33	88-141		IW105	IW40		IW27		IWL5	IW23
	X	X	GM	Ph	core	rim	core	rim	core	core
SiO2 wt%	52.68	53.39	49.52	51.69	52.09	49.81	51.48	51.31	51.32	52.65
TiO2	0.51	0.35	1.27	0.91	0.59	1.19	0.98	0.65	0.98	0.32
Al2O3	2.54	2.87	3.70	2.64	2.74	1.88	1.77	0.99	3.19	3.06
Cr2O3	0.44	0.87	0.14	0.03	0.11	0.00	0.00	0.00	0.69	0.03
FeO*	4.93	4.24	13.58	8.15	7.20	15.69	12.35	19.65	4.10	6.60
MnO	0.15	0.18	0.29	0.20	0.20	0.41	0.31	0.56	0.17	0.23
MgO	17.73	17.52	13.46	16.37	18.17	12.95	14.31	13.66	15.68	17.82
CaO	19.90	19.82	17.85	19.29	17.77	16.78	18.13	12.94	22.28	18.10
Na2O	0.26	0.28	0.40	0.27	0.25	0.31	0.34	0.21	0.31	0.27
Total	99.14	99.52	100.21	99.55	99.12	99.02	99.67	99.97	98.72	99.08
				O = 6						
Si	1.935	1.945	1.871	1.917	1.922	1.919	1.940	1.965	1.905	1.936
Al	0.110	0.123	0.165	0.115	0.119	0.085	0.079	0.045	0.140	0.133
Ti	0.014	0.010	0.036	0.025	0.016	0.034	0.028	0.019	0.027	0.009
Cr	0.013	0.025	0.004	0.001	0.003	0.000	0.000	0.000	0.020	0.001
Fe	0.151	0.129	0.429	0.253	0.222	0.505	0.389	0.629	0.127	0.203
Mn	0.005	0.006	0.009	0.006	0.006	0.013	0.010	0.018	0.005	0.007
Mg	0.970	0.951	0.758	0.905	0.999	0.743	0.803	0.779	0.867	0.977
Ca	0.783	0.773	0.722	0.767	0.703	0.692	0.732	0.531	0.886	0.713
Na	0.019	0.020	0.029	0.019	0.018	0.023	0.025	0.016	0.022	0.019
Total	4.000	3.982	4.023	4.009	4.009	4.016	4.006	4.002	4.139	3.998
Ca	41.1	41.7	37.8	39.8	36.5	35.7	38.0	27.4	47.1	37.7
Mg	50.9	51.3	39.7	47.0	51.9	38.3	41.7	40.2	46.1	51.6
Fe	7.9	7.0	22.5	13.1	11.5	26.0	20.2	32.4	6.8	10.7
Mg-V	0.865	0.881	0.639	0.782	0.818	0.595	0.674	0.553	0.872	0.828

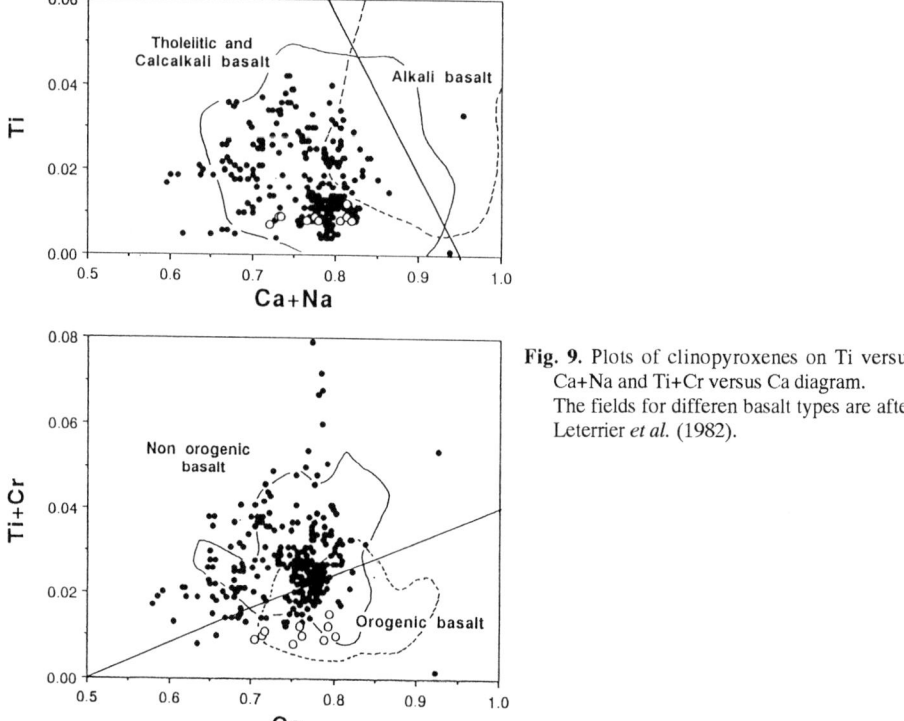

Fig. 9. Plots of clinopyroxenes on Ti versus Ca+Na and Ti+Cr versus Ca diagram. The fields for differen basalt types are after Leterrier et al. (1982).

Ol-Cpx basalt, 60.7 to 56.8 in evolved dolerite, and 55.2 to 52 in most evolved gabbroic dolerite. On the other hand, plagioclases of anorthositic gabbro are high in An contents attaining An88.2 to 82.1.

Clinopyroxene

Clinopyroxenes occur as phenocrysts in some basalts and as poikilitic to ophitic crystals in dolerites and gabbroic dolerites. Needle shaped or dendritic clinopyroxenes appear in the groundmass of basalts and quenched dolerites. Mg-values of clinopyroxenes range from about 0.89 to 0.48. Conspicuous normal zonings are generally observed in poikilitic to ophitic clinopyroxenes. On the other hand, clinopyroxene phenocrysts show complex zoning, indicating complicated crystallization history of the clinopyroxene (Miyashita, 1992).

Cr_2O_3 contents of about 1.5wt% in clinopyroxenes decrease rapidly with decreasing Mg-values of clinopyroxene. Al_2O_3 contents ranging from 1.0 to 5.1 wt%, on the other hand, are different from sample to sample. Xenocryst clinopyroxenes show a narrow range of Al_2O_3, approximately 2.0 to 3.0 wt%. Al_2O_3 contents of poikilitic to ophitic clinopyroxenes from gabbroic dolerites range from 1.0 to 3.0 and decrease with decreasing Mg-values. In dolerites Al_2O_3 contents exhibit a wider compositional range from 1.0 to 4.5 wt%, and also decrease with decreasing Mg-values. The clinopyroxene with the highest Al_2O_3 content occurs as microphenocrysts or quenched crystals in groundmass. Reversed zonings from this pattern of increase in Al_2O_3 with decreasing Mg-value are exclusively found in basalts and the highest Al_2O_3 contents attain 5.1 wt%. Such reversed zoning and high Al_2O_3 contents in clinopyroxenes are probably due to very rapid crystallization which resulted in nonequilibrium incorporation of Al_2O_3.

TiO_2 contents in clinopyroxenes also vary from 0.15 to 1.7 wt%. Phenocrysts and cores of poikilitic clinopyroxenes are usually low in TiO_2, about 0.2 to 0.6 wt%, and increase with de-

Table 4. Selected analyses of Cr-spinel.
FeO and Fe_2O_3 are calculated on the basis of stoichiometry where TiO_2 as ulvospinel.

Sample No	Serpentinite ML7		Basalt					Dolerite and gabbroic dolerite			Andesite
			88-112A	88-141		88W1		88-153	IW15		IW23
	C	M	C	C	M	C	M	C	C	M	C
SiO2 wt%	0.08	0.01	0.00	0.00	0.00	0.00	0.00	0.09	0.07	0.09	0.10
TiO2	1.25	1.24	0.49	0.48	0.52	0.50	0.57	0.51	0.70	1.52	0.30
Al2O3	23.28	22.45	31.02	28.15	29.71	25.21	23.29	25.35	24.81	21.87	13.79
Cr2O3	39.59	41.51	36.14	38.63	35.93	39.96	40.02	39.20	39.25	40.33	54.06
FeO*	16.38	16.50	11.40	12.40	12.77	13.13	14.80	15.21	13.63	14.95	10.16
Fe2O3*	5.58	4.91	3.93	4.09	4.71	5.28	6.67	5.77	6.69	6.75	5.25
MnO	0.25	0.19	0.27	0.34	0.39	0.20	0.22	0.22	0.26	0.22	0.26
MgO	13.06	13.01	16.76	15.74	15.58	14.92	13.65	13.72	14.83	14.00	15.60
CaO	0.01	0.00	0.03	0.00	0.03	0.00	0.00	0.05	0.01	0.13	0.00
Total	99.48	99.82	100.04	99.83	99.64	99.20	99.22	100.13	100.25	99.85	99.51
Si	0.020	0.002	0.000	0.000	0.000	0.000	0.000	0.022	0.017	0.022	0.025
Ti	0.232	0.230	0.086	0.086	0.092	0.091	0.106	0.093	0.127	0.281	0.057
Al	6.762	6.528	8.499	7.862	8.269	7.200	6.760	7.233	7.041	6.332	4.094
Cr	7.715	8.097	6.642	7.237	6.709	7.655	7.793	7.503	7.473	7.833	10.766
Fe3	1.035	0.912	0.688	0.730	0.838	0.963	1.236	1.052	1.212	1.247	0.995
Fe2	3.376	3.405	2.217	2.457	2.522	2.660	3.048	3.080	2.745	3.071	2.140
Mn	0.052	0.040	0.053	0.068	0.078	0.041	0.046	0.045	0.053	0.046	0.055
Mg	4.799	4.785	5.808	5.560	5.485	5.390	5.012	4.952	5.324	5.127	5.858
Ca	0.003	0.000	0.007	0.000	0.008	0.000	0.000	0.013	0.003	0.034	0.000
Total	23.993	23.999	24.000	24.000	24.000	24.000	24.000	23.992	23.994	23.992	23.990

creasing Mg-values. Higher TiO_2 contents exceeding 1 wt% are encountered in rims or in evolved rocks with low Mg-value. However, in the most evolved rock analyzed in which abundant Fe-Ti oxides appear, the TiO_2 contents of clinopyroxenes decrease with decreasing Mg-values, suggesting that the onset of ilmenite decreased TiO_2 in liquid.

Leterrier et al.(1982) proposed discrimination diagrams of basalts using the composition of clinopyroxenes. Fig. 9a shows that the clinopyroxenes of the Shimokawa ophiolite exhibit non-alkaline affinity as well as affinity to basaltic andesite. In a second step of the discrimination (Fig. 9b), however, samples of the Shimokawa ophiolite and basaltic andesite dikes plot in different fields. The former pyroxenes plot in non-orogenic basalt field and the latter in orogenic basalt field. Thus, the compositional characteristics of clinopyroxenes are in good agreement with major and minor element chemistry.

Cr-spinel

Cr-spinels occur as microphe-nocrysts set in the groundmass of less evolved basalts, or inclusions within olivine pseudomorph and large plagioclase crystals. Serpentinites contain a considerable amount of Cr-spinel. Cr-spinel microphenocrysts also appear in less evolved basaltic andesites. Selected analyses of Cr-spinels are shown in Table 4. Although the chemical compositions of Cr-spinels from the Shimokawa basaltic rocks are different from sample to sample, they are within the field of MORBs as shown in Fig. 10. Cr-spinels of serpentinites also have compositions similar to those of basalts though they are slightly richer in Cr_2O_3 and poorer in Al_2O_3 than those

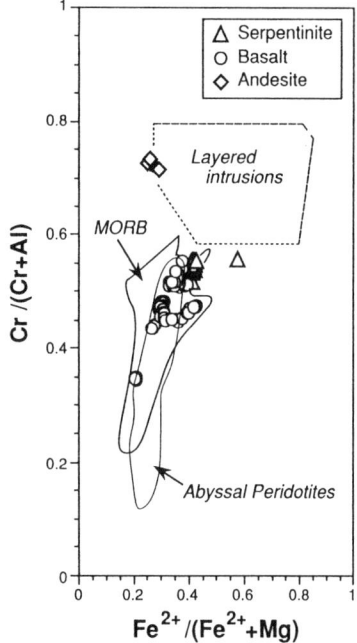

Fig. 10. Plots of Cr-spinels on Cr/(Cr+Al) versus $Fe^{2+}/(Fe^{2+}+Mg)$ diagram.
Composition fields are after Dick and Bullen (1984).

of basaltic rocks (Fig. 10). On the other hand, Cr-spinels of basaltic andesite dikes exhibit distinct compositions, being much higher in Cr_2O_3 and poorer in Al_2O_3 than those of the Shimokawa ophiolite.

Amphiboles

A small amount of brown hornblende, less than 1 modal %, appears in gabbroic dolerite associated with clinopyroxene or chlorite. Although abundant metamorphic amphiboles ranging from actinolite to hornblende in composition occur in the Shimokawa ophiolite, they are bluish green to pale green under the microscope, so they are easily distinguished from each other. The brown hornblendes show strong pleochroism, reddish brown to brown. However, some brown hornblendes grade into the green amphiboles that surrouds them. This is ascribed to the successive formation of amphiboles during the latest igneous to earliest metamorphic stages (Yoshida, 1986). Selected analyses of brown hornblendes are listed in Table 5. TiO_2 contents of hornblendes attain about 2.0 wt%.

Table 5. Representative analyses of brown hornblende from gabbroic dolerite (IW14). Total Fe as FeO.

SiO2	44.86	45.19
TiO2	2.58	1.78
Al2O3	8.71	9.48
Cr2O3	0.00	0.00
FeO	11.85	12.03
MnO	0.17	0.14
MgO	15.18	14.89
CaO	10.84	10.98
Na2O	3.05	2.84
K2O	0.38	0.38
Total	97.62	97.71
Si	6.604	6.633
Al4	1.396	1.367
Al6	0.115	0.273
Ti	0.286	0.196
Cr	0.000	0.000
Fe	1.248	1.257
Mn	0.021	0.017
Mg	3.330	3.257
Fe	0.211	0.220
Ca	1.710	1.727
Na	0.079	0.053
Na	0.791	0.755
K	0.071	0.071
A-site	0.862	0.826

DISCUSSION

Koshimizu et al. (1986) and Bamba (1985) claimed that the Shimokawa ophiolite was generated in island arc environments based on the REE contents and contemporaneous occurrence of terrigenous sediments and MORBs. They considred the andesite dikes as cogenetic and evolved products of the Shimokawa ophiolite. However, our work shows that the andesite dikes intruded after the metamorphism and tectonic emplacement of the ophiolite.

Miyake et al. (1981) and Miyake (1988) emphasized that the Shimokawa complex is tectonically mixed melange whose sediments and basaltic rocks are exotic to each other. Although melange-like lithofacies are locally exposed, the coherent relationship between sediments and basaltic rocks is confirmed by the many lines of evidence we have described. Therefore, the melange-like lithofacies should be interpreted in a different way than typical melange. There are at least two ways to produce such a melange-like occurrence. One way would be collapse of the Shimokawa ophiolite. Such collapse would be encountered at ocean ridges adjacent to transform faults or normal faults in the axial valley along the ocean ridges. Another possibility would be later tectonic movement during the accretionary process. Sinistral deformed structures ascribed to oblique subduction beneath the Eurasian continent are observed around the Shimokawa area (Kimura, 1985; Watanabe and Kimura, 1987; Miyashita et al., 1989). Accordingly, the original structure of the Shimokawa ophiolite could have been modified during the accretionary processes to result in the formation of tectonic melange. The fact that the general stratigraphic succession is traceable along the whole of the complex indicates that the tectonic disturbance was effective only locally.

The appearance of breccias consisting mainly of the same constituents as the Shimokawa ophiolite indicates that the collapse of oceanic crust occurred immediately after the formation of oceanic crust. However, the appearance of exotic pebbles such as limestones suggests that the sedimentation was also occurring. The provenance of the exotic blocks could have been the trench slope of the continent where large limestone blocks might have been exposed, because the western side of the Shimokawa complex is composed of accretionary complexes which include many limestone blocks (Kiminami et al., 1986; Kiyokawa, 1989).

Contemporaneous occurrences of ophiolitic rocks and clastic sediments may be encountered in fore-arc ophiolites (e.g., Bloomer, 1983; Ishii, 1985), however, boninites that are characteristic of the fore-arc ophiolites are not found in the Shimokawa ophiolite, denying the fore-arc origin of the Shimokawa complex.

Contemporaneous occurrence of MORBs and terrigenous sediments would not be peculiar if the ocean ridge were situated near a continent as in present day Gulf of California where thick terrigenous sediments alternate with MORBs (Curray et al., 1982; Moore & Curray, 1982; Griffin et al., 1983). Similar environments also would be expected in the migration of an ocean ridge toward a trench. If ocean ridges arrived near a trench where thick terrigenous clastics were accumulating, MORBs and sediments would occur concomitantly.

Miyashita and Katsushima (1986), Miyashita (1987), Kiminami and Miyashita (1992) and Kiminami et al. (1992) showed many instances of in situ MORBs from the Shimanto to the Hidaka zones. Radiolarian fossil ages of the sediments associated with the MORBs show systematic younging toward east and north, suggesting northward passage of the Kula-Pacific ridge along the Eurasian continent margin (Kiminami and Miyashita, 1989). In this argument, the greenstones of the northern Hidaka zone such as the Tomuraushi and Rurochi complexes accompany Paleocene to Eocene clastic sediments (Kiminami et al., 1990). The geological and petrological features of the Tomuraushi complex, situated at the southern extension of the Shimokawa ophiolite (Fig. 1), are very comparable to those of the Shimokawa ophiolite (Miyashita and Katsushima, 1986) with respect to mode of occurrence, stratigraphic succession, and petrographical features (Miyashita and Yoshida, 1988).

However, the age of the Shimokawa ophiolite remains unsolved. Miyake et al. (1981) reported the metamorphic age as 124Ma on the basis of fission track age of zircon obtained from dolerite. K-Ar using dolerite gave a much younger age, 24.3 ± 7.0Ma (Yoshida and Miyashita, unpublished data), which is ascribed to the later thermal effect. It is well known that ages derived by the fission track method result generally in younger ages than those by K-Ar, due to the difference in closing temperature. Therefore, the Early Cretaceous age appears to be doubtful.

Although the age still remains unsolved, we can conclude that the Shimokawa ophiolite was probably generated at or near a RTT triple junction. If the age of the Shimokawa complex is Early Cretaceous, the ocean ridge might be Fallalon-Izanagi Ridge which is assumed to have passed near Japan margin during Early Cretaceous (Maruyama and Seno, 1986), and the Shimokawa ophiolite was partly collapsed simultaneously or immediately after the formation possibly due to transform faults cutting the ridge or normal faults paralell with the ridges. This ophiolite was partly tectonized during accretion to have produced complicated lithofacies in places. Furthermore, andesite dikes intruded after the tectonic emplacement of the Shimokawa complex, probably during an island arc stage in Hokkaido.

Several specific events accompanying the collision of ocean ridges with trenches have been reported, e.g., near trench magmatism (Marshak and Karig, 1977; Moore et al., 1983; Rogers et al., 1985; Forsythe et al., 1986). However, direct evidence showing the arrival of ocean ridges near the trench has not been clarified. The contemporaneous occurrence of MORBs and trench fill sediments would signify the collision of ocean ridge and trench.

CONCLUSION

The Shimokawa ophiolite, exposed over 20km with width of 1 to 2km in the central part of the northern Hidaka zone, comprises serpentinite, cumulate gabbros, a main body of composite sediments and basaltic rocks, and breccia, in ascending order. The main body is further subdivided into three parts from lower to upper :A, alternation of gabbroic dolerite sheets and clastic sediments (black shale and fine-grained sandstone); B, complicated association of massive and pillow lavas, dolerites, black shale and fine-grained sandstone; and C, thick pillow lavas and subordinate massive lavas with thin intercalations of black shale. Younger basaltic andesite dikes free from metamorphism and metabasalt dikes with primitive features intrude the ophiolites. Contemporaneous occurrence of terrigeneous sediments and basaltic rocks of the main body is revealed by many lines of field evidence, though melange-like bodies are locally observed.

Major and minor element compositions of the basaltic rocks of the main body signify N-MORB features and are clearly distinct from those of other tectonic environments. Mineral compositions of clinopyroxene and Cr-spinel are also comparable with those of MORBs. Basalts are classified into 5 groups based on the phenocryst assemblages, as follows: Ol-Pl(-Cr-Spinel), Ol-Pl-Cpx, Pl-Cpx, Pl and aphyric basalts. General crystallization order is olivine (+ Cr-spinel) - plagioclase - clinopyroxene, which is in good agreement with the low pressure crystallization order in MORBs. However, the occurrence of resorbed clinopyroxenes with high Mg-values in some basalts suggests that the Shimokawa complex underwent polybaric fractional crystallization.

The appearance of breccias consisting mainly of the same constituents as the Shimokawa complex argues that the collapse of oceanic crust happened immediately after the formation of the ophiolite, possibly at transform faults or normal faults. However, the appearance of exotic limestone pebbles in the breccias suggests that the collapse occurred near the trench because the exotic blocks appear to have been derived from the continental-side trench slope where abundant limestone blocks were exposed (Kiminami *et al.*, 1986; Kiyokawa, 1992).

It is concluded that the Shimokawa ophiolite was probably generated at or near a RTT triple junction. Primitive basaltic dikes intruded after the formation of the Shimokawa ophiolite. Subsequently, the ophiolite was accreted and melange-like lithofacies were produced in places. Andesite dikes intruded probably during the island arc stage in Hokkaido. The age of the Shimokawa ophiolite still remains unsolved, but it might have been generated at the Fallalon-Izanagi Ridge or Kula-Pacific Ridge, which was subducting at the trench toward the depth beneath Hokkaido.

Acknowledments: We wish to express our sincere thanks to Profs. T. Bamba and M. Komatsu for their continous encouragements, and to Drs A. Ishiwatari and N. Lindsley-Griffin for critical reading of the manuscript. We are grateful to Drs. K. Kiminami, G. Kimura, Y. Motoyoshi, Messers Yasushi Watanabe, J. Tajika and T. Katsushima for their valuable discussions. We are indebted to Mr. Yamada for preparation of many thin sections. This work was partly supported by Grant-in-Aid from the Japanese Ministry of Education, Science and Culture (No. 02640595) to S. Miyashita.

References

Bamba, T. Implication of the composite mineralization on the massive sulfide deposits of the Shimokawa Mine. *J. Fac. Sci., Hokkaido Univ.*, Ser.IV, **21**, p.363-404 (1985).

Basaltic Volcanism Study Project. *Basaltic volcanism in the terrestrial planets.* Pergamon Press, New York, 1286pp. (1981)

Bloomer, S .H. Distribution and origin of igneous rocks from the landward slopes of the Mariana trench: implications for its structure and evolution. *J. Geophys. Res.*, **88**, 7411-7428 (1983).

Curray, J. P., Moore, D. G. and Kelts, K. Tectonics and geologic history of the passive continental margin at the top of Baja California. *Init. Rep. DSDP*, **64**, 1084-1116 (1982).

Dick, H. J. B. and Bullen, T. Chromian spinels as a petrogenetic indicator in abyssal and alpine-type peridotites and spatially associated lavas. *Contrib. Mineral. Petrol.*, **86**, 54-76 (1984).

Forsythe, R.D., Nelson, E.P., Carr, M.J., Kaeding, M.E., Herve, M., Mpodozis, C., Stoffa, J.M. and Harambour, S. Pliocene near-trench magmatism in southern Chile: A possible manifestation of ridge collision. *Geology*, **14**, 23-27 (1986).

Griffin, B. J., Neuser, R. D. and Schmincke, H. Lithology, petrography, and mineralogy of basalts from DSDP sites 482, 483, 484, and 485 at the mouth of the Gulf of California. *Init. Rep. DSDP*, **65**, 527-548 (1983).

Hart, R. Chemical exchange between sea water and deep ocean basalts. *Earth Planet. Sci. Lett.*, **9**, 269-279 (1970).

Ishii, T. Dredged samples from the Ogasawara fore-arc seamout or "Ogasawara Paleoland" - "Fore-arc ophiolite".

in *Formation of Active Ocean Margin*, eds. Nasu, N. *et al.*, 307-342, TERRAPUB, Tokyo, Japan (1985).
Ishizuka, H. Geochemistry of the Horokanai ophiolite in the Kamuikotan Tectonic Belt, Hokkaido. *J. Geol. Soc. Japan*, **87**, 17-34 (1981).
Kiminami, K., Miyashita, S., Kimura, G., Takika, J., Iwata, K., Sakai, A., Yoshida A., Kato, Y., Watanabe, Y., Ezaki, Y., Kontani, Y. and Katshushima, T. Mesozoic rocks in the Hidaka belt - Hidaka Supergroup. *Monog. Assoc. Geol. Collab. Japan*, 31, 137-155 (1986).
Kiminami, K., Kawabata, S. and Miyashita, S. Discovery of Paleogene radiolarians from the Hidaka Supergroup and its significance with special reference to the ridge subduction. *J. Geol. Soc. Japan*, **96**, 323-326 (1990).**
Kiminami, K., Miyashita, S. Geological consequences of the Kula-Pacific ridge collision and migration of the ridge (Kula-Pacific)-trench-trench triple junction along the Japan margin. *DELP Pub*. 28, 111-114 (1989).
Kiminami, K., Miyashita, S. Occurrence and geochemistry of greenstones from the Makimine Formation in the Upper Cretaceous Shimanto Supergroup in Kyushu, Japan. *J. Geol. Soc. Japan*, **98**, 391-400 (1992).*
Kiminami, K., Kashiwagi, N. and Miyashita, S. Occurrence and significance of in-situ greenstones from the Mugi Formation in the Upper Cretaceous Shimanto Supergroup, eastern Shikoku, Japan. *J. Geol. Soc. Japan*, **98**, 867-883 (1992).*
Kimura, G. The mode of Cretaceous subductiuon in Hokkaido. *Science* (Kagaku), **55**, 24-31 (1985).**
Kiyokawa, S. Cross section of axial zone of Hokkaido with special references to the Idonappu zone. Earth Monthly, 11, 316-322 (1989).**
Kosaka, H. Geochemical characteristics of the Shimokawa diabase sheets, Hokkaido. *Mining Geol.*, **25**, 161-174 (1975).*
Koshimizu, S., Sawai, O. and Bamba, T. REE abundances of the effusive and intrusive rocks from the Shimokawa copper mining area, Hokkaido, Japan. *J. Japan. Assoc. Min. Petr. Econ. Geol.*, **81**, 129-137 (1986).
Leterrier, J., Maury, R. C., Thonon, P., Girard, D. and Marchal, M. Clinopyroxene composition as a method of identification of the magmatic affinities of paleo-volcanic series. *Earth Planet. Sci. Lett.*, **59**, 139-154 (1982).
Mariko, T. Sub-sea hydrothermal alteration of basalt, diabase and sedimentary rocks in the Shimokawa copper mining area, Hokkaido, Japan. *Mining Geol.*, **34**, 307-321 (1984).
Mariko, T., Mochizuki, T. and Horii, M. Overturn of sediments and pillow lavas of the Hidaka series around the Shimokawa ore deposits, Hokkaido. *Mining Geol.*, **32**, 67-72 (1982).*
Maruyama, S. and Seno, T. Orogeny and relative plate motions - an example of the Japanese islands. *Tectonophysics*, 127, 1-25 (1986).
Marshak, R. S. and Karig, D. E. Triple junctions as a cause for anomalously near-trench igneous activity between the trench and volcanic arc. *Geology*, **5**, 233-236 (1977).
Meschede, M. A method of discriminating between different types of mid-ocean ridge basalts and continental tholeiites with the Nb-Zr-Y diagram. *Chem. Geol.*, **56**, 207-218 (1986).
Miyake, T. Geology and mineralization of the Shimokawa Mine: An allochthonous ridge-type massive sulfide ore deposit. *Mining Geol.*, **38**, 215-231 (1988).
Miyake, T., Ochiai, T. and Shikama, M. Metamorphism of Shimokawa ophiolite and ore genesis. *Report of the studies on "Kieslager" in Japan*, Tokyo, p.77-90 (1981).**
Miyashiro, A., Shido, F. and Ewing, M. Metamorphism in the Mid-Atlantic Ridge near 24 and 30 N. *Phil. Trans. Roy. Soc. Lond.*, A, **268**, 589-603 (1971).
Miyashita, S. The greenstones of the Hidaka zone, central Hokkaido. *Prof. Matsui Memorial Vol.*, 215-223 (1987). Institute of Geology, Hokkaido University of Education, Sapporo. **
Miyashita, S. A preliminary study on the zonings of plagioclase and clinopyroxene from the Shimokawa ophiolite, northern Hidaka zone, Hokkaido. *Contrib. Dept. Geol. Min., Niigata Univ.*, 7, 85-101 (1992).*
Miyashita, S. and Katsushima, T. The Tomuraushi greenstone complex of the central Hidaka zone: contemporaneous occurrence of abyssal tholeiite and terrigeneous sediments. *J. Geol. Soc. Japan*, **92**, 535-557 (1986).
Miyashita, S. and Watanabe, Y. Genetic environments of the greenstones in the Hidaka zone, Hokkaido, with special reference to the metallogeny of massive sulfide, and bedded iron and manganese ores. *Mining Geol. Spec. Issue*, 12, 93-104 (1988).
Miyashita, S. and Yoshida, S. Pre-Cretaceous and Cretaceous ophiolites in Hokkaido, Japan. *Bull. Soc. Geol. France.*, (8), IV, 251-260 (1988).
Miyashita, S., Watanabe, Y., Tajika, J. and Kiminami, K. Structural features of the Hidaka zone: sinistral and dextral transpressive zone. *DFLP Pub*., 28, 115-120 (1989).
Moore, J.C., Byrne, Y., Plumley, P.W., Reid, M., Gibbons, H. and Coe, R.S. Paleogene evolution of the Kodiak islands, Alaska: Consequences of ridge-trench interaction in a more southerly latitude. *Tectonics*, **2**, 265-293 (1983).
Mottle, M. J. and Holland, H. D. Chemical exchange during hydrothermal alteration of basalt by seawater. I: Experimental results for major and minor components of seawater. *Geochim. Cosmochim. Acta*, **42**, 1103-1115 (1978).
Mullen, E. D. $MnO/TiO_2/P_2O_5$: a minor element discriminant for basaltic rocks of oceanic environments and its

implications for petrogenesis. *Earth Planet. Sci. Lett.*, **62**, 53-62 (1983).

Nakagawa, M. and Komatsu, M., 1983: Chemical analysis of rocks using fused disk-samples by the X-ray fluorescence method. *Report of research project, Grant-in-Aid for Scientic research from the Ministry of Education,* Niigata Univ., 4-10 (1982).**

Nanayama, F. Sedimentology and sedimentary petrology of the Nakanogawa Group in the Hidaka belt, central Hokkaido, Japan: Three petroprovinces identified in the Paleocene Nakanogawa Group and their geotectonic significance. Dr. Thesis, Hokkaido Univ., pp.130 (1992).*

Pearce, J. A. Trace element characteristics of lavas from destructive plate boundaries. In *Andesite: orogenic andesites and related rocks*, R. S. Thorpe (ed.), 525-548. Chichester: Wiley. (1982).

Pearce, J.A. and Cann, J. R. Tectonic setting of basic volcanic rocks determined using trace element analyses. *Earth Planet. Sci. Lett.*, **19**, 290-300 (1973).

Rogers, G., Saunders, A. D., Terrell, D. J., Verma, S. P. and Marriner, G. F. Geochemistry of Holocene volcanic rocks associated with ridge subduction in Baja California, Mexico. *Nature*, **315**, 389-392 (1985).

Sun, S., Nesbitt, R. W. and Sharaskin, A. Y. Geochemical characteristics of mid-ocean ridge basalts. *Earth Planet. Sci. Lett.*, **44**, 119-138 (1989).

Tamura, Kobayashi, Y. and Shuto, K. Quantitative analysis of the trace elements in silicate rocks by X-ray fluorescence method. *Earth Sci.* (Chikyu-Kagaku), **43**, 180-185 (1989).**

Watanabe, Y. and Kimura, G. Strike-slip fault (Nayorogawa Fault) in northern Hokkaido. *J. Geol. Soc. Japan,* **93**, 1-10 (1987).

Yoshida, A. Geology and metamorphism of the Shimokawa greenstone complex, the Hidaka belt, Hokkaido. Master Thesis of Hokkaido Univ.* (1986).

   * in Japanese with English abstract
   ** in Japanese

Time-Space Distribution and Tectonic Types of Ophiolites in China

WANG Xibin and HAO Ziguo

Institute of Geology, Chinese Academy of Geological Sciences, Beijing 100037, China

Abstract

China's (orogenic) ophiolites and ophiolitic ultramafic bodies are mostly exposed along suture zones and deep fractures, forming 21 ophiolite belts of varying scales and ages. They are mostly Phanerozoic and in rare cases Proterozoic in age.

Paleozoic ophiolites were formed in various periods from the Cambrian to Permian. They do not cluster in some particular periods, though Ordovician ophiolites predominate slightly. The Tethyan ophiolites are mainly of the early Cretaceous and Jurassic age. Some large ophiolite belts are usually multiple belts composed of two or more ophiolite suites of different ages and tectonic types.

To the south of the Tarim and Sino-Korean Platform, ophiolites in Western China have an evolutionary trend of becoming progressively younger from north to south. This time-space distribution might imply that the evolution of the western and southwestern China continent is characterized by the mechanism of southward successive accretion.

On the basis of the types of the ophiolite sections coupled with the degree of maturity (i.e. development stages) of oceanic basins, the authors divide ophiolites of China into four tectonic types: (1) initial oceanic basin-type ophiolite, (2) mature oceanic basin-type ophiolite, (3) island arc-type ophiolite, and (4) residual oceanic basin-type (i.e. non-ophiolite-type) mafic and ultramafic rocks. Ophiolites in China are dominated by the first type. The second and third types are less important, while the fourth type is rare.

Keywords: Ophiolites, Time-Space distributions, Tectonic types

INTRODUCTION

Since the Penrose ophiolite symposium held in 1972 (Coleman, 1977), there has been profound change about the definition of ophiolite. But its basic connotation has still been widely applied by geologists of various countries in the world and constantly enriched and developed, and the broadened definition is more in accord with the geological fact (Moorse, 1982; Coleman, 1984). However, a too broadened definition will cause ophiolite to lose its real connotation. The authors consider that we should rule out beyond the realm of ophiolite all such formations or rock associations that are dominated by flysch with some intercalations of volcanic rocks (including pillow lavas) and radiolarian (either present or absent) cherts but lack other main components of ophiolite. This paper just follows this defined concept.

In the past ten years, unprecedented advances have been made in the study of ophiolites in China and voluminous new data have been accumulated. These results indicate that ophiolites of China not only are widespread and varied in age but also have a great variety of tectonic types. The authors propose the tectonic classification of ophiolites in orogenic belts of the Chinese continent.

For the convenience of correlation, the ophiolite sections studied by the predecessors in China are all expressed by columns in this paper. These columns were constructed according to the sections measured by the original authors (in a few cases according to written description). They are mostly loyal to the original maps, but a few of them were modified slightly.

This paper is not a concluding report, but only sums up and integrates the results obtained by predecessors.

Time-Space Distribution of Ophiolites in China

Ophiolites of China are not only widespread but the ages of their formation range almost throughout the Phanerozoic. Most of them are distributed linearly along suture zones or deep fracture zones in fold belts. The northern, central, and southern nearly E-W-trending gigantic fold systems that traverse the Chinese Continent constitute the basic framework of distribution of ophiolites and ultramafic bodies (Fig. 1). A few ophiolite belts are exposed in the NE-trending tectonic system on the continental margins of eastern China.

According to the latest results of ophiolite research and on the basis of the map* of distribution of basic and ultrabasic rocks in China compiled by the authors in the seventies, 21 ophiolite belts (Fig. 1) and several regions of non-ophiolite type mafic and ultramafic rocks have been distinguished in the paper.

The existing data indicate that ophiolites in most belts were formed in the Phanerozoic (Fig.2), particularly in the Paleozoic, while a few were formed in the Late Proterozoic. At present there are no reliable data about whether there exist still older ophiolites.

Tethyan ophiolite belts in southwestern China

Mesozoic ophiolites are mainly distributed in Tibet and the Sanjiang (the Nujiang River, Lancang River and Jinsha River) area in western Sichuan and Yunnan, it may be forming five major belts: (1) the Yarlung Zangbo ophiolite belt (II_1), (2) the Bangong Lake- Nujiang ophiolite belt (II_2), (3) the Jinshajiang ophiolite belt (II_4), (4) the Ailaoshan ophiolite belt (II_5), and (5) the Ganzi-Litang mafic-ultramafic belt (II_6).

Belt II_1 is distributed in a nearly E-W direction along the Yarlung Zangbo suture zone (Nicolas et al., 1981; Wang X.B. et al., 1984). This belt is the most famous ophiolite belt in China. It is composed of two ophiolite suites of different ages and tectonic types (Wang X.B. et al., 1984,1987; Bao P.S. et al., 1986). One suite is exposed on the northern side of the belt where a relatively complete section is preserved (Wang X.B., 1984,1987). Albite granite in ophiolites has a zircon U-Pb age of 139Ma (Wang X.B. et al., 1987) and

---

*The map of distribution of basic and ultrabasic rocks in China (scale 1:4000000), Wang X.B. et al., 1975.

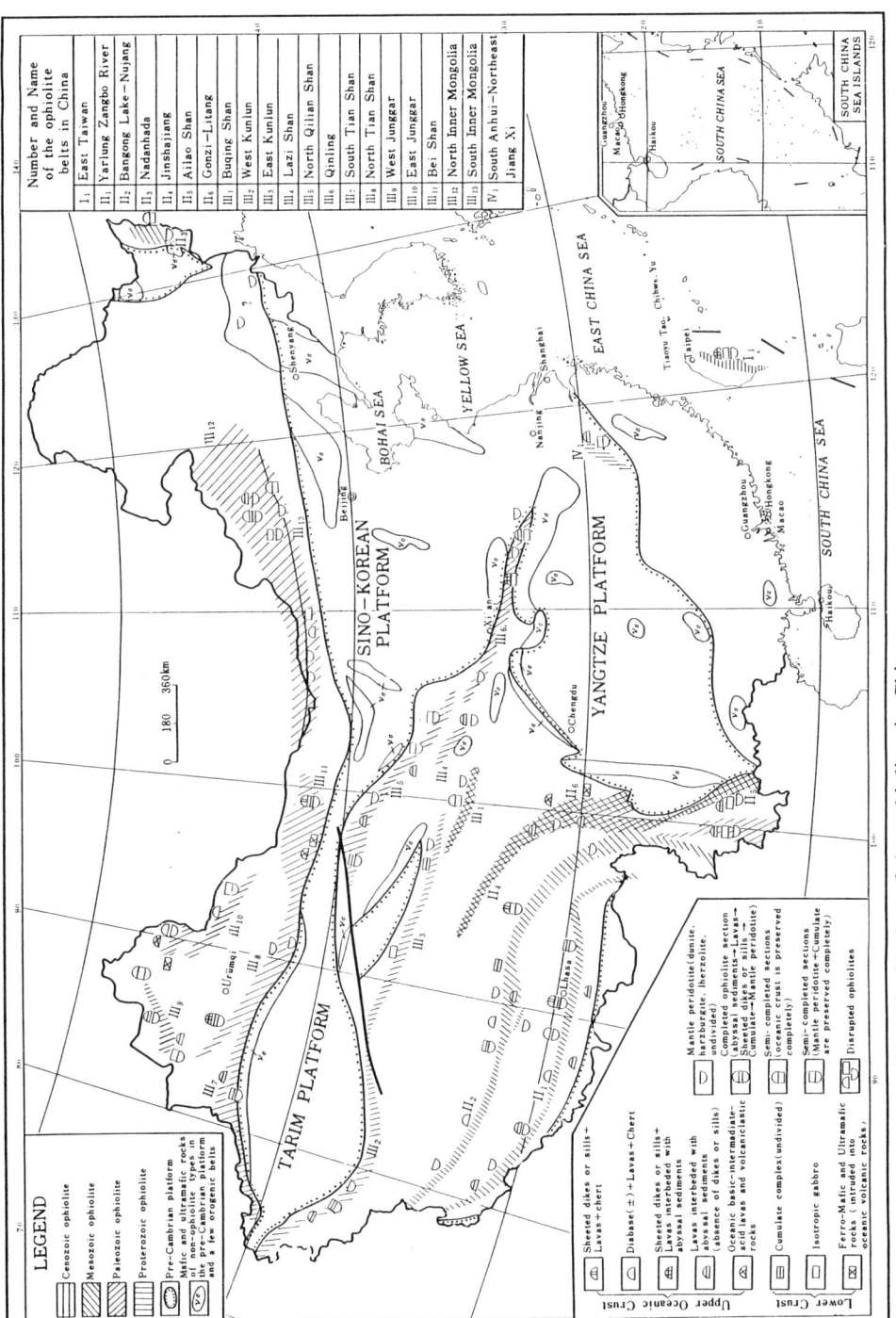

Fig.1 Schematic map of the distribution of the ophiolites in China

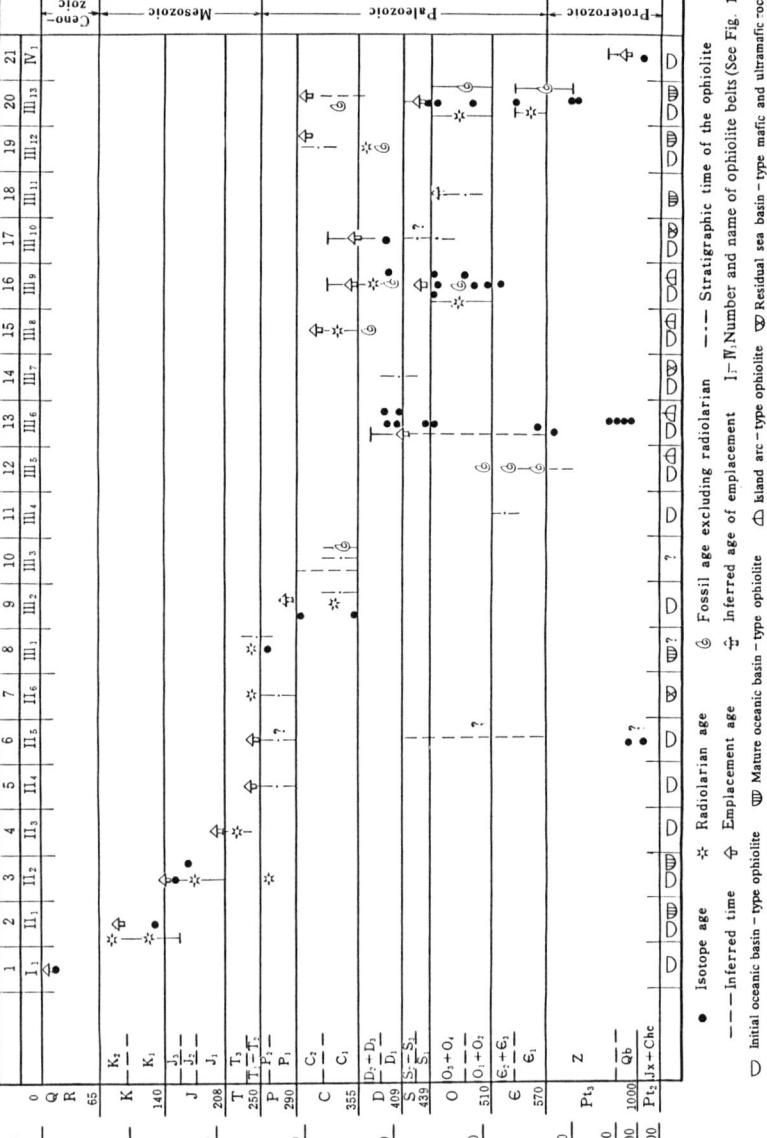

Fig.2 Diagram showing ages of the formation and emplacement and tectonic types the ophiolites in China (the stratigraphic and geologic time of China by Wang Hongzhen et al., 1990)

radiolarians in chert are mostly of Early-Middle Cretaceous age and partly of Late Jurassic age. According to these, the age of this ophiolite suite may be inferred to be Early Cretaceous. The other ophiolite suite is distributed on the southern side of the rock belt. Its age is inferred to be Late Jurassic to Early Cretaceous. According to the fact that hornblende from the metamorphic aureole-- garnet amphibolite -- at the base of the ophiolite yields a K-Ar age of 89Ma, the emplacement time of ophiolites of this belt should be Late Cretaceous.

Located north of belt II_1, belt II_2 is largely distributed along the Bangong Lake-Nujiang River suture zone. Its western end has a tendency to converge with belt II_1 in the Ngari Prefecture of Xizang (Tibet). Westward it enters Kashimir and connects with the Indus Ophiolite belt (Asrarullah et al., 1979). Eastward it extends along the Nujiang River valley into northwestern Yunnan, where it turns south. In the light of the identification of radiolarian fossils and other fossils, ophiolites of this belt originated in the Jurassic. The tectonic emplacement time of the ophiolites is considered to be before the Early Cretaceous (Wang X.B. et al., 1984). According to Zhang Qi (1985), the eastern sector of this belt might be represented by Late Permian ophiolite in western Yunnan.

Belt II_4 is mainly distributed along both sides of the Jinshajiang fracture (Zhang Z.M. et al.,1979). This fracture represents a plate subduction zone whose activity ended in the terminal Triassic (Zhang Z.M. et al.,1981). Two spilite-keratophyre series are developed on both sides of the subduction zone. Spilite-Keratophyre on its western side form ophiolite with gabbro and harzburgite. Spilite occurs in the Permian Gajin Snow Mountain Group and the early Middle Zhongxinrong Group (Zhang Z.M., 1981), and the latter is in turn overlain unconformably by the Upper Triassic Jiapila Formation. Ophiolites of this belt originated in the Permian to Early and Middle Triassic, and might have emplaced before the Late Triassic.

Belt II_5 mainly extends in a NW-SE direction along the western side of the Mount Ailao fracture (Duan X.H.,1981; Yang J.R. ,1986). Its northwestern sector might be connected to belt II_4, while its southeastern sector extends southward and enters Vietnam (Duan X.H., 1981).

There has been much controversy over the age of the ophiolite of this belt because of no reliable age data. There are three opinions about its age: the first opinion is that it is tentatively classified as the Early Paleozoic product (Yang J.R., 1986); the second opinion proposes that it is the 900- 1100Ma ancient oceanic crust(Fan C.J., 1986);the authors of this paper hold that this belt has many similarities to the belt II_4 in respect to the volcanic-sedimentary formation. Therefore it is inferred that it might be the Permian to Early-Middle Triassic (Zhang Q. et al., 1988) product and have been emplaced prior to the Late Triassic.

Lying east of belt II_4, belt II_6 is distributed largely in a N-S direction. Divergent views exist as to whether this belt is an ophiolite belt. Some geologists consider that this belt is a volcanic-matrix ophiolitic mélange zone (Wang L.C. et al., 1985; Chen B.W, 1983), because this belt consists mainly of a sequence of basaltic volcanic rocks(intercalated with radiolarian siliceous rocks). Besides, small amounts of ferro-ultramafic rocks[*] have been reported. Hence, in this paper this belt is tentatively regarded as a non-ophiolite type mafic-ultramafic belt. Its age might be Permian to Early Triassic (Wang L.C. et al., 1985).

---

[*]Unpublished

To sum up, the five rock belts in western China become younger successively from east to west, i.e. their ages are Permian-Early and Middle Triassic→Jurassic→Early Cretaceous (and Late Jurassic to Early Cretaceous) (Fig.2).

Ophiolites distributed along the southern margin of the Tarim and Sino-Korean platform in central China

Traversing central China and stretching generally in a nearly E-W direction, the Qinling-Qilian-Kunlun Mountain System constitutes the important boundary between the northern and southern platform and the Yangtze platform) (Ren, 1990), and ophiolites distributed along this mountain system form an ophiolite belt dominated by Paleozoic ophiolite. From west to east there are following six ophiolite belts (Fig.1): (1) the West Kunlun ophiolite belt (III_2), (2) the East Kunlun ophiolite belt (III_3), (3) the Buqinshan Mountain ophiolite belt (III_1),(4) the Lajishan Mountain ophiolite belt(III_4),(5) the North Qilian ophiolite belt (III_5), and (6) the Qinling- Tongbai ophiolite belt (III_6). Besides, there are also two possible greenstone. belts (the northern Qaidam belt and the south Qinling Belt).

Belt III_1, III_2 and III_3 are all exposed in the Kunlun fold system and distributed along the southern margins of the Tarim platform and Qaidam land mass.

Belt III_2 is exposed on the south of the West Kunlun northern margin fracture (Jiang C.F., 1992). Its succession has been mostly destructed, but semi-complete ophiolite sections may be seen locally. The basalt in ophiolite has Rb-Sr isochron ages of 359Ma and 297Ma and the age of radiolarian fossils is not earlier than Carboniferous. According to these the age of ophiolite of this belt is determined to be Early Carboniferous (Jiang C.F. et al.,1992). It is inferred that its tectonic emplacement might take place in the Late Carboniferous to Early Permian.

Belts III_1 and III_3 both occur in the East Kunlun fold system, but the former is mainly distributed in a NE-SE direction along the southern side of the East Kunlun southern deep fracture, while the latter mainly stretches in an E-W direction along the middle fracture in the East Kunlun (Jiang C.F., 1992).

With regard to the age of the ophiolite of belt III_3, the Rb-Sr isochron age of basalt is 260Ma and the age determined by radiolarian fossils is Late Permian to Middle Triassic(Jiang C.H., 1992). At present there are two opinions about the age of formation of belt III_3: one suggests that the ophiolite is of Early Carboniferous age (Jiang C.H., 1992), and the other considers that the ophiolite was emplaced during the closing of the Kunlun paleo-oceanic basin in the terminal Caledonian period and that the age of its formation should be earlier than this period (Gao Y.L., 1988).

The North Qinlin ophiolite belt (III_5) is located on the southern margin of the Tarim and Sino-Korean platform and in the North Qilian Caledonian fold belt. Ophiolite is best develop here in the Qinling-Qilian-Kunlun fold system. Ophiolites are exposed as NW-SE-trending bodies. Volcanic rocks are well developed in these ophiolites, which consist of a spilite- keratophyre sequence of Sinian, Middle Cambrian, Lower Ordovician and Upper Ordovician ages. (Xiao et al., 1978; Wang Q. et al., 1976).

Situated southeast of Belt III_5, Belt III_4 belong to the Qilian Caledonian fold belt. The ophiolite is Middle-Late Cambrian in age.

Belt III_6 is connected to belt III_5 on the west, and its eastward end is at the west of Xinyang. The internal composition, tectonic deformation and metamorphism of the Qinling orogenic belt are all very complex. This belt is controlled by the Shang-Xian fracture and the Shang-Dan fracture and may be divided into the northern and southern subbelts (Zhang G.W. et al., 1988; Sun Y. et al., 1988). Rocks have been commonly undergone greenschist facies and even amphibolite facies metamorphism. Locally (in Guojiagou) sheeted dike swarms are seen (Zhang G.W., 1988; Sun Y. et al., 1988). Terrigenous fragments accompany the ophiolite section. Such kind of ophiolite is impossible to be of abyssal origin. Most geologists believe that the ophiolite is Early Paleozoic (Zhang G.W. et al., 1988; Xu Z.Q., 1988), and some geologists consider that its age is terminal Ordovician to initial Devonian (Sun Y. et al., 1988; Wang R.M. et al., 1990). However, some geologists consider that it originated in Late Proterozoic (as exemplified by the Erlangping ophiolite). The reported isotopic ages may fall into three groups: the first group includes 397Ma (Rb-Sr age), 402.6 ± 17.4Ma (Sm-Nd age), 440Ma (U-Pb age), 447.8 ± 41Ma (Rb-Sr age) (Zhang G.W. et al., 1988), and 391Ma and 410Ma (Rb-Sr age) (Wang R.M. et al., 1990); the second group includes 575.5 ± 65Ma (Rb-Sr age) (Wang R.M. et al.,1990) and 575.5 ± 79.6Ma (Lu X.X., 1988); the third group comprises 761 ± 87Ma (Rb-Sr age), 799Ma (Rb-Sr age), 884Ma (K-Ar age) and 920Ma (K-Ar age) (Hu S.X. et al.,1988). The above-mentioned isotopic ages suggest that there are at least three phases of ophiolite in this belt: (1) Late Ordovician to Early Devonian, (2) Early Cambrian and (3) Late Proterozoic. Tectonic emplacement might have taken place in the terminal Caledonian to Early Variscan (Xu Z.Q. et al., 1988).

Ophiolites distributed along the northern margin of the Tarim and Sino-Korean platform in northwestern and northern China

Situated north of the Tarim and Sino-Korean platform, the Tianshan-Mongolia-Xing'an, fold system represents a tectonic junction zone between the Siberian platform and the Sino-Korean paraplatform (Ren et al., 1990). The age of ophiolite belts distributed in this fold system correspond to those on the south of the Tarim and Sino-Korean platform (Fig 1). There are seven major ophiolite belts in the region: (1) the South Tianshan ophiolite belt (III_7), (2) the North Tianshan ophiolite belt (III_8), (3) the West Junggar ophiolite belt (III_9), (4) the East Junggar ophiolite belt (III_{10}), (5) the Beishan ophiolite belt (III_{11}), (6) the Northern Inner Mongolian ophiolite belt (III_{12}), and (7) the Southern Inner Mongolian ophiolite belt (III_{13}).

The ophiolite belts in the Tianshan Mountains are largely parallel to several large-scale shear fracture belts. The ophiolite distributed along the northern edge of the Tianshan main fracture forms the North Tianshan ophiolite belt (III_8). The ophiolite of this belt is remnants of Early Carboniferous-Early Middle Carboniferous oceanic crust (Wang Z.X., 1990). Siliceous rocks (chert) yield Early Carboniferous radiolarians and Late Devonian conodonts (Wang Z.X. et al., 1990; Xiao et al., 1991). Tectonic emplacement took place prior to the Middle Carboniferous.

Belt III_7 comprises two subbelts: one is located on the northern margin of the South Tianshan Mountains, and the other largely extends along the fracture at the main ridge of Halik Mountain. Mafic and ultramafic rocks form an ophiolite belt together with Late Silurian to Early Devonian volcanic rocks and siliceous rocks (chert), so the ophiolite in this belt might be formed in the Late Silurian to Early Devonian (Zhang Z.X. et al., 1990).

In addition, on the eastern sector (Huangshan area) of this belt there also exists a residual oceanic basin non-ophiolite type mafic and ultramafic formation associated with sulfide Copper-Nickel deposits, which indicate that this belt is different from typical geosynclinal fold belts in respect the tectonic nature.

The East and West Junggar ophiolite belts (III_9 and III_{10}) lie north of the Tianshan fold belt, on the eastern and western sides of the Junggar basin.

Belt III_9 includes six subbelts, most of which had undergone strong deformation and metamorphism so as to form ophiolitic mélange, but relatively complete sections may be observed in a few places (e.g. at Honggulelang). On the basis of radiolarian, brachiopod and coral fossils and isotopic dating, they are at least three phases of ophiolite in the belt: (1)Early-middle Devonian (Feng et al.,1991; Xiao et al.,1991), with an isotopic age of 395 ± 1.2Ma (Sm-Nd isochron age) (Zhang C. et al.,1992); (2) Ordovician, with isotopic age of 444 ± 27Ma, 447 ± 56Ma (Sm-Nd isochron age) (Zhang C. et al., 1992); (3) Late Cambrian - Middle Ordovician, with isotopic age of 489 ± 53Ma (Sm-Nd isochron age) (Zhang C. et al., 1992), 480-520Ma (Pb-Pb age) (Tilton et al., 1986) and 508 ± 20Ma (Pb-Pb age) (Feng et al., 1991; Xiao et al., 1991). Silurian ophiolite might also exist in this belt (Feng et al., 1991; Zhang C. et al., 1992).

Belt III_{10} is composed of two parallel subbelts--the northern and southern subbelts, which extend in a NW-SE direction. Its geological features are similar to those of belt III_9; ophiolite therein mostly occurs in the form of mélange. According to microfossils and isotopic ages, the age of the ophiolite is Middle Devonian or even earlier. The isotopic age (gabbro K-Ar age) is 388-392Ma (Li J.Y., 1991), and gabbro and ultramafic bodies are found to be overlain unconformably by the Lower Carboniferous Nanmingshui Formation with definite fossils (ammonites--Gastrioceras sp.) at Nanmingshui. Its tectonic emplacement should have happened before the Early Carboniferous (Li J.Y., 1991; Xiao et al., 1991). Besides, siliceous rocks (chert) with Ordovician radiolarians have been observed to be contained in Lower Devonian basal conglomerate in other places of the belt. Therefore it is reasonable to infer that there might exist Ordovician to Middle Silurian ophiolite in this belt (Li J.Y., 1991).

The eastern and western parts of the Beishan ophiolite belt (III_{11}) discontinuously connect with the Inner Mongolian ophiolite belt and the Tianshan ophiolite belt, respectively. This belt has not yet been studied intensively. According to a report (Zuo et al., 1987), there occurs Late Middle Ordovician ophiolite in the belt and the section is relatively complete, which is seldom seen in Paleozoic ophiolite belts.

Located on the northern margin of the North China platform and in the eastern sector of the Mongolia-Xing'an fold belt, Inner Mongolia is an important region for Paleozoic ophiolite in China. In terms of its spatial distribution, it may be divided into the northern and southern belts.

The northern Inner Mongolian ophiolite belt (III_{12}) includes ophiolite blocks at Mount Hegengshan. Alten Gol Sum and Manlai Sum, and westward it enters the People's Republic of Mongolia, where it is connected to the Mount Zuolun ophiolite bodies(Xiao et al., 1991). Ophiolite in this belt is concentrated in the vicinity of Mount Hegengshan, where a relatively complete section has been preserved. The age of ophiolite as determined according to the radiolarian fossils in chert and coral fossils in limestone (intercalated in volcanic rocks) is Late Early Devonian to Early-Middle Devonian (Tang K.D. et al., 1991).It is noteworthy that according to field investigations conducted by the authors (1991), there also occurs basalt of the ophiolite suite of Early Carboniferous age (in Hebai and Mount Bayinshan) in the Hegengshan area in addition to the above-mentioned Devonian ophiolite. According to the fact that ultramafic rocks are unconformably overlain by the

Lower Permian Zhesi Formation, the emplacement of the ophiolite should have occurred before the Early Permian.

The southern Inner Mongolian ophiolite belt (III_{13}) comprises the Mount Solonshan subbelt in the western sector, the Ondor Sum subbelt in the central sector and the Xilamulun River subbelt in the eastern sector.

The Mount Solonshan subbelt in the west is located on the Sino-Mongolia border. The ophiolite is mainly marked by exposed ultramafic rocks. Relevant age data are limited. It has been determined to be of Middle Carboniferous age according to the occurrence of Middle Carboniferous small fusulinid fossils in the limestone intercalation in basalt of the ophiolite (Wang Q. et al., 1991). As the Lower Permian Zhesi Formation overlies the ultramafic body unconformably, its emplacement should have taken place before the Early Permian.

The Ondor Sum subbelt in the central sector is a relatively deeply metamorphosed ophiolite belt. Examination of microfossils and isotopic data indicate that the ophiolite of this sector is a polyphase product. There are three phases: (1) Cambrian to Ordovician, (2) Early Cambrian or slightly earlier (Wang Q. et al., 1991; Tang K.D. et al.,1991), and (3) Precambrian (Wang Q. et al.,1991). There are two groups of isotopic ages: one is 446Ma (plagiogranite Sm-Nd age) and 444-492Ma (Pb-Pb age) (Xiao et al.,1991); the other is 630Ma (Rb-Sr age) (Peng L.H., 1984). The former age group should be considered to be Ordovician, while the latter age Late Precambrian (Late Proterozoic) to Early Cambrian. Tectonic emplacement took place before the Late Silurian (Wang Q. et al.,1991; Tang K.D., 1991).

The Xilamulun River subbelt in the east is mainly exposed along the northern bank of the Xilamulun River, including the Balengshan, Erbadi and Kedanshan ophiolite bodies. The ophiolite of this sector is a polyphase product too. One phase involves Cambrian and Ordovician, and the other terminal Proterozoic to Early Cambrian (Li J.Y.,1987; Wang Q. et al., 1991). The ophiolite emplacement occurred prior to the Late Silurian (Tang K.D., 1991).

Ophiolites on the continental margin of eastern China

Less ophiolites are exposed in eastern China. They are scattered in a few tectonic units of differing natures and ages, all extending in a NE direction. Three ophiolite belts may be distinguished: the Eastern Taiwan ophiolite belt (I_1), Nadanhada ophiolite belt (II_3) and Southern Anhui-northeastern Jiangxi ophiolite belt (IV_1) (Fig. 1).

Belt I_1 is located on the western side of the Coastal Range in eastern Taiwan. The ophiolite is exposed as mélange in Pliocene Strata. It was formed in the Miocene (Suppe et al., 1981), at about 15Ma (Jahn B.M., 1986), and emplaced in the Pliocene, at about 4-5Ma. So it is known as the youngest ophiolite in the world (Jahn B.M., 1986; Ishiwatari,1990).

Lying on the Sino-Russian border in northeastern Heilongjiang province, belt II_3 is separated from Russia by a river on its east. Tectonically this belt is a part of the Sikhote-Alin belt (Ren J.S. et al., 1980). Identification of radiolarian fossils from silicious rocks (chert) suggests that the upper limit of the ophiolite age of this belt is Middle- Late Triassic (Kang B.X. et al.,1990).The ophiolite might have emplaced in the Late Triassic to Early Jurassic (Kang B.X. et al., 1990).

Situated on the southern margin of the Yangtze platform and in the Jiangnan anteclise, the belt IV1 is the known oldest ophiolite belt in China now. The age of ophiolite formation is 1024 ± 30Ma (whole-rock-mineral interior isochron Sm-Nd age) (Zhou X.M. et al., 1989). It might have emplaced prior to 769Ma, i.e. during the Late Xuefeng movement (Zhou X.M. et al., 1989).

In summary, the following speculations may be gained on the temporal-spatial distribution of China.

1. The history of the genesis and obduction emplacement of ophiolite of China range throughout the whole Phanerozoic, or from Early Paleozoic to Cenozoic. It seems not to cluster in some particular periods (Fig.2) as suggested by Abbate et al., (1985); only the Ordovician ophiolite predominates slightly. Tethyan ophiolites in southwestern China might have formed mainly from Jurassic to Early Cretaceous and partly in the Early to Middle Triassic. They might have emplaced mainly in the Early to Late Cretaceous and Late Triassic.

2. With the southern margin of the Tarim and Sino-Korean platform, ophiolites in western and southwestern China have an evolutionary trend of becoming progressively younger from north to south, for example, from Early Paleozoic (Cambrian, Ordovician) of the North Qilian and Qinling belts (III_5 and III_6)→Late Paleozoic and Early Mesozoic (C_1, P, T_{1-2}) of the East and West Kunlun belts (III_1, III_2, III_3)→late Late Paleozoic- early Early Mesozoic (P, T_{1-2}) of the Jinsha River belt (II_4)→Jurassic (J) of the Bangong Lake-Nujiang River Belt (II_2)→Early Cretaceous (J_3-K_1) of the Yarlung Zangbo belt (II_1). This feature might imply that the evolution of the western and southwestern continent is characterized by the mechanism of southward successive accretion.

3. Voluminous data about the ages of ophiolites and the ages of their tectonic emplacement indicate that the oceanic crust of ophiolite was all formed in a relatively short period of time, then followed immediately by tectonic emplacement. This shows that most ophiolites in orogenic belts of China represent remnants of oceanic crust that was disrupted soon after its formation.

4. Some large-scale ophiolite belts usually show the features of multiple origins. They are commonly composed of ophiolite suites of two or more phases and several different tectonic types. Such kind of ophiolite usually represents the position of a suture zone as well.

Types of Chinese Ophiolite Sections

Section types

Large quantity of ophiolite data have so far shown the diversity of ophiolites (Moorse, 1982; Coleman, 1984; Ishiwatari, 1990), and Chinese ophiolites show more diversity, which is not only manifested by their igneous assemblages, but also by the internal structure of the sections, i.e. the stratigraphic section types. Ophiolite sections of China may be summarized into the following four types.

Type 1 section (Fig.3)

This type of section has a relatively complete sequence, consisting in ascending order of mantle peridotites (usually dominated by harzburgite or its altered products) $\pm$ cumulate $\pm$ sheeted dikes (or sills) (or massive diabases) $\pm$ volcanic rocks. The uppermost part of

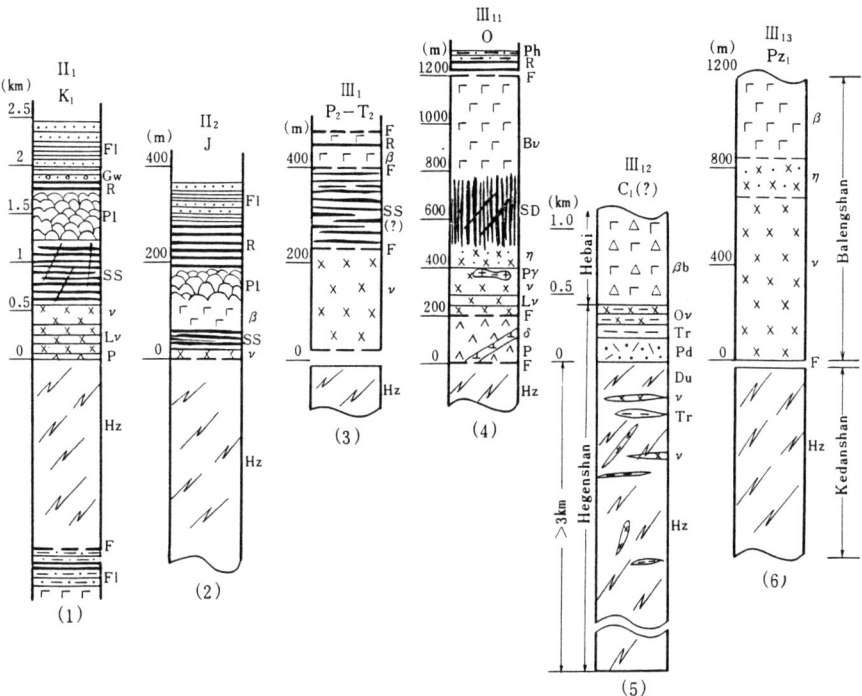

Fig.3 Columnar sections of type 1 ophiolite

1--Flysch, Ph--Phylites, Gw--Graywackes, R--Radiolarian cherts or siliceous rocks, Pl--Pillow lavas, β--Massive basalts, βb--Breccia basalts, Bv--Basic volcanic rocks, SD--Sheeted dikes, SS--Sheeted sills, η--Diabases, δ--Diorites, Pγ--Plagiogranites, ν--Homogeneous gabbros, Oν--Olivine-gabbros, Lν--Layered gabbros Tr--Troctolites, P--Pyroxenites and/or wehrlites, Pd--Plagioclase-bearing dunites, Du--Mantle dunites, Hz--Harzburgites, F--Faults, II_1--III_{13}--Number and name of the ophiolite belts (see Fig.1), K_1--Pz_1--The ages of the ophiolites. (1)--Jiding ophiolite section (after Wang X.B. et al., 1984), (2)--Dingqing ophiolite section (after Zheng H.X., 1983 Fig.1), (3)--Delistangou ophiolite section (after Jiang C.F. et al., 1992, Fig.4-18), (4)--Yueyashan ophiolite section (after Zuo G.C. et al., 1987, Fig.7), (5)--Hegenshan-Hebai section (this paper), (6)--Balongshan-Kedanshan section (after Li J.Y., 1987, Fig.2).

intercalated in the upper part of the sequence. That is to say, the section is simple in structure composed all of mafic and ultramafic rocks. This type of section has typical features of oceanic crust. The examples are the Early Cretaceous ophiolite section (Wang X. B. et al, 1984, 1987) at Xigaze, and the Jiding section (Fig. 3-1) of belt II_1, the Dingqing section (Fig. 3-2) of belt II_2, the Delistangou section (Fig. 3-3) of belt III_1, the Yueyashan section (Fig. 3-4) of belt III_{11}, the Hegengshan-Hebai section (Fig. 3-5) and the Balengshan-Kedanshan section (Fig. 3-6) of belt III_{13}.

Type 2 Section (Fig. 4 and Fig.5)

The constituent rocks of this type of section are primarily similar to those of type 1 sections, but the internal structures of the section are different: there are more intercalations of abyssal and bathyal deposits in the upper part of the sequence of type 2, and even alternate with each other. This type can be subdivided into three subtypes.

The first subtype (Fig.4-1-3): The upper sequence of this subtype is characterized by the occurrence of alternating beds of volcanic rocks and the presence of sheeted dikes. It can be represented by the Kunshan section (Fig.4-1) of belt I_1, the Bayingou section of belt III_8 (Fig.4-2) and the Guojiagou section of belt III_6 (Fig. 4-3).

The second subtype (Fig.4-1-3): This kind of section differs from the first subtype by lacking the dike (or sill) swarms and having generally smaller cumulate complex with less developed layered structure, which are often dominated by homogeneous gabbro. Such kind of section exist widely in various orogenic belts. The examples are the Saga section of belt II_1 (Fig. 4-4), the Loubuzhong section of belt II_2 (Fig. 4-5), the Raohe section of belt II_3 (Fig. 4-6), the Shuangqou section of belt II_5 (Fig. 4-7), the Hegengshan-Jidong section of belt III_{12} (Fig. 4-8), the Huolashan section of belt III_7 (Fig. 4-9), the Yushigou section of belt III_5 (Fig. 4-10) and the Wulangou section of belt III_{13} (Fig. 4-11).

The third subtype (Fig. 5): Apart from large quantities of abyssal deposits intercalated in the upper sequence of this subtype of section, there also occur a lot of pyroclastics-tuff, which alternate with lava and radiolarian chert. It lacks dike swarms. This subtype can be represented by the Kudi section of belt III_2 (Fig. 5-1), sections in the East and West Junggar ophiolite belt, such as the Sartuohai section of belt III_9 (Fig. 5-2) and the Fuchuan section of belt IV_1 (Fig. 5- 3). In addition, there are also some ophiolite section in the Qinling belt (III_6) that are interacted with a large amount of tuffaceous components (as exemplified by Erlangping ophiolite).

Type 3 Sections(Fig. 6)

The structure of sections of this type is similar to that of type 1, but the uppermost volcanic lavas are characterized by the occurrence of andesite or andesitic basalt and intermediate-acid volcanic rocks, while the cumulate complex shows distinct features of polyphase magmatism (Laurent et al, 1993). Sections of this type are represented by the Hongguleleng section (Fig. 6) of the west Junggar belt (III_9).

Type 4 Sections (Fig. 7)

This type of section is composed mainly of a sequence of marine intermediate-basic volcanic rocks. Siliceous rocks and calc-alkali volcanic and pyroclastics sometimes in the upper sequence. It lacks other major elements of an ophiolite suite (mantle peridotite,

Time-space distribution and tectonic types of ophiolites in China

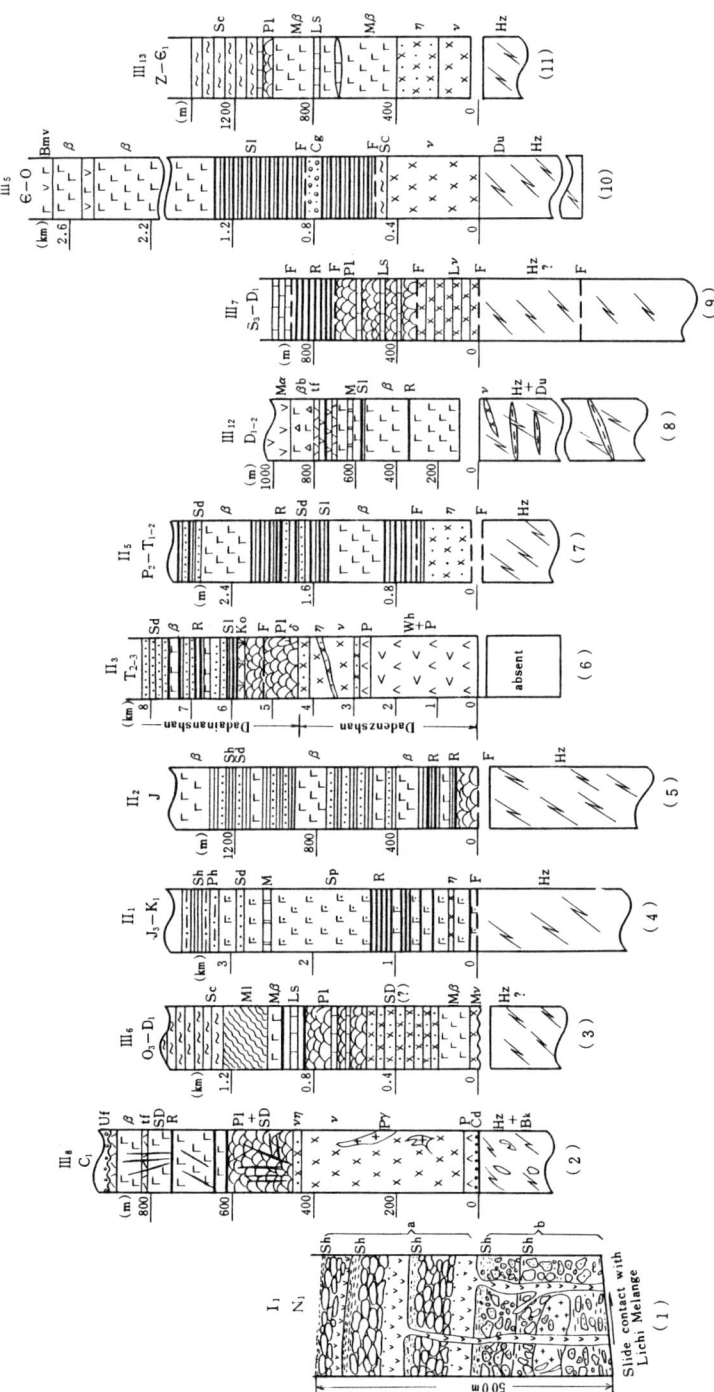

Fig.4 Columnar sections of type 2 ophiolites (subtype 1 and subtype 2)

Sd--Sandstones, Cg--Conglomerates, Sh--Shales or red shales, Ls--Limestones, M--Marble, Sc--Schists, Ml--Mylonites and mylonitic granites, tf--Tuffs, Mα--Meta-andesites, Mβ--Metabasalts, Sp--Spilites, Bmv--Basic-intermediate volcanic rocks, Ko--Komatiite, Bk--Blocks, Mν--Metagabbros, Wh--Wehrlites, Cd--Cumulate dunites, Uf--Unconformity, others are same as Fig.3. (1)--Columnar section of Kuanshan ophiolite (after Liou et al., 1979, Fig.7; Suppe et al., 1981), a--Extrusive sequence: composed of glassy pillow lavas, pillow breccias and some massive flows, b--Plutonic sequence: composed of angula diabase-, gabbro-, pyroxenite-, plagiogranite-, and harzburgite-bearing breccias. (2)--Columnar section of Bayinggou ophiolite (after Xiao X.C., 1991, Fig.11), (3)--Columnar section of Guojiagou ophiolite (after Zhang G.W. et al., 1988, Fig.2), (4)--Columnar section of Saga ophiolite (after Mei H.J. et al., 1981, Fig.6-5), (5)--Columnar section of Luobuzhong ophiolite (after Wang X.B. et al., 1984, 1987, Fig.II-1-8), (6)--Columnar section of Raohe ophiolite (after Kang B.X. et al., 1990, Fig.1 and 2), (7)--Columnar section of Shuanggou ophiolite (after Yang J.R., 1986, Fig.4), (8)--Columnar section of Hegenshan-Jidong ophiolite (this paper), (9)--Columnar section of Huolashan ophiolite (after Wang Z.X., 1990, Fig.5-10A), (10)--Columnar section of Yushigou ophiolite (after Xiao X.C. et al., 1978, Fig.2), (11)--Columnar section of Wulangou ophiolite (after Wang Q. et al., 1991, Fig.27).

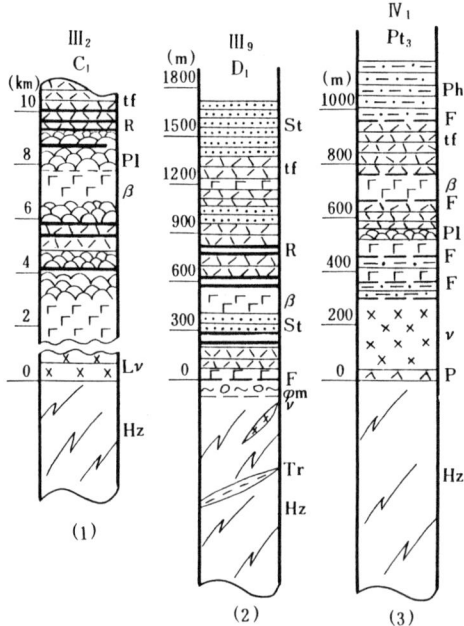

Fig.5 Columnar section of type 2 ophiolite (subtype 3)

St--Siltstones, φm--Ophiolitic mélange, others are same as Fig.3 and Fig.4. (1)--Columnar section of Kudi ophiolite (after Jiang C.F., 1992, Fig.4-2) (2)--Columnar section of Sartuohai ophiolite (after Zhu B.Q., 1987, Fig.9) (3)--Columnar section of Fuchuan ophiolite (this paper).

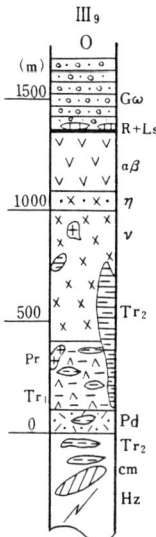

Fig.6 Columnar section of type 3 ophiolite (Hongguleleng section)

$\alpha\beta$--Andesite-basalts, Tr_1--Leuco-troctolites, Tr_2--Melano-troctolites, Cm--Cumulate complex: Layered plagioclase-bearing dunites + plagioclase - bearing lherzolites + bistagites, others are same as Fig.3 and Fig.4.

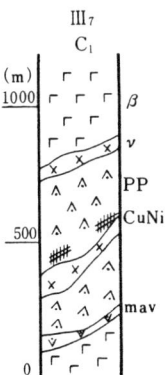

Fig.7 Columnar section of type 4 "ophiolite" (Tudun section) (after Wang Z.X.,1990, Fig.5-22).

PP--Plagioclase lherzolite, mav--Intermediate acid volcanic rocks, CuNi--Copper-Nickel sulfides, others are same as Fig.3 and Fig.4.

layered cumulate and sheeted dikes). However, this marine volcanic sequence is intruded by mafic and ultramafic rocks of ferruginous series, forming something looking like an "Ophiolite suite". The authors do not believe that this suite of complex belongs to ophiolite, but consider that it represents mafic and ultramafic rocks of non- ophiolite type. However, its generation has certain associations with the evolution in the terminal closing stage of oceanic basin. Therefore, it also differs from mafic and ultramafic assemblages in a Precambrian craton. Section of the type can be represented by Tudun section (Fig. 7), and some sections in the Ganzi-Litang rock belt (II_6).

Some Features of Chinese Ophiolite sections

The main features are summarized as follows.

1. The thickness of ophiolite crust is not big, mostly less than 2km, and even a few hundreds of meters, only in a few cases, it thickens up to 2-3km. Compared with mean oceanic crust, the total thickness of ophiolite crust is clearly inadequate.

2. The upper sequence (upper crust) of ophiolite often shows alternating beds of volcanic and abyssal deposits, and is even frequently intercalated with terrigenous clastic. There are appreciable difference between them and the typical mid-oceanic ridge ophiolite.

3. Sheeted dike swarms (or sill swarms) are usually less developed or absent.

4. In terms of the REE geochemical features, most ophiolite volcanic (at the sections of type 2,3 and 4) show a distribution pattern of REE enrichment type (P type), while a few of them (type 1 section) show LREE depletion type and flat type (N type and T type).

5 Magma chambers represented by cumulates are usually relatively small in size (mostly not thicker than 1 km). There are mainly two types of assemblages of cumulates complexes (Wang X. B. et al, 1987). One type is the dunite_-pyroxenite (or wehrlite)- -gabbro assemblage (the crystallization order of the cumulate minerals is Ol-Cpx-Pl). The other is the feldspar-bearing dunite--troctolite--olivine gabbro assemblage(the mineral crystallization order is Ol-Pl-Cpx). There is an additional type that have both types of cumulates mentioned above (Bao et al, 1984). though it is not common. The cumulate of the Ol-Opx-Cpx assemblage (Ishiwatari, 1990), is not found yet in China.

6. Mantle peridotite at the base of ophiolite is dominated by harzburgite with subordinate dunite of varying amounts, rare lherzolite remains in harzburgite as nodular residue. Mantle peridotite can be divided into three melted residue type according to their melting degrees (Wang X. B. et al, 1989): highly melted residue type, moderately melted residue type, and sightly melted residue type. According to the amount of segregated mafic material within the mantle peridotite, there are two types of rock bodies: one type contains no or few segregated mafic material, which is called "pure" mantle peridotite; the other contains more or even large amounts of troctolite--gabbro, and in some cases it even forms a conspicuous crust-mantle transition zone (Wang X.B. et al.,1989) below and above the Moho (e.g. Hegengshan and Hongguleleng sections). The former rock type is associated with chrome-rich podiform chromite, while the latter with aluminum-rich podiform chromite (Bao et al, 1990).

Tectonic Types of China's Ophiolites

So far, classifications of ophiolite have been made from different view-points by some geologists (Miyashiro, 1975; Moorse, 1982; Coleman, 1984; Zhang Q., 1990). In this paper, an attempt is made to propose a scheme of classification on the basis of the structural types, of the ophiolite sections and the characters of their associated deposits, combined with an analysis of the maturity of the ocean basin development, i.e., its development stages.

It is known to all that ophiolite represents a part of the oceanic lithosphere, which can be produced in various environments. The authors believe that the tectonic setting at the time of ophiolite formation, i.e. its tectonic type, is tightly controlled the maturity of the ocean basin development.

Thus, tectonic types of China's ophiolites can be tentatively classified as follows:

1. Initial ocean basin-type ophiolite

This type of ophiolite is characterized by the development of type 2 sections. That is to say, it has three primary features: a) sheeted dikes are less developed or lacking; b) volcanic rocks in the upper sequence contain more abyssal and bathyal deposits or terrigenous clastic deposits; and c) magma chambers are usually small and layered structures are mostly less developed. These primary features might reflect that the ocean basin was at the initial development stage, i.e. it was not mature yet, with a low spreading rate and a relatively high deposition rate. By that time the mantle below the crust was also at a low melting degree, and a spreading center had not been formed yet. The basaltic magma that melted out was relatively enriched in alkalis, and such incompatible elements as LREE; as a result, volcanic rocks of this type of ophiolite commonly shows an enriched type (P type) distribution pattern. Apparently this type of ophiolite is limited in size as it is constrained by immature development of the ocean basin.

The tectonic locations of its formation are mostly restricted to various small-sized basins on paleocontinental margins (including all kinds of basins in the island-arc system). The crust represented by the section of this kind of ophiolite is not the real oceanic crust, but mostly shows the feature of transitional crust. This type of ophiolite is the main type of China's orogenic ophiolite (Fig. 2).

2. Mature ocean basin type-ophiolite

This type of ophiolite has the features of type 1 sections. Its main features are the following: a) the upper sequence of the ophiolite section contains nearly no or few intercalations of abyssal deposits, indicating no introduction of terrigenous fragments, and the section structure is simple with a clear sequence; b) sheeted dike swarms (or sill swarms) are generally developed to various extent; and c) pillow lava and cumulate have a consistent LREE-depleted (N-type) distribution pattern. These features show the maturity of the ocean basin, the increase of spreading rate and the formation of a continuous spreading center.

This type of section occurs mainly in the Tethyan ophiolite belts (belts II_1, and II_2). As their thicknesses are relatively small, their formation environment might still be restricted to small ocean basins on paleocontinental margins, though the crust represented by them may represent real oceanic crust. This type of ophiolite is more in accord with the ophiolite definition given in the Penrose symposium.

3. Island-arc type-ophiolite

Macroscopically, the section sequence and structure of this type of ophiolite show no significant difference from those of the above mentioned, ocean basin type ophiolite. Except for the occurrence of andesite and andesite basaltic rocks in the upper sequence, it has all the features of the ocean basin type ophiolite, and likewise also contains the abyssal deposits, such as radiolarian chert. This shows that both the island-arc type ophiolite and ocean basin type ophiolite have certain genetic relations. The oceanic island- arc formation also involves a process of generation, development and maturation. The island-arcs represented by those ophiolites are all immature ones at their youth stage, and have been developed on the base of oceanic crust. For example,the ophiolites at the Hongguleleng section of belt III_9 and the Yushigou section of belt III_5 can also be regarded as island- arc type ophiolite.

4. Residual sea basin-type, i.e. non-ophiolite type mafic and ultramafic rocks

This type of "ophiolite" has the assemblage features of type 4 section. It lacks nearly all the features of an ophiolite suite, except for the occurrence of marine volcanic and the possible appearance of siliceous rocks.However, its formation has certain relations with the terminal stage, i.e. the closing stage, of the ocean basin development. The crust at this time had already entered into the stage of development of the continental crust, though sea water still existed by then.

Conclusion

In summary, the following conclusions may be gained on the Chinese ophiolites:

1. Ophiolites of China are widespread, and exposed in four areas: (1) Tethyan ophiolites in southwestern China; (2) Qinlin-Qilian-Kunlun ophiolites distributed along the southern margin of the Tarim and Sino-Korean platform in central China; (3) Tianshan-Inner Mongolia-Xing'an ophiolite distributed along the northern margin of the Tarim and Sino-Korean platform in northwest and northern China; (4) Ophiolites on the continental margin of eastern China.

2. The history of the genesis and obduction emplacement of ophiolite of China ranges from Late Proterozoic to Cenozoic, they are mostly of Paleozoic and rarely of Proterozoic. Phanerozoic ophiolites do not form and cluster in some particular periods, though Ordovician ophiolites slightly predominate. The Tethyan ophiolites are mainly of early Cretaceous and Jurassic, and partly in Early to Middle Triassic.

3. To the south of the Tarim and Sino-Korean platform, ophiolites in southwestern and western China have an evolutionary trend of becoming progressively younger from north to south. This time-space distribution clearly indicates progressive accretion of the Chinese continents toward the southwest through Paleozoic and Mesozoic.

4. Some large ophiolite belts are usually multiple belts composed of two or more ophiolite suites of different ages and tectonic types.

5. On the basis of the types of the ophiolite sections integrating with the maturity (i.e. development stage) of oceanic basins, the authors divide ophiolites of China into four tectonic types: (1) Initial oceanic basin-type ophiolite, (2) Mature oceanic basin-type ophiolite, (3) Island are-type ophiolite, and (4) Residual oceanic-type (i.e. non-ophiolite-type) mafic and ultramafic rocks.

Ophiolites in China are dominated by the first type. The second and third types are less important, while the fourth type is rare.

6. The synthetic study on ophiolites of China reveals that the total thickness of the oceanic crust at all ophiolite sections in China is notably insuffient usually less 2-3km as compared with the average thickness of the present oceanic crust (5-7km). It is evident that the great majority of the ophiolite sections in China are the product of small oceanic basins.

Acknowledgement

This research is a part of the subject "The Evolution of the Continental Lithosphere of China and Its Adjacent Areas" (directed by Professor Ren Jishun), a project supported by the National Natural Science Foundation of China. We acknowledge here the supports and valuable suggestions and opinions about the writing of this paper given by Professor Xiao Xuchang and Professor Ren Jishun and also the drawing of all figures in this paper done by Ms. Dong Xiaojing.

References

E. Abbate, V. Bortolotti, P. Passerni, and G. Principi. The Rhythm of Phanerozoic Ophiolites. Ofioliti, 6, 109-125 (1985).
Asrarullah, Z. Ahmad, S. G. Abbas. Ophiolites in Pakistan: An Introduction. In: Geodynamics of Pakistan. A. Farah and K. A. Dejong (Eds) PP.101-104, Geological Survey of Pakistan, Queta (1979).
Bao Peisheng and Wang Xibin. Evolution of oceanic crust of Mesozoic Tethys in the light of two suites of volcanic rocks in the Yarlung Zangbo Ophiolite belt. Sci. Sinica (Series B), XXIX, 1317-1329 (1986).
Bao Peisheng and Wang Xibin. The cumulate complex of the ophiolite suite in the N. Xizang (Tibet) (in Chinese). In: Himalayan Geol., Li G. C. and J.C. Mercier (Eds), II, 149-161, Geological Publishing House, Beijing (1984).
Bao Peisheng and Wang Xibin. Geochemistry evidence for the genesis of Xigaze ophiolite, Xizang (Tibet) (in Chinese). In: Himalayan Geol. Li G.C. and J. L. Mercier (Eds), Geological Publishing House, Beijing (1984).
Bao Peisheng, Wang Xibin, Hao Ziguo and Peng Genyeng. A new idea about the genesis of the aluminium-rich podiform chromite deposit--with the Sartuohai chromite deposit of Xinjiang as an example (in Chinese). Mineral. Deposits. 2, 97-110 (1990).
R. G. Coleman. The diversity of ophiolites. Geol. Mijnb., 63, 267-283 (1984).
--------------. Ophiolites, Ancient Oceanic Lithosphere? Springer-Verlag, Berlin, Heidelberg, New York (1977).
Chen Bingwei. Some new observations on the tectonic development of Sanjiang region, East Xizang (Tibet) (in Chinese). Contrib. Geol. Qinghai-Xizang (Tibet) Plateau, 12, 165-175 (1983).
Duan Xinhua and Zhuo Hong. The Ailaoshan-Tengtiaohe fracture--the subduction zone of an ancient plate (in Chinese). Acta Geol. Sinica 55, 258-265 (1981).
Feng Yimin et al.. Tectonic evolution of the West Junggar, Xinjiang, China (in Chinese). In: Tectonic evolution of the southern margin of the Paleo-Asian composite megasuture. Xiao X.C. and Tang Y.Q. (Eds) 13, 66-88 (1991). Beijing Scientific and Technical Publishing House, Beijing (1991).
Fan Chenjun. The tectonic-metamorphic belt of Mountain Ai-Lao in Yunnan Province (in Chinese). Yunnan Geol. 5, 281-289 (1986).
Gao Yanlin, Wu Xiangnong and Zuo Guochao. The characters and tectonic significance of ophiolite first discovered in the east Kunlun area (in Chinese). Bull. Xi'an Inst. Geol. Min. Res., Chinese Acad. Geol. Sci. 21, 17-28 (1988).
Hu Shouxi and Lin Qianlong et al.. The geology and metallogeny of the amalgamation zone between ancient North China plate and South China plate (in Chinese). Nanjing University Press, Nanjing (1988).

Hao Ziguo et al.. Geological characteristics and genetic study on ophiolites of the two types in the western Junggar, Xinjiang (in Chinese) Acta Petrol. Mineral. 8, 299-309 (1989).

A. Ishiwatari. Time-space distribution and petrologic diversity of Japanese ophiolites. Ophiolite Genesis and Evolution of Oceanic Lithosphere. 723-743 (1990).

Jiang Chunfa et al.. Opening and Closing Tectonics of Kunlun Mountains (in Chinese). Geological Publishing House, Beijing (1992).

Jahn Bor-ming. Mid-ocean ridge or marginal basin origin of the East Taiwan ophiolite:Chemical isotopic evidence.Contrib. Mineral. Petrol., 92, 194-206 (1986).

Kang Baoxiang et al.. Raohe ophiolite and its geological significance in Nadanhadaling. Helongjiang Geol. 1, 3-16 (1990).

Liu Baotian, Jiang Yaoming and Qu Jingchuan. The discovery of a Paleooceanic crust strip along the line from Litang to Ganzi Sichuan and its significance on Plate tectonics (in Chinese). Contrib. Geol. Qinghai-Xizang (Tibet) Plateau, 12, 119-124 (1983).

Li Jinyi. On evolution of Paleozoic plate tectonics of east Junggar, Xinjiang, China (in Chinese). In: Tectonic Evolution of the Southern Margin of the Paleo-Asian Composite Megasuture. Xiao X.C. and Tang Y.Q. (Eds). 13, 92-105, Beijing Scientific and Technical Publishing House (1991).

----------. Essential characteristics of early Paleozoic ophiolites to North of Xila Mulun river. Eastern Inner Mongolia and their plate tectonic significance (in Chinese). Contrib. Project Plate Tect. North China 2,136-148 (1987).

J. G. Liou, and W. G. Ernst. Oceanic ridge metamorphism of the East Taiwan ophiolite. Contrib. Mineral. Petrol., 68, 335-348 (1979).

Lu Xinxiang, Geochemistry of M-type granitoid rocks in ophiolite suite in Eastern Qinling orogenic belt (in Chinese) In: Formation and Evolution of the Qinling Orogenic Belt. Zhang G.W.(Ed.) PP.149-161. Northwestern University Press, Xi'an (1988).

Li Chunyu. A preliminary study of plate tectonics of China (in Chinese). Bull. Chinese Acad. Geol. Sci. Series 1, Vol. 2, 1, 11-18 (1980).

R. Laurent, Wang Xibin and Bao Peisheng. The Hongguleleng island-arc ophiolite, A sequence of multiple intrusions in the Paleozoic ophiolites of Xinjiang, China (in press,1993).

Mei Houjun and Lin Xuenong. Xizang (Tibet) Ophiolite. In: Magmatism and Metamorphism in Xizang (Tibet) (in Chinese) Zhou Y.S. et al (Eds). PP. 147-200, Chap. 6, Scientific Publishing House, Beijing (1981).

E. M. Moorse. Origin and emplacement of ophiolites. Reviews of Geophysics and Space Physics, 20, 735-760 (1982).

A. Miyashiro. Classification, characteristics and origin of ophiolites. J. Geol., 83, 249-281 (1975).

A. Nicolas et al.. The Xigaze ophiolite (Tibet): a peculiar oceanic lithosphere. Nature, 294, 414-417 (1981).

Peng Lihong. The ages and tectonic significance of the ophiolite in southern zone of Onder Sum formation, Inner Mongolia (in Chinese). Kexue Tongbo, 2, 104-107 (1984).

Ren Jishun. Jiang Chunfa, Zhang Zhengkun and Qin Deyu, The Geotectonic Evolution of China (in Chinese). Scientific Publishing House, Beijing (1980).

J. Suppe, J. G. Liou and W. G. Ernst. Paleogeographic origins of the Miocene East Taiwan ophiolite, Am. J. Sci., 281: 228-246 (1981).

Sun Yong, Yu Zaiping and Zhang Guowei. Geochemistry of the Eastern Qinling ophiolites (in Chinese). In: Formation and Evolution of the Qinling Orogenic Belt. Zhang G. W. (Ed.). PP.65-73 Northwestern University Press, Xi'an (1988).

Tang Kedong and Zhang Yunping. Tectonic evolution of Inner Mongolian suture zone (in Chinese). In: Tectonic evolution of the southern margin of the Paleo-Asian composite megasuture. Xiao X.C. and Tang Y.Q. (Eds), Beijing Scientific and Technical Publishing House, Beijing (1991).

G. R. Tilton, S.T. Kwon, R.G. Coleman and Xiao X.. Isotopic studies from the western Junggar Mts, NW China. Geol. Soc. Amer., Abstr. Programs, 18, 773 (1986).

Wang Xibin and Bao Peisheng. Types of melting residue and structural deformation of the mantle peridotite bodies in orogenic belts of China. Progress in Geosciences of China (1985-1988), Papers to 28th IGC, 13-66 (1989).

Wang Xibin, Bao Peisheng, Deng Wanming and Wang Fangguo. Xizang (Tibet) Ophiolites (in Chinese). Geology Publishing House, Beijing (1987).

Wang Xibin, Bao Peisheng and Zheng Haixiang. A structurally disrupted ophiolite in the lake area of northern Xizang (Tibet) and geochemistry (in Chinese). In: Himalaya Geol., Li G. C. and J. L. Mercier (Eds). II, PP. 115-138. Geological Publishing House, Beijing (1984).

Wang Xibin, Cao Yougong and Zheng Haixiang. Ophiolite assemblage of the middle Yarlung Zangbo river in Xizang (Tibet) and a model of oceanic crust evolution (in Chinese). In: Sino-French cooperative investigation in Himalayas. Li G.C. and J.L. Mercier (Eds). PP.181-204, Geological Publishing House, Beijing (1984).

Wang Quan, Liu Xueya and Li Jinyi. Plate tectonics between Cathaysia and Angaraland in China (in Chinese). Beijing (Peking) University Press, Beijing (1991).

Wang Quan and Liu Xueya. Paleo-ocean crust of the Chilienshan (Qilianshan) region, western China and its tectonic significance (in Chinese). Scientia Geol. Sinica, 1, 42-54 (1976).

Wang Renming, Chen Zhenzhen, Li Pingfan and Su Shangguo. Tectonic environment and crustal evolution of the Kuanping Group and the Erlangping Group in Tongbai area, Henan Province (in Chinese). In: Geol. Memoirs of the Qinling-Daba Mountains. Liu C.H. and Zhang S. G. (Eds), 1, PP. 99-108, Beijing Scientific and Technical Publishing House, Beijing (1990).

Wang Zuoxun et al.. Polycycle Tectonic Evolution and Metallogeny of the Tianshan Mountains (in Chinese). Scientific Publishing House, Beijing (1990).

Wang Hongzhen and Li Guangcen. Correlation table of stratigraphical subdivision (in Chinese). Geological Publishing House, Beijing (1990).

Wang Lienchang, Li Dazhou, Zhang Qi and Zhang Kuiwu. Ophiolitic mélange in Litang, Sichuan Province (in Chinese). Acta Petrol. Sinica. 2, 15-26 (1985).

Xiao Xuchang et al.. On tectonic evolution of the southern margin of the Paleoasian composite megasutrue zone (in Chinese). In: Tectonic Evolution of the Southern Margin of the Paleo-Asian Composite Megasuture. Xiao X.C. and Tang Y. Q. (Eds), Beijing Scientific and Technical Publishing House, Beijing (1991).

Xiao Xuchang, Chen Guoming and Zhu Zhizhi. A Preliminary study on the tectonics of ancient ophiolites in the Qilian mountain, Northwest China (in Chinese). Acta Geol. Sin. 52, 281-294, (1978).

Xu Zhiqin, Lu Yilun, Tang Yaoqing and Zhang Zhitao. Formation of the Composite East Qinling Chains (in Chinese). Environmental Science Publishing House, Beijing (1988).

Yang Jiarui. The features of ophiolitic sequence and their geological significance in Shuang Gou area, along the middle Ailao mountain (in Chinese), Yunnan Geol., 5, 292-300 (1986).

Zuo Guochao et al.. A discovery of nape structure and ophiolitic mélange in the early Paleozoic ophiolite zone at Baiyunshan Xichangjing district of Beishan range (in Chinese). Contrib. Porj. Plate Tectonic in Northern China. 2, 51-57 (1987).

Zhu Baoqing, Wang Laisheng and Wang Lianxiao. Paleozoic era ophiolite of southwest part in western Junggar of Xinjiang, China (in Chinese). Bull. Xi'an Inst. Geol. Min. Res., Chinese Acad. Geol. Sci., 17, 3-30 (1987).

Zheng Haixiang, Pan Guitang, Xu Yaorong and Wang Peisheng. Some new information about the ultramafics along Nujiang tectonic belt -- a complete ophiolite suite (in Chinese). Contrib. Geol. Qinghai-Xizang (Tibet) Plateau, 13,191-192 (1983).

Zhou Xinmin, Zhou Haibo, Yang Jiedong and Wang Yinxi. Sm-Nd isochron age and its geological significance of Fuchuan ophiolite in the Shexian area, Anhui province (in Chinese). Kexue Tongbao, 16, 1243-1245 (1989).

Zhang Qi. Classification of ophiolites. Scientia Geol. Sinica, 1, 54-60 (1990).

Zhang Qi, Li Dazou and Zhang Kuiwu. A preliminary study on Tongchangjie ophiolite mélange from Yun County, Yunnan Province (in Chinese). Acta Petrol. Sinica, 3, 1-12 (1985).

Zhang Qi, Zhang Kuiwu, Li Dazou and Wu Haiwei. A preliminary study of Shuanggou ophiolite in Xinping County, Yunnan Province (in Chinese). Acta Petrol. Sinica. 4, 37-48 (1988).

Zhang Guowei, Sun Yong and Yu Zaiping. The ancient continental margin of the northern Qinling Mountains (in Chinese). In: Formation and Evolution of the Qinling Orogenic Belt, Zhang G. W. (Ed.), PP.48-61. Northwestern University Press, Xi'an (1988).

Zhang Chi and Huang Xuan. The ages and tectonic settings of the ophiolites in the western Junggar, Xinjiang (in Chinese), Geol. Review, 6, 509-522, (1992).

Zhang Zhimeng and Jin Meng. Two kinds of mélange and their tectonic significance in Xiangcheng-Derong area, Southwestern Sichuan (in Chinese). Sci. Geol. Sinica. 3, 205-213 (1979).

Zhang Zhimeng, Wang Zhongshi and Zheng Yumin. Petrographic characteristics and tectonic significance of the spilite-keratophyric rocks in the Xiangcheng-Derong area of southwestern Sichuan (in Chinese). Acta Geol. Sinica. 55, 179-193 (1981).

The Hongguleleng island-arc ophiolite, a sequence of multiple intrusions in the Paleozoic ophiolites of Xinjiang, China

R. LAURENT[1], X. WANG[2] and P. BAO[2]

[1] *Department of Geology, Laval University, Quebec, CANADA, G1K 7P4*
[2] *Institute of Geology, Chinese Academy of Geological Sciences, Beijing, CHINA*

ABSTRACT --The Hongguleleng ophiolite is located at the northern margin of the Junggar basin of Xinjiang, northwest China, and is part of a Paleozoic broad suture zone between the Siberian and Tarim-China-Korea plates. The ophiolite is closely associated with thick sequences of andesitic lavas and volcaniclastic Flysch of Siluro-Devonian age. Our study is concentrated on the plutonic rocks of the ophiolite. We distinguish: a) a mantle unit consisting of mantle lherzolite, harzburgite, dunite with chromite ore, and of xenolithic lenses of an iron-rich variety of lherzolite with olivine Fo 71-78; b) a complex crustal sequence defining distinct magmatic events. The crustal plutonic rocks consist of a sequence of deformed meta-gabbros which are intruded first by dunite and undeformed layered troctolites and, secondly, by plutons of poikilitic wehrlite and lherzolite. These poikilitic ultramafites are also intrusive into the mantle peridotites and the troctolitic series. Finally, dykes and plugs of calc-alkaline diorite intrude both the mantle peridotites and all the crustal plutonic rocks. Spinels of each unit are distinctive in composition. Spinel chemistry, the observed magmatic relations and the close association of the plutonic rocks studied with a volcanic association of island-arc character strongly suggest that the ophiolite was formed in the suprasubduction zone environment of a Silurian island-arc within the Junggar ocean basin.

INTRODUCTION

This paper presents a detailed study of the Hongguleleng ophiolite which is located at the northern margin of the Junggar basin of Xinjiang, northwest China (figure 1). The ophiolite is the largest body within the northeast trending Hoboksar ophiolite belt of Late Paleozoic age. This belt defines a broad zone of suture between the Siberian and Tarim-China-Korea plates (Feng et al., 1989). Convergence of these two plates during the Paleozoic progressively destroyed the Junggar ocean basin, where the ophiolite was formed, and led to the tectonic emplacement of the ophiolite by southward thrusting in Devonian time. The ophiolite occurs in close association with thick sequences of volcaniclastic Flysch and andesitic lavas of Siluro-Devonian age. Huo (1987) suggested that the ophiolite was formed in a Devonian back-arc basin. However, and though the precise age of formation of the ophiolite is not known, a hornblende K-Ar age of 400-420 Ma has been obtained on the youngest group of andesite porphyry dykes that intrude the plutonic rocks at Hongguleleng (Bai et al., 1986). This result suggests that the ophiolite is probably of Silurian age or older. Still recently, the Hongguleleng ophiolite was considered to be a remnant of oceanic lithosphere consisting of a mantle peridotite unit and of crustal rocks forming a single differentiation sequence (Bai and Zhou, 1988; 1991). Our field work in 1991 completed by analyses of the samples collected now allows us to put in evidence the structural and petrological complexity of this ophiolite.

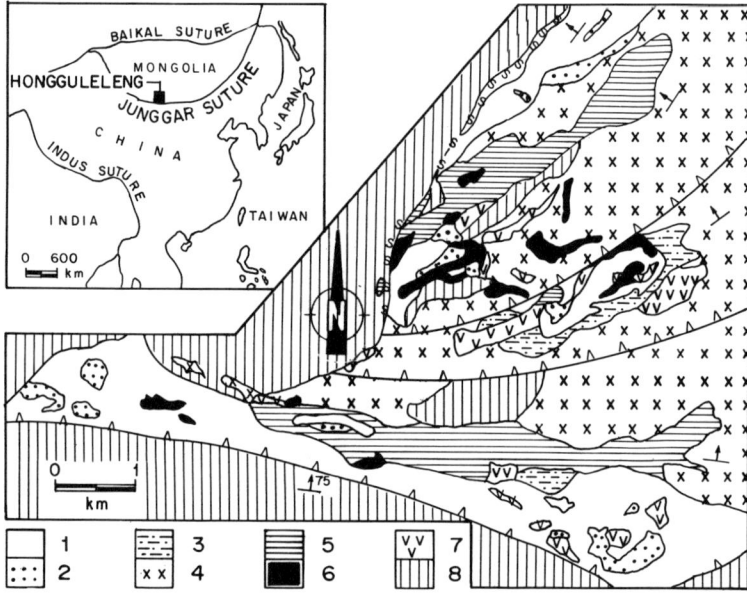

Figure 1: Geological map of the Hongguleleng ophiolite (modified from Bai and Zhou, 1988). Hongguleleng is located at 86°26'27" East and 46°47'40" North. Units: 1) mantle harzburgite; 2) dunite; 3) mantle lherzolite; 4) metagabbro; 5) troctolitic series; 6) poikilitic ultramafites; 7) diorite intrusions; 8) Siluro-Devonian sediments and volcanic rocks.

FIELD RELATIONS

On the south and west, the plutonic rocks of the Hongguleleng ophiolite (figure 1) rest in tectonic contact on Siluro-Devonian sedimentary rocks consisting of thick sequences of shales and sandstones interbedded with volcaniclastic tuffs. Locally and mainly in the northeast, the plutonic rocks of the ophiolite are topped by thrust-sheets of basaltic and andesitic lavas and by Silurian volcaniclastic tuffs and breccias. The plutonic sequence is structurally divided into three major slices by thrust faults dipping towards the north and the northwest. Shards of country-rocks and blocks of metagabbro, peridotite and plagiogranite occur along the tectonic contacts and fault zones.

The plutonic sequence includes mantle and ocean-crust rocks. The mantle rocks form the base of the ophiolite slices. They are made up of a complex assemblage of residual harzburgite and dunite, of chemically less depleted plagioclase-bearing lherzolite and of iron-rich lherzolite. The bulk of the mantle peridotites consists of harzburgite and dunite.

The plagioclase lherzolites occur as discrete bodies within the harzburgite and dunite at the top of the mantle rocks near their contact with the overlying cumulate rocks of the crustal sequence. The iron-rich lherzolite is found as small xenolithic lenses included in the mantle peridotites. The lenses are only a few meters long and less than one meter in width. They are massive showing no tectonite fabrics and they rest in sharp contact with the country-rock. This iron-rich lherzolite either represents segmented dykes or is a tectonic inclusion. The problem has not yet been solved by our field observations, but the chemistry and mineral composition discussed further on show that the iron-rich lherzolite is a distinct unit which may be correlatable with the late poikilitic ultramafic bodies intrusive into the cumulate rocks of the crustal sequence. Finally, the mantle peridotites also contain dykes of troctolite and dykes and plugs of younger hornblende diorite.

The plutonic sequence of crustal rocks above the mantle peridotites is not simple and does not derive from a single major magmatic event but from several ones. On the basis of cross-cutting relatioships observed in the field, we can distinguish four major magmatic events which, from old to young, are: 1) metagabbros, 2) fresh troctolitic series, 3) intrusive poikilitic ultramafites, and 4) late hornblende diorite plutons and andesite porphyry dykes. The first oldest crustal sequence of gabbros is strongly deformed and metamorphosed in the amphibolite and greenschist facies. After deformation and metamorphism, this metagabbroic sequence has been intruded by the troctolitic series which are neither significantly deformed nor metamorphosed. The troctolitic series are bedded and well layered and they grade in composition from dunite and plagioclase dunite at the base to troctolites upwards. The third sequence consisting of poikilitic ultramafites is intrusive into the two first ones. These ultramafic rocks occur as strings of small bodies cross-cutting the metamorphic foliation of the metagabbros as well as the bedding of the fresh troctolites, or they also occur as sill-like bodies a few centimeters to several meters thick within the troctolitic series. These rocks are massive, undeformed wehrlite and lherzolite characterized by a poikilitic texture. Their occurrence, structural features and composition are very similar to the ultramafic bodies intrusive into the crustal sequences of the Mesozoic ophiolites from Cyprus (Benn and Laurent, 1987; Laurent et al., 1991) and Oman (Lippard et al., 1986; Juteau et al., 1988). The youngest intrusive rocks are diorites and andesite porphyries of calc-alkaline parentage. They are likely to be the magmatic feeders of the basaltic and andesitic lavas that rest in tectonic contact on the plutonic rocks of the Hongguleleng ophiolite.

PETROGRAPHIC DESCRIPTION AND CHEMISTRY

1. Mantle peridotites

For comparative purposes compositions of pyroxenes, olivine and plagioclase are summarized in figure 2, and selected data on the mineral chemistry of pyroxenes and spinels are given in tables 1 and 2.

The mantle peridotites consist mainly of harzburgite associated with chromite-bearing dunite. They are characterized by porphyroclastic textures and by metamorphic foliations and lineations induced by high-temperature plastic deformation (Nicolas et al., 1980).

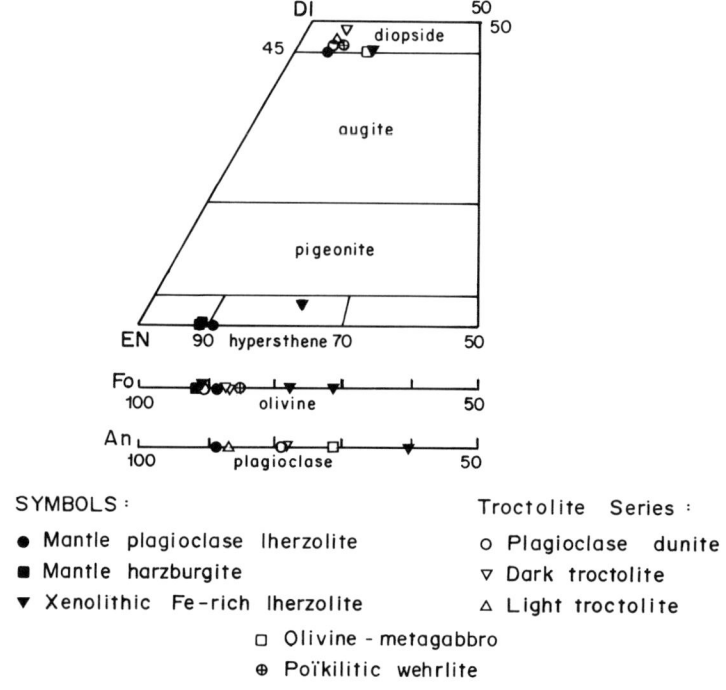

Figure 2. Average composition of pyroxenes, olivine and plagioclase in main rock units of the Hongguleleng ophiolite.

Olivine is forsterite 90-91 with NiO contents of 0.3 to 0.7 wt.%. Orthopyroxene is enstatite 91-92 with about 1 wt.% of CaO and relatively high contents of chromium (0.65 to 0.75 wt.% Cr_2O_3). Accessory spinels are rich in chromium and have Cr# (100 Cr/Cr+Al) varying between 50 and 60. The chromite ore, massive or disseminated, which occurs in locally discordant dunite bodies within the mantle harzburgite is also rich in chromium with an average Cr# of 60 (figure 3).

The lherzolite has a more fertile composition than the harzburgite and dunite. The rock contains clinopyroxene and some plagioclase as major components besides olivine and orthopyroxene. Olivine is forsterite 89 with NiO contents of 0.2 to 0.5 wt.%. Orthopyroxene is enstatite 90 with 1.3 wt.% CaO and about 0.50 wt.% Cr_2O_3. Clinopyroxene is a diopside ($En_{49}Fs_5Wo_{46}$) rich in chromium (about 1 wt.% Cr_2O_3). Plagioclase is anorthitic (An_{89-90}). The accessory spinels are rich in aluminum compared to the harzburgite and dunite spinels and they have Cr# varying between 20 and 30.

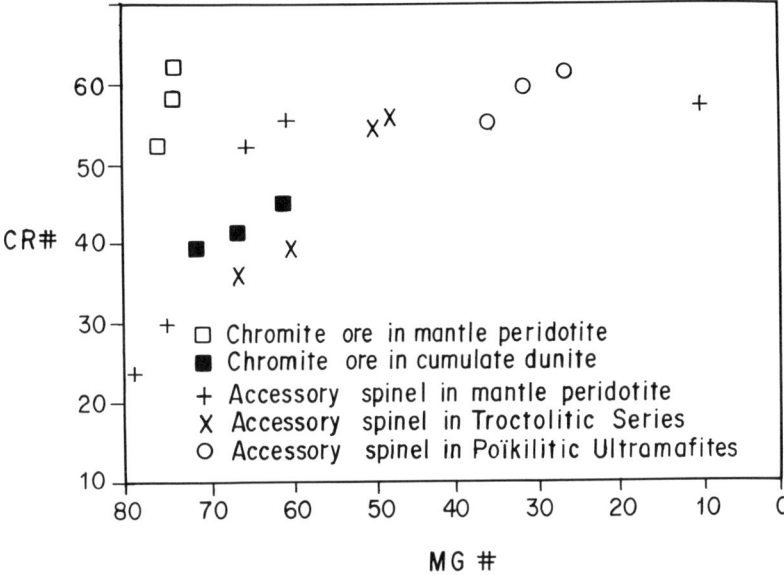

Figure 3. Cr# versus Mg# for chromite and spinel from the Hongguleleng peridotites and gabbroic rocks.

We see that compositions of the mantle peridotites grade from residual harzburgite to relatively fertile lherzolite (Ishiwatari, 1985). The origin of this compositional heterogeneity may be inherited from the partial melting process if the percentage of melt formed and then extracted has been controlled by local and significant variations in the physical conditions. Alternatively, the heterogeneity could result from a post-melting episode of metasomatism. Some of the residual harzburgites may have locally been impregnated by basaltic melts and turned at relatively low pressure into plagioclase lherzolite. Because of the position of the lherzolite at the top of the mantle peridotites, we consider that the second explanation is more probable than the first one.

The last type of peridotite we have documented in the Hongguleleng mantle sequence is an iron-rich lherzolite only found as small xenolithic lenses. The rock has a magmatic porphyritic texture with undeformed pyroxene phenocrysts set in a protogranular matrix of smaller olivine grains and spinels. Compared to the other peridotites the three major mineral components of this lherzolite have much lower Mg# (100 Mg/Mg+Fe). Olivine is chrysolite (Fo_{71-78}) with NiO contents of 0.05 to 0.24 wt.%. Orthopyroxene is bronzite ($En_{77}Fs_{20}Wo_3$) with an average Cr_2O_3 content of 0.26 wt.%. Clinopyroxene is a diopsidic augite ($En_{44}Fs_{12}Wo_{44}$) with Cr_2O_3 contents of 0.4 to 0.8 wt.%. Accessory spinels are low in both aluminum and chromium (9.0 wt.% Al_2O_3 and 19.0 wt.% Cr_2O_3), but they are rich in iron (about 60.0 wt.% $FeO+Fe_2O_3$) and titanium.

TABLE 1: Selected microprobe analyses of pyroxenes

WT.%	1	2	3	4	5	6	7	8	9	10
SiO_2	52.01	52.50	52.61	50.55	51.98	51.52	56.38	55.06	56.54	52.82
TiO_2	0.47	0.32	0.61	0.97	0.24	1.13	0.16	0.08	0.00	0.27
Al_2O_3	3.77	2.08	2.05	3.11	4.52	3.21	1.67	2.63	2.08	1.80
FeO	2.78	3.43	6.72	3.81	2.64	7.17	5.77	6.58	5.47	13.06
MnO	0.04	0.20	0.16	0.22	0.13	0.20	0.09	0.11	0.09	0.31
MgO	16.94	16.61	15.23	16.06	16.44	13.88	33.83	33.29	33.59	27.31
CaO	21.83	23.83	21.41	22.95	21.14	21.38	1.05	1.34	1.19	2.64
Na_2O	0.56	0.22	0.20	0.39	nd	0.56	nd	0.05	nd	0.40
K_2O	0.03	0.03	nd	nd	nd	nd	nd	nd	0.02	nd
Cr_2O_3	1.09	0.41	0.19	0.83	1.08	0.48	0.68	0.54	0.72	0.26
TOT.	99.52	99.62	99.18	98.89	98.17	99.53	99.63	99.68	99.44	99.87
EN	49.5	46.6	44.7	48.0	49.6	44.0	91.2	89.8	91.5	76.5
FS	4.6	5.4	10.80	6.0	4.5	12.0	8.8	10.2	8.5	20.0
WO	45.9	43.0	44.5	46.0	45.9	44.0	-	-	-	3.3
Mg#	91.6	89.6	80.2	88.3	91.8	77.6	91.3	90.0	91.7	78.9

nd = not determined,   EN = % enstatite,   FS = % ferrosilite,   WO = % wollastonite

SAMPLES: # 1 to 6 are **clinopyroxenes** from:

1) Plagioclase dunite;   2) Dark troctolite (1 + 2 = Troctolitic Series);   3) Olivine-metagabro;
4) Poïkilitic wehrlite;   5) Mantle plagioclase lherzolite;   6) Xenolithic Fe-rich lherzolite.

SAMPLES: # 7 to 10 are **orthopyroxenes** from:

7) Clinopyroxene-bearing Mantle harzburgite;   8) Mantle plagioclase lherzolite;   9) Mantle harzburgite;   10) Xenolithic Fe-rich lherzolite.

All samples, except nb. 4, have been analyzed with a JXA-733 automated wavelength-dispersion electron-microprobe at the Institute of Geology, Beijing. Sample 4 has been analyzed with an ARL wavelength-dispersion electron-microprobe at Laval University, Quebec.

TABLE 2: Selected microprobe analyses of chromite and spinel

Wt. %	7	11	15	19	23	24	28	29	33
TiO_2	0.13	0.29	0.10	0.43	0.52	0.28	0.11	0.02	0.09
Al_2O_3	22.52	26.07	20.37	33.86	31.94	28.70	23.77	27.11	42.19
Fe_2O_3*	4.95	4.49	4.88	5.52	6.35	7.76	6.47	6.02	4.53
FeO	9.91	8.99	9.76	11.03	12.69	15.52	12.95	12.04	9.06
MnO	0.19	0.15	nd	0.14	0.22	nd	1.09	0.27	0.35
MgO	15.52	16.04	15.23	15.52	13.95	13.16	10.94	12.80	15.51
Cr_2O_3	47.18	43.69	49.61	31.91	33.57	34.85	44.25	41.20	26.11
NiO	0.17	0.09	nd	0.30	0.02	nd	0.12	0.10	0.35
Mg#	73.6	76.1	73.4	71.5	66.2	60.2	60.1	65.4	75.3
Cr#	58.4	52.9	62.0	39.5	41.4	44.9	55.5	50.5	29.3

	36	38	39	42	43	47	48	50	51
TiO	0.26	5.89	0.54	1.31	0.37	1.11	1.97	1.05	1.15
Al_2O_3	47.04	9.60	36.67	19.76	33.01	20.88	15.07	17.38	20.07
Fe_2O_3*	4.30	20.12	6.07	10.10	7.36	9.66	14.41	12.25	11.65
FeO	8.60	40.23	12.14	20.24	14.73	19.31	28.82	24.49	23.30
MnO	nd	0.47	nd	0.20	0.58	0.48	nd	0.78	0.95
MgO	18.04	2.40	13.52	10.16	12.38	10.21	5.64	6.03	7.14
Cr_2O_3	21.49	19.03	30.66	37.39	30.67	37.98	34.60	37.75	35.39
NiO	nd	0.32	nd	0.37	0.04	nd	nd	nd	nd
Mg#	78.9	9.6	66.5	47.2	60.0	48.5	25.9	30.5	35.3
Cr#	23.5	57.1	35.9	55.9	38.4	55.0	60.7	59.3	54.2

* calculated from FeO total   nd = not determined

<u>SAMPLES:</u>

Nos 7, 11, 15, chromite ore in Mantle dunite. Nos 19, 23, 24, Chromite ore in cumulate dunite. Nos 28, 29, Spinel in Mantle harzburgite. Nos 33, 36, Spinel in Mantle lherzolite. No. 38, Spinel in Xenolithic Fe-rich lherzolite. Nos 39 to 47, Troctolitic Series: 38 is plagioclase dunite, 42, 43 and 47 are troctolites. No. 48 is poïkilitic wehrlite. Nos 50 and 51 are poïkilitic lherzolites.

All samples have been analyzed with the microprobe of the Institute of Geology, Beijing (same as table 1), except the samples 15, 24, 36, 47, 48, 50 and 51 which were analyzed with a JEOL automated scanning electron-microscope at Laval University, Quebec.

With the exception of the iron-rich lherzolite, the mantle peridotites were or still are residues of melting. Therefore, the highly incompatible trace elements should also be depleted relative to chondritic composition. For example, the rare earth element (REE) abundances in alpine peridotites are characterized by a Light REE depletion ranging from about 0.01 to 0.6 x chondritic values and Heavy REE abundances that are 1 to 2 x chondritic values (Frey, 1982). The mantle peridotites of Hongguleleng (figure 4 and table 3) show a different pattern. In these rocks, the LREE have abundances that are 1 to 2 x chondritic values while Eu and the HREE are depleted ranging from about 0.05 to 0.4 x chondritic values. This geochemical signature suggests that the Hongguleleng peridotites may have been partially melted at low pressure and, later, metasomatically enriched in LREE. We also note that the plagioclase lherzolite is richer in HREE than the harzburgite. In contrast, the relatively compatible transition elements show a normal distribution (table 3); their abundances, except for Cr and Ni, are lower than average values for undepleted lherzolites and peridotites (Basaltic Volcanism Study Project, 1981). The iron-rich lherzolite is characterized by higher contents of Sc, Ti and V than the other Hongguleleng mantle peridotites.

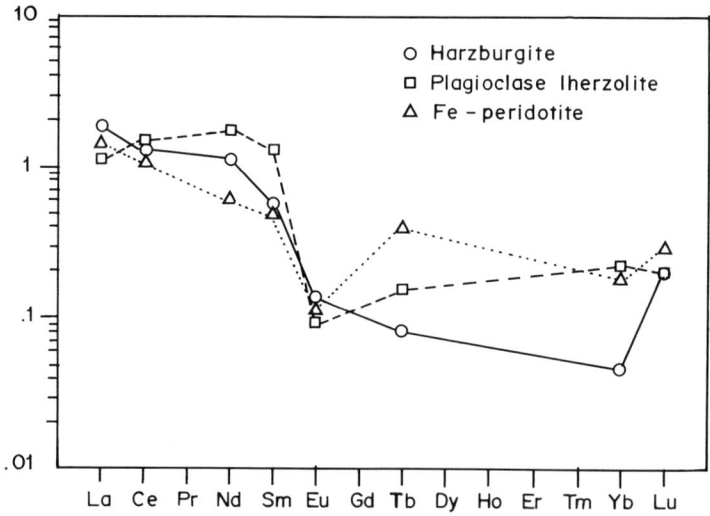

Figure 4- Chondrite normalized rare earth element abundances in peridotites from Hongguleleng.

Table 3: Bulk-rock compositions major, transition and rare-earth elements

Wt. %	1	2	3	4	5
SiO_2	43.96	42.44	45.07	45.33	-
TiO_2	0.15	0.08	0.02	0.24	-
Al_2O_3	22.95	8.22	0.77	2.49	-
Fe_2O_3	0.81	3.28	2.86	4.17	-
FeO	4.35	6.20	5.54	4.93	-
MnO	0.06	0.14	0.11	0.17	-
MgO	8.96	31.69	43.87	38.83	-
CaO	17.98	7.52	1.33	3.07	-
Na_2O	0.88	0.22	-	-	-
K_2O	0.11	0.05	-	0.15	-
P_2O_5	0.05	0.05	-	-	-
TOTAL	100.26	99.89	99.57	99.38	-
Mg#	75.9	86.1	90.6	88.9	-
Cr^{ppm}	315	1863	1346	1157	1159
Ni	123	1363	1360	1556	987
Co	26	72	64	94	82
Sc	34	8	16	12	31
V	121	20	60	34	140
Cu	118	120	84	3	137
Zn	52	56	63	75	133
La	0.48	0.42	0.59	0.41	0.46
Ce	1.51	1.90	1.14	1.31	0.89
Nd	1.40	0.60	0.70	1.09	0.37
Sm	0.34	0.43	0.11	0.26	0.10
Eu	0.21	0.02	0.01	0.01	0.01
Tb	0.10	0.01	0.00	0.01	0.02
Yb	0.17	0.02	0.01	0.04	0.04
Lu	0.01	0.01	0.01	0.04	0.01

SAMPLES:
1) Olivine meta-gabbro; 2) Dark troctolite; 3) Mantle harzburgite; 4) Mantle plagioclase lherzolite; 5) Xenolithic Fe-rich lherzolite.
The five samples were analyzed at the Institute of Geology, Beijing. Major elements were analyzed by X-ray fluorescence with separate titration for $FeO-Fe_2O_3$. The analyses of the REE were done by plasma mass spectrometry (ICPMS) and the other trace elements by plasma optical emission spectrometry (ICPOES). The analytical error of these trace element data is smaller than 10%.

2. Metagabbros, cumulates and poikilitic ultramafites

The oldest plutonic rocks of the crustal sequence are deformed and altered gabbros resting locally in fault contact directly on the mantle peridotites. The metagabbro has cataclastic textures and its primary minerals are replaced to variable extents mainly by secondary green and brown hornblendes or by greenschist mineral assemblages. Some fresh samples can be found showing that the protolith of this unit is olivine gabbro. Olivine is chrysolite (Fo_{86}) with a NiO content of 0.17 wt.% and plagioclase is labradorite-bytownite (An_{62-80}). The clinopyroxene, a diopsidic augite ($En_{45}Fs_{11}Wo_{44}$) with 2.05 wt.% Al_2O_3 and 0.19 wt.% Cr_2O_3, has probably recrystallized in equilibrium with secondary brown hornblende. The troctolitic cumulates are intrusive into the base of the metagabbroic sequence; they are not deformed nor metamorphosed though olivine is partly serpentinized. The sequence is bedded and layered and grades in composition from plagioclase dunite and dark olivine-rich troctolite to light olivine-poor troctolite and gabbro. The troctolites contain some clinopyroxene, but always in small amounts, as well as accessory spinels. This suite generated by the cotectic crystallization of olivine and plagioclase is not strongly fractionated as shown by the chemistry of its main mineral components (figure 2 and tables 1, 2). Olivine varies from Fo_{91} (with 0.28 wt.% NiO) in dunite to Fo_{87} (down to 0.15 wt.% NiO) in troctolites. Plagioclase varies from a maximum value of An_{91} to a minimum value of An_{80}. Clinopyroxenes are diopsidic augites with an average composition of $En_{49}Fs_6Wo_{45}$ and Cr_2O_3 contents varying from 1.1 to 0.4 wt.%. Accessory spinels are as rich in chromium and aluminum as the chromite ore in cumulate dunite but values of their Mg# vary much more from a maximum of 67 to a minimum of 47 (figure 3).

Metagabbros and troctolitic cumulates are intruded by small bodies of poikilitic wehrlite and lherzolite. In these rocks olivine (Fo_{85} with 0.23 NiO wt.%) and spinels are the cumulus phases while the clinopyroxene is the main intercumulus phase (oikocryst) and its composition is diopsidic ($En_{48}Fs_6Wo_{46}$ with 0.83 wt.% Cr_2O_3). Orthopyroxene and plagioclase occur in variable amounts but they are extensively altered. Accessory spinels have lower Mg# than any other spinels from the troctolites, from the chromite ores or from the mantle peridotites. The only exception are the iron-rich spinels from the xenolithic lenses of "magmatic" lherzolite in the mantle peridotites (figure 3). REE data on one sample of olivine-rich troctolite and of metagabbro are compared graphically in Figure 5. In both samples abundances of LREE are equal to 1 to 2 x chondritic values. However the troctolite is strongly depleted in Eu and HREE compared to the metagabbro. Abundances of the transition elements are also quite distinct (data in table 3). They are close to chondritic values in the troctolite while the metagabbro shows significant enrichment of Sc, Ti and V followed by depletion of Cr, Mn, Co and Ni. These distinct geochemical signatures, particularly the REE, confirm that these two units derive from different magmas.

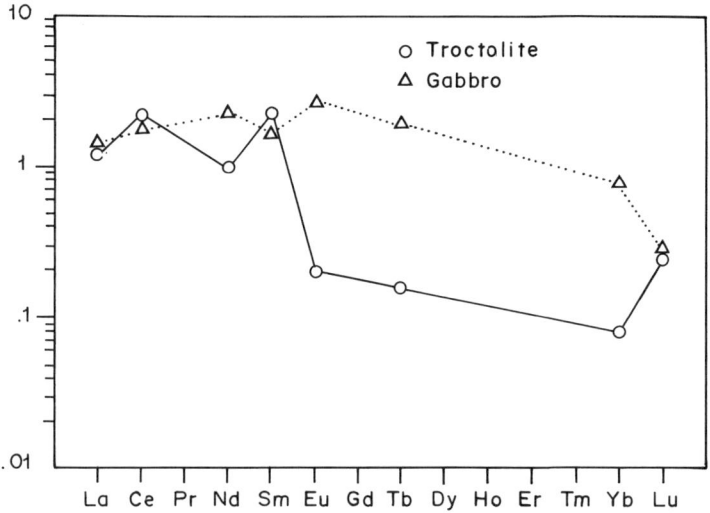

Figure 5- Chondrite normalized rare earth element abundances in gabbroic rocks from Hongguleleng.

COMPOSITION OF SPINELS

Spinels are more resistant to alteration than most of the major silicates of peridotites and gabbros and they are also sensitive petrogenetic indicators of the composition of the system as well as of the physical conditions that exist at the time of their crystallization (Irvine, 1967). Now we will use the composition of spinels to better define the different units recognized in the field and whose petrographic descriptions have been given above.

Spinels in the Hongguleleng ophiolite are found as concentrations of chromite ore and as accessory phases. The concentrations of chromite ore are massive or disseminated but they always occur in dunites from either the mantle sequence or from the "stratiform" cumulates at the base of the crustal sequence. Accessory spinels are distributed in all peridotites and troctolites. The ore-bearing mantle dunite occurs as lenses and boudinage dykes within the upper part of the mantle harzburgite; its contact is sharp and locally discordant with the layering or the foliation and lineation of the enclosing harzburgite. The dunite and its associated ore show at variable degrees evidences of high temperature solid-state deformations: plastic deformations of olivine and cataclastic "pull-apart" deformations of chromite. The basal dunite cumulates also contain chromite ore showing structural and textural magmatic features similar to the deposits of stratiform complexes, such as the Bushveld complex, where they have not been significantly deformed. However, most basal dunite cumulates of ophiolites are deformed and their deformation has usually been annealed by recrystallization. These circumstances make difficult the distinction between mantle and cumulate dunites. For this problem the field observations are critical. At Hongguleleng, the mantle harzburgite grades upwards into lherzolites so that the dunites overlying the mantle peridotites are interpreted to be of cumulate origin. The contact between the two units is faulted and the "cumulate"dunites with their chromite ore are deformed but they do not grade laterally into harzburgite and they are overlain by meta-gabbros or troctolitic rocks.

In figure 3 and table 2 our data on the chemistry of the Hongguleleng spinels are compiled in function of Mg# and Cr#. Chromite ore in mantle dunite is richer in Cr_2O_3 and poorer in Al_2O_3 than chromite ore in cumulate dunite (figure 3). We also notice that compositions of accessory spinels from the mantle harzburgite are close to the field of mantle dunite chromite ore like the compositions of accessory spinels from the troctolitic cumulates overlap the field of cumulate dunite chromite ore. The Mg# of the two ore types is also slightly different, the mantle one is close to 75 while the Mg# of cumulate ore varies between about 70 and less than 60. Since the Mg/Fe ratio of spinels may in part be temperature dependent (Irvine, 1967), the observation suggests that the mantle chromites have crystallized at significantly higher temperatures than the cumulate chromites. The higher Cr# of mantle ore is also consistent with the experimental data of Roeder and Reynolds (1991) which indicate that Cr# in chromite increases with increasing temperature. Fe# (the trivalent cation ratio $Fe^3/Fe^3+Al+Cr$) versus Mg# does not allow a clear separation of the two types of chromite ore because the contents of iron in the cumulate chromite ore are, on the average, only about 2 wt.% higher than those from the mantle chromite ore.

Figure 3 also shows that the accessory spinels of the units previously defined are chemically distinct. Accessory spinels of mantle lherzolite and harzburgite have same Mg# (and Fe#) but very different Cr#. Xenolithic iron-rich lherzolite spinels on the other hand are characterized by low Mg# (and high Fe#). Finally, spinels from the poikilitic ultramafites can be distinguished from the most mafic troctolitic cumulates by higher Cr# and very significant low Mg# values.

CONCLUSIONS

Spinels found in volcanic rocks in different geological settings have distinctive compositions. Roeder and Reynolds (1991) have shown that spinels in basalts from different oceanic environments have different Cr# and Fe#. This is ascribed to differences in the composition of the melt from which the spinels crystallized. The Al/Cr appears to be proportional to that of coexisting melt and the chromites that have crystallized early from basaltic melts have Fe# below 15 (Roeder and Reynolds, 1991). On the base of the Mg# and Cr# (figure 3) we concluded above that the mantle chromite ore crystallized at higher temperatures than the cumulate chromite ore. Those two ore-types also formed from melts of distinct compositions with the mantle melt having a lower Al/Cr. However both ores could be comagmatic. The parental magma rising through the upper mantle would originally have a low Al/Cr. Fractionation of olivine and chromite (with high Cr# and Mg#) at high temperature and the accretion of these minerals as dunite dykes and chromitite within the mantle harzburgite would change the magma composition and strongly increase its Al/Cr. The cumulate chromite ore formed later, at lower temperatures, in subcrustal conditions from this evolved magma would

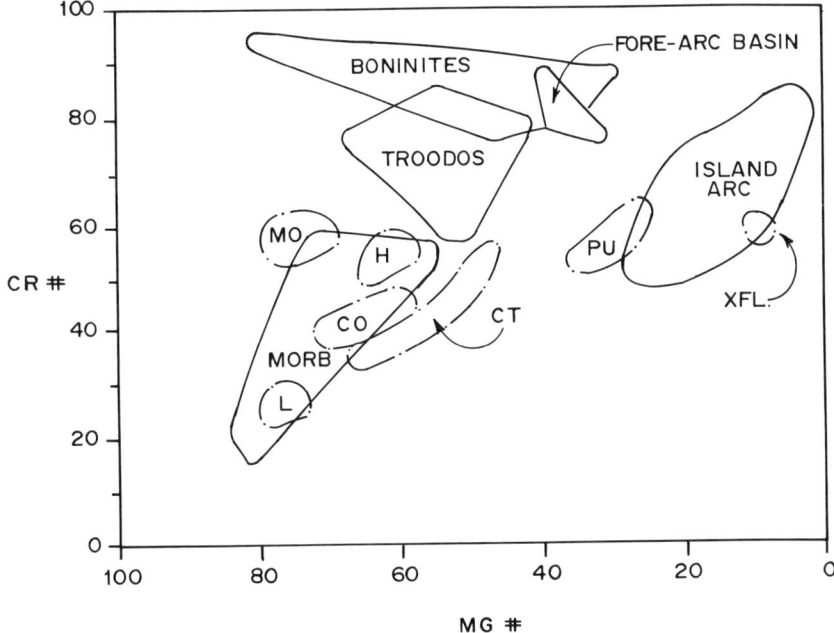

Figure 6- Cr# versus Mg# for chromite and spinel from the Hongguleleng peridotites and gabbroic rocks compared with MORB, island-arcs and the Troodos ophiolite (data sources compiled by Kepezhinskas et al., Ms). Hongguleleng units: MO, mantle chromite ore; CO, cumulate chromite ore; L, mantle lherzolite; H, mantle harzburgite; XFL, xenolithic Fe-rich lherzolite; CT, troctolitic series; PU, poikilitic ultramafites.

necessarily have lower Cr# and Mg# alike those of our cumulate samples. Similar relations have been well documented in Oman where Brown (1980) has shown that Cr# of chromite increases significantly downward as a function of the depth of chromite within a thickness of 6.5 km of mantle peridotites.

The Hongguleleng ophiolite is closely associated with basaltic and andesitic lavas and its plutonic mantle and crustal rocks are intruded by dykes and small bodies of hornblende diorite and andesite porphyry. Earlier work (Huo, 1987) and our (unpublished) analyses of lavas and dykes indicate that these rocks are of calc-alkaline parentage. Their REE and other incompatible elements have a typical signature of island-arc volcanic rocks.

The chemistry of the spinels analyzed in this study also suggests that the plutonic rocks evolved in a subduction zone environment. In figure 6 the spinel data are compiled and compared with MORB, island-arcs and the suprasubduction zone Troodos ophiolite of Cyprus. We notice that most of the Hongguleleng crustal rocks are spread across a field connecting MORB with island-arc spinels. Only spinels from the mantle harzburgites and lherzolites and the chromite ores plot within the MORB field with, however, the exception of the xenolithic iron-rich lherzolite which plots in the Kamchatka island-arc field defined by Kepezhinskas et al. (Ms). So our present observations are interpreted to mean that the Hongguleleng ophiolite was formed in an island-arc which is a tectonically very complex environment. The complicate history of several magmatic events separated by an episode of deformation and metamorphism as documented by the Hongguleleng ophiolite fits a subduction setting better than any other oceanic environment.

ACKNOWLEDGEMENTS

This work was carried on as a cooperative exchange program between the Institute of Geology in Beijing, China, and the Geology Department of Laval University, Canada. Financial support for the stay of R. Laurent in China, for the field work in Xinjiang and the laboratory work in Beijing was provided by the Chinese Ministry of Geology and Mineral Ressources. The analytical work and the stay of P. Bao at Laval University was supported by the Natural Sciences and Engineering Research Council of Canada. Thoughtful reviews by Mei-Fu Zhou and A. Ishiwatari improved the content of the paper.

REFERENCES

BAI, W.-J., & ZHOU, M.-F., 1988. Variations in chemical composition of chrome spinels from the Hongguleleng ophiolite, Xinjiang, China. Acta Mineralogica Sinica, 8, 313-323 (in chinese).

-------------------------- , 1991. Geochemical character and origin of cumulates of the Hongguleleng ophiolite, Xinjiang, China. Bulletin of the Chinese Academy of Geological Sciences, 24, 63-78 (in chinese).

BAI, W.-J., REN Y., WANG S., LE BEL L., OHNENSTETTER D., & OHNENSTETTER M., 1986. Petrology and chrome mineralization of the Hongguleleng ophiolite, Xinjiang, China. Bulletin of the Chinese Academy of Geological Sciences, 14, 1-50 (in chinese).

BASALTIC VOLCANISM STUDY PROJECT, 1981. Basaltic volcanism on the terrestrial planets. Pergamon Press Inc., New York, 1286 p.

BENN, K., & LAURENT, R., 1987. Intrusive suite documented in the Troodos ophiolite plutonic complex, Cyprus. Geology, 15, 821-824.

BROWN, M., 1980. Textural and geochemical evidence for the origin of some chromite deposits in the Oman ophiolite. In: A. Panayiotou, ed., Ophiolites - Proceedings International Ophiolites Symposium, Cyprus 1979, Geological Survey Department Cyprus, 714-721.

FENG, Y., COLEMAN, R.G., TILTON, G., & XIAO, X., 1989. Tectonic evolution of the West Junggar region, Xinjiang, China. Tectonics, 8, 729-752.

FREY, F.A., 1982. Rare earth element abundances in the upper mantle rocks. In: P. Henderson, ed., Rare earth geochemistry. Elsevier, Amsterdam, 153-203.

HUO, Y., 1987. Petrochemistry and geological evolution of island arc and back-arc basin, Hoboksar of Junggar, Xinjiang. Bulletin of the Xinjiang Institute of Geol. Min. Res., Chinese Acad. Geol. Sci., 15, 57-68 (in chinese).

IRVINE, T.N., 1967. Chromian spinel as a petrogenetic indicator. Part 2, Petrologic applications. Canadian Journal of Earth Sciences, 4, 71-103.

ISHIWATARI, A., 1985. Igneous petrogenesis of the Yakuno ophiolite (Japan) in the context of the diversity of ophiolites. Contribution to Mineralogy and Petrology, 89, 155-167.

JUTEAU, T., ERNEWEIN, M., REUBER, I., WHITECHURCH, H., & DAHL, R., 1988. Duality of magmatism in the plutonic sequence of the Sumail Nappe, Oman. Tectonophysics, 151, 107-135.

KEPEZHINSKAS, P.K., TAYLOR, R.N., LEDNEVA, G.V., & TANAKA, H., Ms. Zoned spinels in ultramafic-mafic plutons from the North Kamchatka arc (Russia): petrogenetic implications and comparison with ophiolite plutonic complexes. Submitted to Mineralogical Magazine.

LAURENT, R., DION, C., & THIBAULT, Y., 1991. Structural and petrological features of peridotite intrusions from the Troodos ophiolite, Cyprus. In: Tj. PETERS et al., ed., Ophiolite genesis and evolution of the oceanic lithosphere. Kluwer Academic Publ., Dordrecht, 175-194.

LIPPARD, S.J., SHELTON, A.W., & GASS, I.G., 1986. The ophiolite of Northern Oman. Geological Society, London, Memoir 11, 178 p.

NICOLAS, A., BOUDIER, F., & BOUCHEZ, J.L., 1980. Interpretation of peridotite structures from ophiolites and oceanic environments. American Journal of Science, 280A, 192-210.

ROEDER, P.L., & REYNOLDS, I., 1991. Crystallization of chromite and chromium solubility in basaltic melts. Journal of Petrology, 32, 909-934.

Petrology and Tectonic Settings of the Neyriz Ophiolite, Southeastern Iran

K. SARKARINEJAD

Department of Geology, College of Sciences, Shiraz University, Shiraz, Iran

Abstract: The ophiolitic rocks exposed over 407 km^2 were mapped on a scale of 1:25,000, and this showed that predominant rock unit in the area is harzburgite and serpentinized harzburgite. The harzburgite has fairly uniform mineral proportions of $OL_{84}OPX_{16}$. The average mineral chemistries of olivine ($Fo_{91.2}$) and enstatite ($En_{90.5}$) are considered to be representative of depleted residual mantle. The Neyriz harzburgite which resulted from partial melting of the mantle is depleted in Na, K, Ca, Ti and enriched in refractory elements such as Mg.

The present study has established that the magmatic events within the Neyriz ophiolite were produced by three stages of harzburgite and lherzolite partial melting in three different types of geological environments. The lower pillow lavas with MORB affinities have been derived by partial melting of harzburgite under an oceanic spreading axis. The upper pillow lavas are also derived from the same source with a more depleted nature and higher degree of partial melting of depleted mantle.

The reconstructed paleospreading ridge is based on the sheeted dyke orientations, and shows that Neyriz ophiolite is closely related to the spreading ridge in the southern part of the Tethyan Ocean.

Keywords: Neyriz ophiolite Iran Tethys Ocean harzburgite

INTRODUCTION

The current understanding of the Neyriz ophiolite in terms of petrology and structural geology is limited to some reconnaissance surveys. Haynes and McQuillan [6] pointed out that the ophiolite consists of lenticular pods of harzburgite, pyroxenite and dunites and these were emplaced as a series of cold intrusions in the form of thrust slices. Stonely [19] suggested that the peridotites are made up predominantly of mantle harzburgite tectonites with minor spinel lherzolite and chromitite. Hall [5] suggested that the Neyriz ultramafic rocks are typically serpentinized harzburgites and small podiform chromitite bodies.

GEOLOGICAL AND STRUCTURAL SETTINGS

Field mapping of the Neyriz ophiolite on the scale of 1:25,000 [14] has shown that it consists of the following rock sequences:

Tectonized harzburgite and lherzolite unit. Harzburgite and serpentinized harzburgite are the dominant ultramafic rock types in the Neyriz ophiolite (Fig. 1) and occupies an area of approximately 54.2 km^2. The harzburgite is foliated. This foliation is defined by the preferred orientation of bladed elongate orthopyroxenes. In this oriented harzburgite, the modal composition ranges from $OL_{70}OPX_{30}$ to $OL_{85}OPX_{15}$.

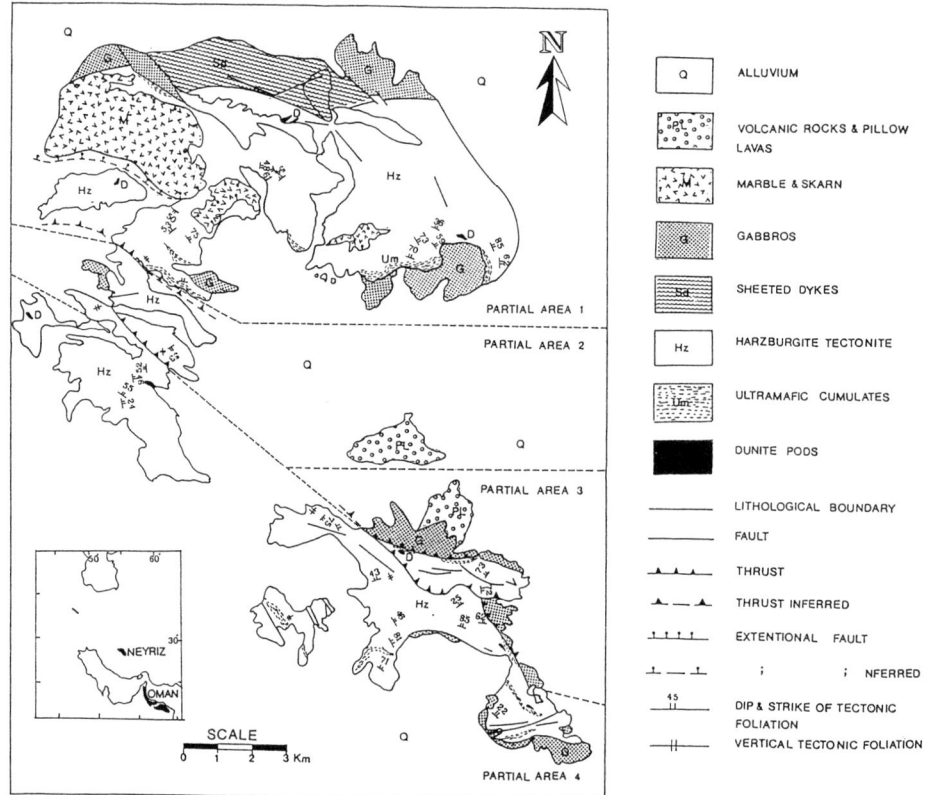

Figure 1. Geological sketch map of the Neyriz ophiolite

The orientation of poles to foliation in the three largest partial areas(Fig. 2) are presented as eigenvectors V_1, V_2 and V_3 after Woodcock [20], three eigenvectors V_1, V_2 and V_3 are defined, where V_1 is the mean orientation of the data, V_3 is pole to the best fit girdle and V_2 is perpendicular to V_1 and V_3. The V_1, V_2, V_3 and K parameter of the orientation of harzburgite foliation in the three largest area are as follows:

Location	V_1 axis	V_2 axis	V_3 axis	$I_n(s_1/s_2)$	$I_n(s_2/s_3)$	K
partial area 1	082°/43°	200°/27°	311°/35°	0.4	0.06	0.6
partial area 2	229°/32°	106°/41°	342°/32°	1.48	0.57	2.59
partial area 3	197°/40°	089°/19°	340°/43°	0.81	0.3	2.7

K-parameter [20] of harzburgite foliation orientation from partial area 1 shows that the distribution of harzburgite foliation poles is a great circle griddle, whilst the K-parameter foliation poles from partial areas 2 and 3 are clusters and the orientation in partial areas 2 and 3 are similar. Harzburgite foliation of partial area 1 is folded possibly during and after emplacement. The mean attitude of harzburgite fold axes of partial area 1 is 311°/35°. Ultramafic cumulate units composed of interlayered dunite, chromite, websterite and clinopyroxenite sits on top of the harzburgite.

The ultramafic units transit to a mafic cumulate unit is composed of low–level layered gabbros. The low–level gabbros unit consists of metagabbro, troctolite, anorthosite and gabbro. Rhythmic and microrhythmic layering and wispy structure are well developed and preserved in the gabbro unit. The orientation of poles to gabbro layering in the three largest areas of the map are presented as eigenvectors V_1, V_2 and V_3 and they may be summarized as follow:

Location	V_1 axis	V_2 axis	V_3 axis	$I_n(s_1/s_2)$	$I_n(s_2/s_3)$	K
Partial area 1	205°/22°	297°/5°	039°/68°	0.78	0.149	5.38
Partial area 3	151°/67°	241°/4°	289°/17°	0.45	1.06	0.24
Partial area 4	196°/21°	296°/252°	071°/56°	0.45	0.50	0.9

The K–parameter of gabbro layering from partial area 1 is 5.83 indicating a cluster distribution. K–parameter of gabbro layering from area 3 and 4 indicating a great circle girdles. The gabbros from partial area 3 and 4 is folded with fold axes of 289/17 and 071/56, respectively.

The sheeted dyke complex on top of the low–level gabbro occupies an area of approximately 7.12 km² and the main sheeted dykes are situated in the northern part of the ophiolite sequence (Fig. 1); the complex has a fault contact with serpentinite and low–level gabbros. Sheeted dykes appear as small outcrops in the central part of the ophiolite sequence.

Statistical treatment of field measurements of sheeted dyke orientation from partial area 1 shows the following results:

Location	V_1 axis	V_2 axis	V_3 axis	$I_n(s_1/s_2)$	$I_n(s_2/s_3)$	K
Partial area 1	079°/50°	166°/40°	206°/28°	0.66	0.42	1.57

The K–parameter of sheeted dyke orientation is a cluster distribution. The V_2 axis in sheeted dyke orientation is 166°/40°, striking NW–SE and dipping NE, and V_1 is interpreted as the spreading direction in which the Neyriz ophiolite was formed, if we assume that no rotation of the sheeted dykes occurred during emplacement.

Pillow lavas occur over an area of approximately 1.87 km² mainly in the central and south east of the southern ophiolite sequence. There are also some outcrops above the sheeted dyke complex in northern and central part of the northern main gabbro unit. Massive blocky lavas, 220 m thick, are also present, overlaying pillow lavas and radiolarite.

The Neyriz colour melange was not studied in detail in the present work, but three traverses through the melange have shown that the melange consists of radiolarite, tuffs, pillow lavas with lesser amounts of serpentinite and exotic blocks of limestone.

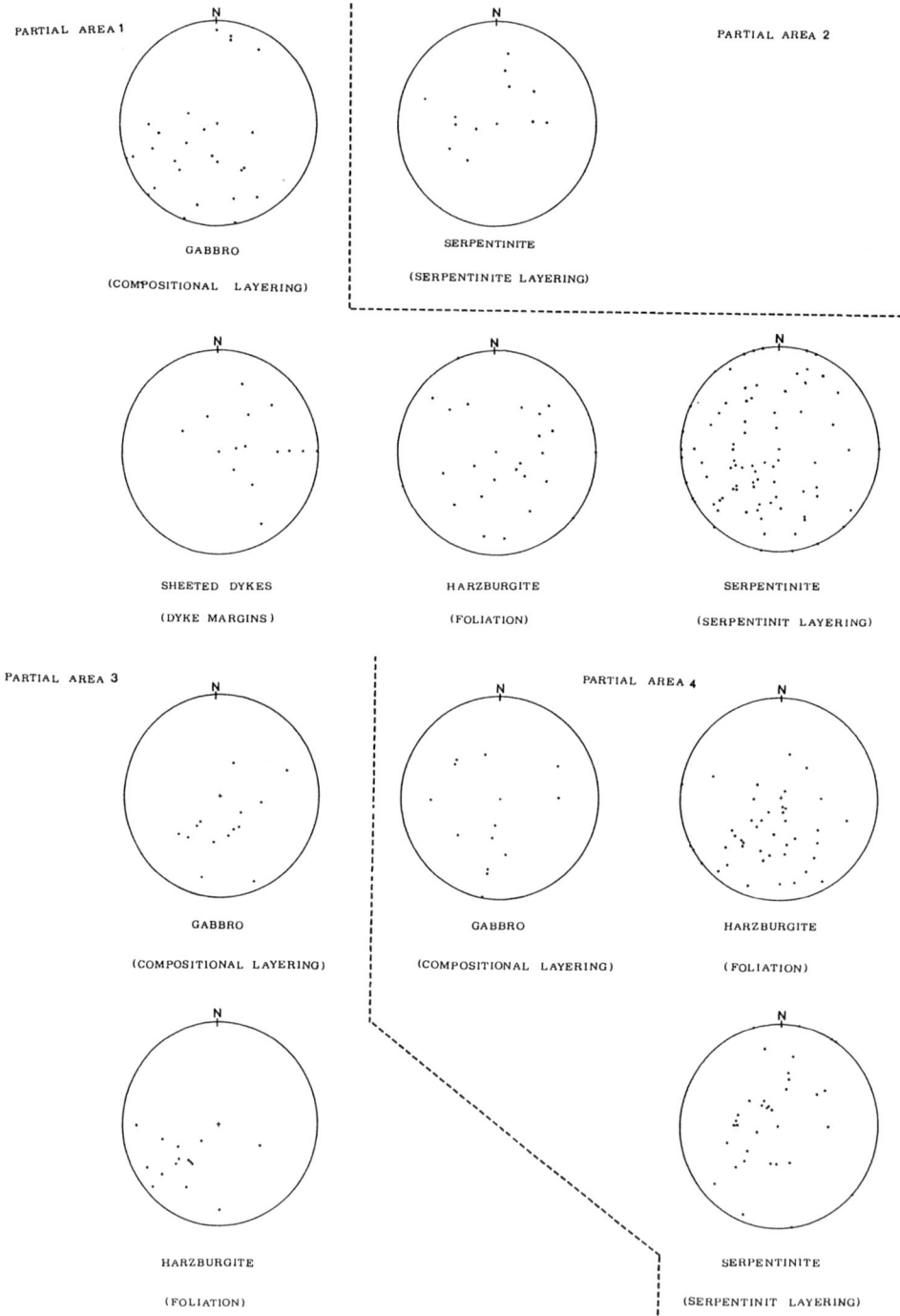

Figure 2. The orientation poles of harzburgite foliation, serpentinite layering, gabbro compositional layering and sheeted dyke margin orientations from three partial areas of Fig. 1.

MINERAL CHEMISTRY OF THE HARZBURGITE

The average olivine Fo content from 24 analyses of the Neyriz harzburgite (Table 1) is $Fo_{91.21\pm0.4}$. Analyses of olivines from widespread localities within the Neyriz ophiolite show insignificant variations of Fo content. The harzburgite with olivine content of $Fo_{91.21\pm0.4}$ is seen to be typical of the alpine-type peridotites or ophiolites (Table 2). The NiO content of olivine is $0.32\pm0.05\%$.

Table 1. Electron microprobe analysis and structural formulae of olivines of the harzburgite tectonite.

	01200G	01200H	Co-011	Co-011	Co-012	Co-014A	Co-015
	Weight percent						
SiO_2	41.113	41.305	40.121	41.167	41.173	40.590	40.609
MgO	50.166	49.589	50.271	50.781	50.287	50.135	50.510
FeO	8.140	8.688	7.952	7.765	7.947	8.709	8.476
NiO	0.140	0.257	----	----	----	----	----
MnO	----	----	----	----	----	0.146	0.153
Cr_2O_3	----	----	----	----	0.142	----	----
Total	99.756	99.480	98.343	99.71	99.550	99.570	99.757
	Number of ions on the basis of 4 oxygens						
Si	1.002	1.008	0.992	1.00	1.004	0.992	0.990
Mg	1.823	1.803	1.852	1.839	1.827	1.826	1.835
Fe	0.166	0.177	0.164	0.157	0.162	0.177	0.172
Ni	0.007	0.005	---	---	---	---	---
Mn	---	---	---	---	---	0.003	0.003
Cr	---	---	---	---	0.162	---	---
Total	2.998	2.992	3.008	2.998	2.995	2.998	3.000

Table 2. olivine composition of alpine-type peridotites (tectonites)

Peridotite Body	Location	Composition (%Fo)	Data Source
Troodos	Cyprus	90.7–91.3	Greenbaum [4]
Burro Mtn.	California	91.1–91.4	Loney et al. [7]
Papua	Papua	91.6–93.6	Menzies and Allen [10]
Othris	Greece	90 –92	Menzies [9]
Oman	Oman	90.2–91.7	Gass and Brown [3]
Neyriz	Iran	90.2–92.1	Present study

The pyroxene in harzburgite is dominantly enstatite and clinopyroxene appears as an accessory phase or as individual grains. Harzburgite enstatite has the composition $En_{90.52\pm0.82}Wo_{1.21\pm0.8}Fs_{8.27\pm0.26}$ (Table 3).

Table 3. Electron microprobe analysis and structural formulae of orthopyroxenes of the harzburgite tectonite.

	3226E	779F	779G	779I	115
SiO_2	57.477	57.297	57.033	57.821	56.999
Al_2O_3	1.095	1.073	1.269	1.083	0.945
Cr_2O_3	0.455	0.688	0.631	0.496	0.539
FeO	5.643	5.531	5.315	5.478	5.509
MgO	34.169	34.740	33.849	34.516	34.539
CaO	1.116	0.689	1.781	0.631	0.589
MnO	0.113	----	0.134	----	0.137
Total	100.016	100.012	100.125	99.255	99.287
Number of ions on the basis of 6 oxygens					
Si	1.978	1.971	1.967	1.982	1.975
Al^{IV}	0.022	0.029	0.033	0.018	0.025
Al^{VI}	0.022	0.014	0.019	0.026	0.014
Cr	0.012	0.019	0.017	0.013	0.015
Fe^{2+}	0.181	0.159	0.153	0.157	0.160
Mg	1.725	1.781	1.74	1.769	1.784
Ca	0.041	0.025	0.066	0.023	0.022
Mn	0.003	----	0.004	----	0.004
Total	3.994	3.998	3.999	3.989	3.998

TRACE ELEMENT CHARACTERISTICS OF THE SHEETED DYKES, PILLOW LAVAS, AND COLOUR MELANGE BASALTS

The major and trace elements of 35 unaltered samples from the Neyriz sheeted dykes, pillow lavas, tuffs and colour melange pillow lavas and massive basalts were analyzed by XRF (Table 4, 5, 6 and 7). The most important trace elements discussed in this section are Ti, Zr and Y. When these trace elements plotted on the ternary Ti–Zr–Y diagram (Fig. 3), 50% of the Neyriz samples, mostly ophiolitic pillow lavas, fall within the field of ocean floor basalts, 25% of the samples of sheeted dykes and ophiolitic pillow lavas fall within the field of low-K tholeiites and 25% of the samples, mostly colour melange basalts fall within the field of within plate basalts.

The Ti–Zr diagram was devised by Pearce [12] for paleotectonic reconstruction of various volcanic rocks. Using Pearce's diagram for discrimination of the Neyriz basaltic rocks, most of ophiolitic pillow lavas and colour melange pillow lavas have a basic composition with MORB characteristics whilst some colour melange basalts fall in the fields of within plate and island-arc volcanics (fig. 4A). In general, the Ti against Zr diagram (fig. 4A) is particularly useful for showing the dividing line between basic and intermediate rock types and fractionation trend. Alabaster et al. [1] suggested that the transition from basic to intermediate character is marked by the incoming of Fe–Ti oxides as a cumulus phase which caused sudden fall in the Ti/Zr ratio of residual magma. Thus, it is apparent that the Neyriz ophiolite and colour melange, basic rocks predominate but intermediate and even acidic rocks occur. Basaltic rocks with high Ti and Zr (upper part of MORB field) represent the lower unit of the pillow lavas. Basaltic rocks with low Ti and Zr (lower part of the MORB field) represent the upper pillow lavas unit (Fig. 4A).

Table 4. Whole rock analyses of Lower Pillow Lavas

	10-60	11-49	10-103A	14-134	10-103B	CM412	CM417
Weight percent (wt.%)							
SiO_2	53.39	54.81	50.33	55.17	54.80	49.73	50.34
TiO_2	2.12	0.73	2.13	1.11	2.03	0.86	1.81
Al_2O_3	14.16	15.55	15.3	14.56	10.45	14.07	15.26
Fe_2O_3	3.62	1.99	3.63	2.92	3.53	2.72	3.96
FeO	11.17	6.22	10.06	5.0	8.59	5.80	8.41
MgO	5.29	3.63	6.34	5.24	7.21	9.34	4.05
CaO	7.35	11.48	9.49	12.65	11.72	12.57	8.86
Na_2O	1.78	5.22	2.19	1.16	1.36	2.16	6.04
K_2O	0.24	0.13	0.21	0.81	0.09	1.86	0.04
MnO	0.27	0.10	0.15	0.12	0.10	0.20	0.17
P_2O_5	0.168	0.13	0.168	0.52	0.13	0.09	0.21
S	0.01	0.04	0.01	0.02	0.01	n.a.	n.a.
Total	100.0	100.03	99.99	100.0	100.02	99.4	99.21

Part per million (ppm) key for tables: n.a.- not analysed, n.d.- not detected

Ba	35	49	26	390	40	220	14
Ca	n.a.	n.a.	n.a.	n.a.	n.a.	n.a.	9
Co	n.a.	n.a.	n.a.	n.a.	n.a.	19	38
Cr	71	n.a.	73	180	227	497	36
Cu	33	n.a.	34	49	137	14	10
Ga	n.a.	n.a.	n.a.	n.a.	n.a.	7	17
La	n.a.	n.a.	n.a.	n.a.	n.a.	4	n.d.
Ni	16	n.a.	19	91	116	62	9
Nb	3	3	3	42	3	3	4
Pb	n.a.	n.a.	n.a.	n.a.	n.a.	8	2
Rb	3	25	2	20	5	3	1
Sr	167	307	98	426	115	173	80
Th	n.a.	n.a.	n.a.	n.a.	n.a.	n.d.	n.d.
U	n.a.	n.a.	n.a.	n.a.	n.a.	n.d.	n.d.
V	n.a.	n.a.	n.a.	n.a.	n.a.	110	306
Y	39	22	42	19	41	14	32
Zn	86	18	79	109	71	74	95
Zr	96	5	104	117	114	44	105

	PL41	PL42	PL44	PL45	PL47	PL405	PLL1
wt.%							
SiO_2	57.29	56.19	57.43	57.26	49.02	49.44	56.30
TiO_2	1.92	1.96	1.99	1.78	1.11	1.19	1.92
Al_2O_3	12.99	12.70	13.37	12.61	14.17	14.69	13.50
Fe_2O_3	3.93	4.06	3.08	3.88	2.80	2.95	3.84
FeO	8.36	8.64	8.05	8.28	5.96	6.23	8.17
MgO	4.04	4.18	3.62	4.17	7.75	6.45	4.09
CaO	7.48	7.79	6.23	6.24	14.52	13.88	7.62
Na_2O	3.52	3.92	4.77	4.58	3.65	3.83	3.86
K_2O	0.11	0.16	0.41	0.67	0.67	0.84	0.17
MnO	0.11	0.15	0.14	0.16	0.15	0.15	0.14
P_2O_5	0.15	0.15	0.15	0.14	0.16	0.15	0.15
S	n.a.	n.a.	n.a.	n.a.	n.a.	n.a.	n.a.
Total	99.90	99.90	100.04	99.86	99.96	99.80	99.80

Table 4. (Continued)

ppm	PL41	PL42	PL44	PL45	PL47	PL405	PLL1
Ba	9	19	22	49	19	130	108
Ca	17	16	18	9	15	20	17
Co	33	33	35	31	38	27	32
Cr	54	44	71	44	61	276	295
Cu	45	41	47	15	43	62	79
Ga	18	19	15	14	19	13	15
La	n.d.	n.d.	n.d.	n.d.	n.d.	n.d.	n.d.
Ni	12	11	12	10	10	96	89
Nb	4	4	3	4	4	4	4
Pb	4	3	3	6	10	3	n.d.
Rb	n.d.	3	5	9	1	7	11
Sr	113	93	94	163	96	229	258
Th	1	n.d.	n.d.	n.d.	4	n.d.	n.d.
U	n.d.	1	n.d.	n.d.	n.d.	n.d.	n.d.
V	404	411	414	369	411	206	204
Y	40	41	37	37	40	25	28
Zn	75	76	81	59	84	60	72
Zr	121	122	117	111	120	88	107

Table 5. Whole rock analyses of upper pillow lavas.

wt.%	13–132	46–155	CM402	CM404	CM408	CM410	CM416	CM460
SiO_2	46.55	51.00	48.75	50.36	53.72	57.34	62.01	49.85
TiO_2	0.90	0.54	1.059	1.10	1.15	0.67	0.38	2.06
Al_2O_3	15.32	9.69	13.84	14.51	17.19	13.37	6.93	15.65
Fe_2O_3	2.46	2.06	2.91	3.01	3.19	2.23	2.11	3.07
FeO	7.68	6.44	6.22	6.41	6.79	4.77	4.49	6.54
MgO	13.26	9.66	7.83	8.0	3.18	9.17	4.145	4.23
CaO	12.19	18.77	14.55	10.97	7.37	6.88	16.19	11.0
Na_2O	1.21	4.06	3.39	3.79	4.45	5.13	1.73	5.31
K_2O	0.05	2.19	0.59	0.64	1.28	0.04	0.14	0.38
MnO	0.27	0.19	0.16	0.15	0.15	0.27	0.17	0.09
P_2O_5	0.08	0.29	0.13	0.14	0.15	0.06	0.70	0.5
S	0.04	0.01	n.a.	n.a.	n.a.	n.a.	n.a.	n.a.
Total	100.01	99.99	99.389	99.08	98.52	99.86	98.995	99.13
ppm								
Ba	29	83	103	103	186	47	580	111
Ce	n.a.	n.a.	14	11	26	21	n.a.	31
Co	n.a.	n.a.	33	38	23	27	19	28
Cr	n.a.	n.a.	387	375	66	403	497	66
Cu	n.a.	n.a.	74	63	245	6	14	28
Ga	n.a.	n.a.	14	13	20	9	7	14
La	n.a.	n.a.	n.d.	n.d.	n.d.	n.d.	4	7
Ni	n.a.	n.a.	127	134	11	176	62	25
Nb	3	3	4	4	4	2	3	18
Pb	n.a.	n.a.	1	1	n.d.	n.d.	n.d.	n.d.
Rb	n.a.	n.a.	4	n.d.	6	3	8	2
Sr	58	36	8	8	11	n.d.	3	13
Th	211	366	217	224	375	73	173	616
U	n.a.	n.a.	n.d.	n.d.	n.d.	n.d.	n.d.	n.d.
V	n.a.	n.a.	191	204	288	149	110	264
Y	22	17	24	25	28	15	14	19
Zn	165	59	67	68	62	45	74	65
Zr	57	35	84	84	86	56	44	152

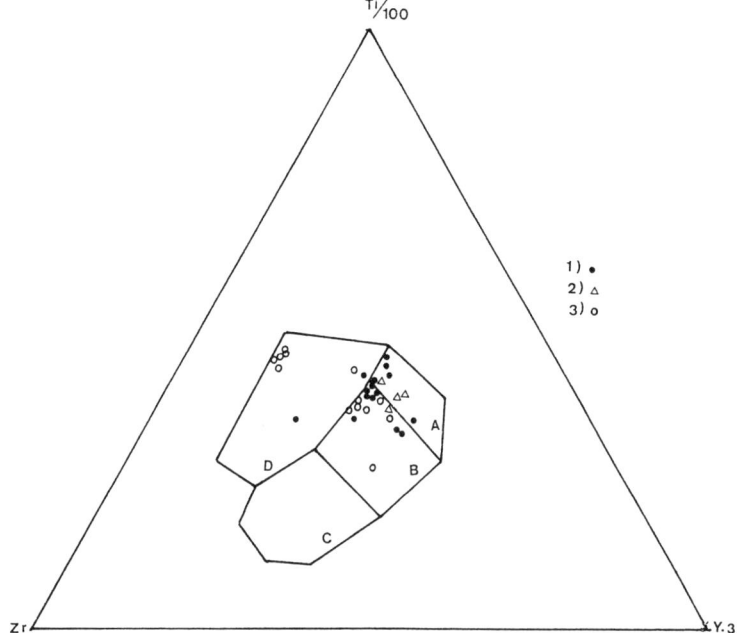

Figure 3. Ti, Zr and Y ternary diagram "within plate" basalt, i.e. ocean island and continental basalts, plot in field D, ocean-floor basalts, OFB in field B, Low-potassium tholeiites (LKT) in field A and B, symbols are as follows: 1. Neyriz ophiolitic pillow lavas; 2. sheeted dykes; 3. colour melange pillow lavas and basalts.

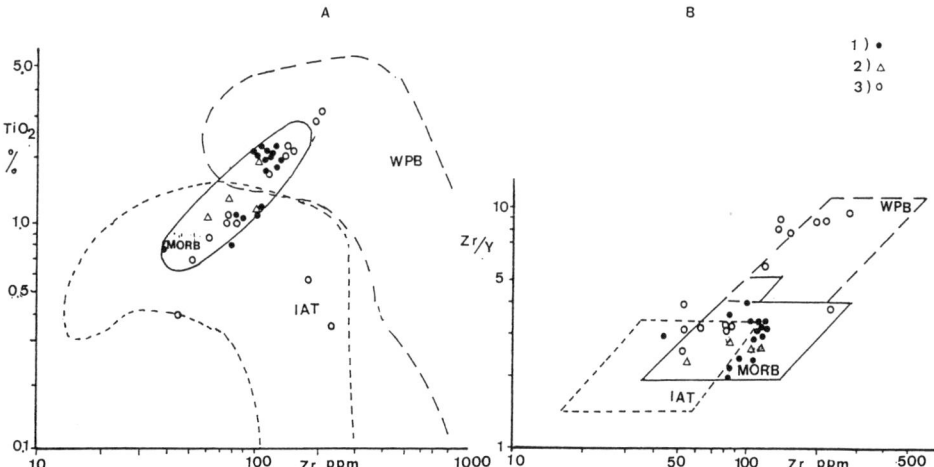

Figure 4A. Ti-Zr diagram for the Neyriz ophiolitic pillow lavas and basaltic rocks. Symbols are the same as in Fig. 2. **4B.** Discrimination diagram of Zr/Y ratio against Zr showing the mid-ocean ridge basalt affinities of the ophiolitic pillow lavas and island-arc and within plate affinities of the colour melange pillow lavas and basaltic rocks symbols are the same as in Fig. 3.

Table 6. Whole rock analysis of within plate massive basalts.

	221	CM401	CM409	CM414	CM470
wt.%					
SiO_2	55.29	46.25	69.97	47.39	50.37
TiO_2	2.01	2.92	0.62	3.09	2.05
Al_2O_3	14.01	15.37	9.12	15.58	17.31
Fe_2O_3	2.68	3.69	1.48	2.94	2.34
FeO	8.36	7.02	3.15	6.26	4.98
MgO	5.29	2.88	4.49	2.18	2.28
CaO	7.45	14.0	8.48	12.94	12.5
Na_2O	4.32	5.57	1.30	5.88	5.94
K_2O	0.23	1.105	1.06	1.28	2.07
MnO	0.18	0.161	0.05	0.141	0.12
P_2O_5	0.12	0.66	0.10	0.70	0.41
S	0.11	n.a.	n.a.	n.a.	n.a.
Total	100.05	99.226	99.97	99.381	99.92
ppm					
Ba	44	67	147	107	283
Ce	n.a.	56	49	46	26
Co	n.a.	26	17	27	37
Cr	64	20	394	17	62
Cu	50	18	16	15	22
Ca	n.a.	16	10	17	15
La	n.a.	13	19	12	16
Ni	14	8	162	8	49
Nb	3	30	11	36	17
Pb	n.a.	4	10	9	4
Rb	13	15	40	17	34
Sr	197	292	134	332	656
Th	n.a.	n.d.	13	4	n.d.
U	n.a.	n.d.	2	n.d.	n.d.
V	n.a.	141	82	123	113
Y	42	25	20	26	20
Zn	96	75	45	75	94
Zr	104	213	197	226	164

The upper unit of pillow lavas is mostly that of the colour melange. Smewing [16] showed that there are contrast between the high–Ti axis sequence (lower pillow lavas and sheeted intrusive complex) and the unconformably overlaying, low–Ti upper pillow lavas in the Troodos ophiolite.

The discrimination diagram of Zr/Y against Zr (Fig. 4B) [12] shows that the most of the Neyriz ophiolitic pillow lavas fall within the MORB field. Some of the colour melange pillow lavas and basaltic rocks fall with transitional island–arc tholeiite and MORB and there are some colour melange basalts which fall in the WPB field. The presence of the three types of basalt in the Neyriz area can be interpreted as follow:

a) MORB/IAT types can be found in some marginal basin setting [13].

b) MORB/WPB types can be found in diffuse spreading centers such as Iceland and Afar [12].

The present study shows that the different types of basalts, notably, lower pillow lavas and upper pillow lavas, possibly have been produced by different degrees of partial melting of similar source. The same origin is also suggested for Oman lavas [1]. The lower pillow lavas with a high Zr content and a fairly high Zr/Y ratios may have originated by initial magmatism related to ocean floor spreading.

The upper pillow lavas with lower Zr content, fairly low Zr/Y ratios and lower Ti contents originated in magmatism in an embryonic arc associated with syn–collision settings [1]. Geochemical data plus field investigations in the colour melange support the embryonic arc or immature island–arc setting.

Table 7. Whole rock analysis of sheeted dykes.

	322-8	76	35-60	32-23
wt.%				
SiO_2	58.28	49.67	48.20	55.60
TiO_2	1.98	1.39	1.05	1.80
Al_2O_3	13.87	15.12	12.86	14.44
Fe_2O_3	2.42	2.87	2.66	2.63
FeO	7.56	8.94	8.31	8.23
MgO	4.87	7.67	7.70	4.49
CaO	6.06	10.06	15.68	6.91
Na_2O	3.41	3.81	3.13	5.05
K_2O	0.63	0.19	0.18	0.49
MnO	0.17	0.17	0.08	0.16
P_2O_5	0.13	0.11	0.09	0.18
S	0.04	0.01	0.04	0.01
Total	99.96	100.01	99.92	99.99
ppm				
Ba	34	184	28	40
Nb	4	4	3	4
Rb	35	50	62	27
Sr	242	539	207	157
Y	42	35	27	46
Zn	32	39	62	12
Zr	110	82	58	114

A SUGGESTED MODEL FOR THE TECTONIC SETTING OF THE NEYRIZ OPHIOLITE

Field work in the Neyriz ophiolite has shown that the sheeted dike complex is exposed over 7.12 km² in fault contract with serpentinite. Moores et al. [9] suggested that well developed sheeted dyke complexes in ophiolites indicate that they have formed at a spreading center.

Stereographic plots of sheeted dyke margin orientation show a mean orientation of 166°/40° or general NW–SE direction.

As stated before, the whole Neyriz ophiolite has undergone post–emplacement folding. The present sheeted dyke orientations must be rotated to their original positions about horizontal axis to give an original orientation of 90°/156° (Fig. 5A).

This orientation of 90°/156° is interpreted as a paleospreading ridge orientation for the Neyriz ophiolite. Using this in a palaeogeographic reconstruction, transform fault zones mostly are perpendicular to the Neyriz spreading ridges direction with an orientation of 245°. A transform fault separates the Neyriz and Oman spreading ridges (Fig. 5B).

The Neyriz paleospreading ridge direction shows a slight difference in orientation with the Oman paleospreading ridge. Paleomagnetic data is needed to confirm this. Smewing [16] suggested that Oman spreading ridge direction is N-S based on sheeted dyke orientations. Boudier and Coleman [2] proposed that the Oman and Neyriz spreading have a 148 orientation. Moores et al. [11] based on a regional model for whole Middle East ophiolite believed that the Oman ophiolite may originally have been a N-S trending spreading center that underwent little rotation during spreading. The present study on the sheeted dykes supports the above suggestions and it is estimated that the Neyriz sheeted dykes possibly underwent about 10° anticlockwise rotation due to emplacement and post-emplacement folding. The Neyriz ophiolite is estimated to have undergone a higher degree of rotation compared with the Oman ophiolite due to its present position between Oman and Troodos. Luyendyk and Day [8], using paleomagnetic data, suggested that the Oman ophiolite formed near the equator and was transported northward to its present position, where as the Troodos ophiolite originally formed at about 20°N. Paleomagnetic data also indicates a 60° to 90° anticlockwise rotation of the Troodos ophiolite [15].

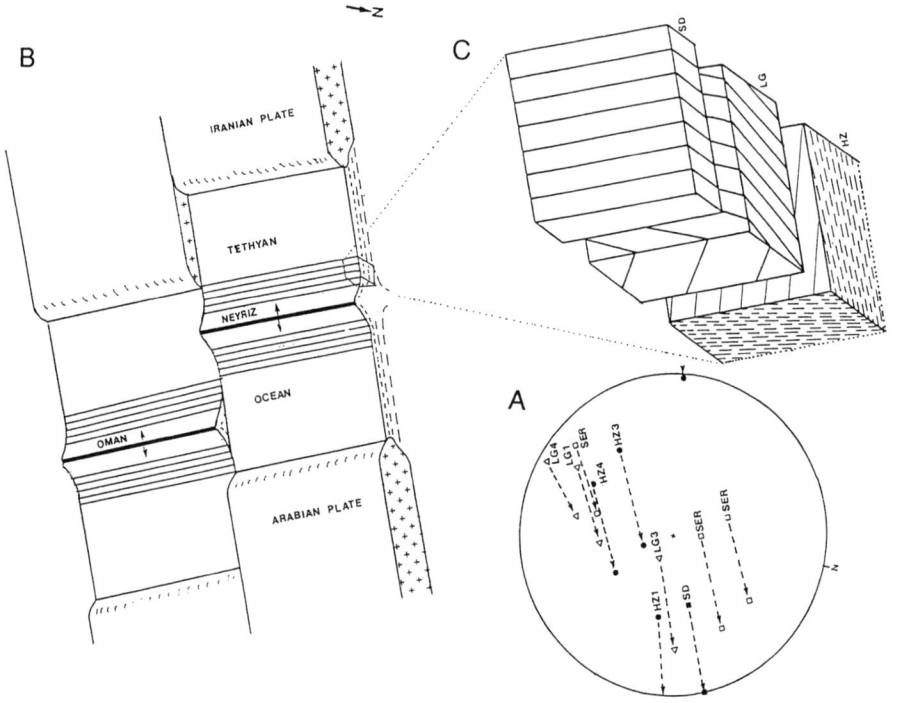

Figure 5A. Stereogram of the mean plane of sheeted dyke margins, gabbro compositional layering and harzburgite foliation of the whole Neyriz ophiolite sequence. To remove post- emplacement folding. These structural elements are rotated to original positions by rotating the dykes to vertical. **5B.** A model for the tectonic setting of the Neyriz spreading ridge and transform fault zone. **5C.** Block diagram illustrating structural relationships of sheeted dykes, gabbro layering and harzburgite foliation orientations.

The present gabbro compositional layering and harzburgite foliation directions of partial area 3 (Fig. 1) are 199°/15° and 315°/54°, due to post-emplacement folding, the gabbro layering and harzburgite foliation directions have been rotated their present orientations. The rotated attitude of gabbro layering and harzburgite foliation are 165°/60° and 248°/20°. The harzburgite foliation of the partial area 3 (20° to SE) is perpendicular to the sheeted dyke orientation (Fig. 5C). This indicate that plastic flow plane of harzburgite was perpendicular to the spreading ridge direction, and this possibly indicates that deformation has occurred after the harzburgite moved away from spreading ridge.

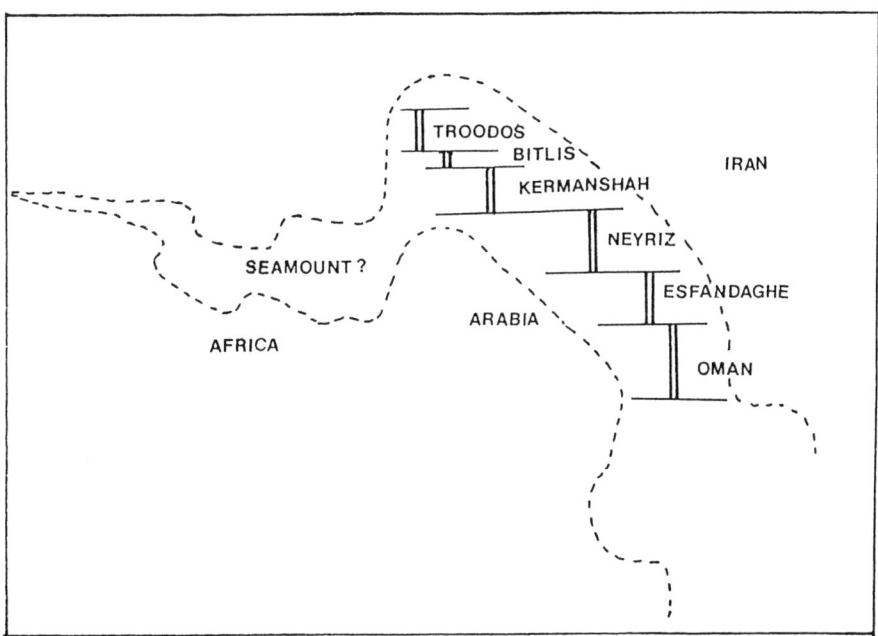

Figure 6. Hypothetical scheme for generation of the Neyriz ophiolite in the south of the Tethyan ocean. This scheme is developed from a model originally proposed by Moores et al. [11].

CONCLUSIONS

The Neyriz ophiolitic pillow lavas and colour melange pillow lavas and massive basalts exhibit considerable geochemical variations resulting from their generation in three different tectonic environments and they are as follows:
a) The lower pillow lavas are derived from mantle source from fractional crystallization of a magma chamber beneath the oceanic ridge.
b) The upper pillow lavas are also derived from the same source with a more depleted mantle wedge beneath an island–arc are in an embryonic arc environment.
c) The massive lavas originated from partial melting of enriched mantle source in a syn-collision environment.

The Neyriz spreading ridge is closely related to the Tethyan spreading ridge of Oman and Troodos. These were formed in the south of a narrow ocean, Tethys. According to Smith [18], the Tethyan ocean apparently formed during Triassic to late Cretaceous rifting and spreading between Eurasian and African plates that extended from Gibraltar to the Himalayas.

The reconstruction of the Neyriz Paleospreading ridge suggests that it was formed in the southern Tethyan ocean and separated from Oman, and other Iranian ridge such as Isfandaghe to the east of the Neyriz spreading and Kermanshah, Biltis to the west of the Neyriz spreading ridge by transform faults (Fig. 6). During ophiolite emplacement and post–emplacement Tertiary folding the Neyriz transform ridge has undergone 10 degrees of anticlockwise rotation. This rotation is greater than that of the Oman ridge–transform and less than that of the Troodos ridge–transform system.

Acknowledgment: The financial support of the Research Council of the Shiraz University is gratefully acknowledged.

References:

1. T. Alabaster, J.A. Pearce, and J. Malpas. The volcanic stratigraphy and Petrogenesis of the Oman ophiolite complex. *Contrib. Mineral. Petrol.*, **81**, 168–183. (1982).
2. F. Boudier, and R.G. Coleman. Cross section through peridotite in the Semail ophiolite south eastern Oman Mountains. *J. Geophys. Res.*, **86**, 2573–2592. (1981).
3. I.G. Gass and M.A. Brown. Nature and composition of the Oman ophiolite harzburgite. Unpublished Report, Leeds university. (1980).
4. D. Greenbaum. The Geology and evolution of the Troodos plutonic complex and associated chromite deposits, Cyprus. Unpublished Ph.D. thesis, Department of Earth Science, Leeds Univ. (1972).
5. R. Hall. Ophiolite–related contact metamorphism in skarn from Neyriz, Iran. *Proceed. Geol. Assoc.*, **92**, 231–240. (1981)
6. S.J. Haybes and H. McQuillan. Evolution of the Zagros suture zone, southern Iran. *Geol. Soc. Am. Bull.*, **92**, 739–747. (1974).
7. R.A. Loney, G.R. Himmelberg and R.G. Coleman, R.G. Structure and petrology of the Alpine–type peridotite of Burro Mountain, California, U.S. *J. Petrology,* **12**, 245–310. (1971).
8. B.P. Luyendyk and K. Day. Paleomagnetism of the Semail ophiolite, Oman. The Wadi Kadir gabbro section. *J. Geophys. Res.* **87**, 10903–10917. (1982).
9. M. Menzies. Mineralogy and partial melt textures within an ultramafic body, Greece. *Contrib. Mineral. Petrol.* **42**, 273–385. (1973).
10. M. Menzies and C. Allen. Plagioclase lherzolite–residual mantle relationship within two Eastern Mediterranean ophiolites. *Contrib. Mineral. Petrol.* **45**, 197–213. (1974).
11. E.M. Moores, P.T. Robinson, J. Malpas and C. Xenophonotos. Model for the origin of the Troodos massif, Cyprus, and other mideast ophiolites. *Geology*, **12**, 500–503. (1984).
12. J.A. Pearce. Geochemical evidence for the genesis and eruptive setting of lavas from Tethyan ophiolites. In: *Ophiolites (Proceedings of the Troodos'79 International Ophiolite Symposium)*. A. Panayiotou (ed.). pp. 261–272. Geological Survey of Cyprus, Nicosia. (1980).
13. A. Reay, J.M. Rooke, R.C. Wallace and P. Whelan. Lavas from Nivfo'ou island, Tonga, resemble ocean floor basalts. *Geology*, **2**, 605–606. (1974).
14. K. Sarkarinejad. The geology and tectonic settings of the ophiolites and associated rocks in the Neyriz area, southeastern Iran. Unpublished Ph.D. thesis, University of Wales (356 pp.) (1985).
15. A.W. Shelton and I.G. Gass. Rotation of Cyprus microplate. In: *Ophiolites (Proceedings of the Troodos'79 International Ophiolite Symposium)*, A. Panayiotou (ed.). pp. 61–65. Geological Survey of Cyprus, Nicosia (1980).
16. J.D. Smewing, K.O. Simonian and I.G. Gass. Metabasalts from the Troodos massif, Cyprus: Genetic implication deduced by petrography and trace element geochemistry. *Contrib. Mineral. Petrol.* **51**, 49–64. (1975).
17. J.D. Smewing. An upper Cretaceous ridge–transform intersections in the Oman ophiolite. In: *Ophiolites (Proceedings of the Troodos'79 International Ophiolite Symposium)*. A. Panayiotou (ed.). pp. 407–413. Geological Survey of Cyprus, Nicosia. (1980).
18. A.G. Smith. Alpine deformation of the oceanic areas of the Tethys, Mediterranean and Atlantic. *Geol. Soc. Am. Bull.* **82**, 2039–2070. (1971).
19. R. Stoneley. The geology of the Kuh–e Dalneshin area of southern Iran, and its bearing on the evolution of southern Tethys. *J. Geol. Soc. London*, **138**, 509–526. (1981).
20. N.H. Woodcock. Specification of fabric shapes using eigenvalues method. *Geol. Soc. Am. Bull.* **88**, 1236. (1977).

Polychronous Ophiolite Belts of Central Kazakhstan and Their Evolution.

A.S.YAKUBCHUK

Geological Faculty, Moscow State University,
Leninhills, Moscow 119899, RUSSIA

Abstract

Central Kazakhstan ophiolites belong to numerous sutures between Early Paleozoic island arcs and Precambrian (1100 Ma) continental terranes. Ophiolites are presented as melanges (dominant) and dismembered (common) or undismembered (rare) bodies occurring in nappe-piles within sutures. Age data, which are obtained recently using conodont remnants from interlava layers, prove that emplacement of many ophiolites took place after their birth only within 20-30 Ma. With regard to ages of adjacent arc volcanics there are three major age generations of ophiolites: 1) Vendian-Early Cambrian pre-island arc ophiolites (rare), 2) Early Cambrian-Middle Ordovician syn-island arc ophiolites (dominant) and 3) Middle Devonian syn-orogeneous ophiolites (only in Junggar).

Vendian-Early Cambrian pre-island arc ophiolites occur in the Arkalyk and Jalair-Naiman sutures; in the East Erementau zone the basalts of this age are preserved only. Basalts of East Erementau zone and Jalair-Naiman sutures show E-MORB (early stages) and N-MORB (late stages) affinity. Early Cambrian ophiolites formed during earliest stages of the Boshekul-Chinghiz arc terrane evolution. Ordovician ophiolites form synchronous pairs with island-arc terranes. These pairs are regarded as the fragments of arc - back-arc systems. The pairs of the Stepnyak arc terrane of the Middle Cambrian-Ordovician and the Chistopol - Aksu-Iradyr sutures with ophiolites of the Early Ordovician, the Boshekul-Chinghiz arc terrane of the Early Cambrian-Middle Ordovician and the Maikain-Balkybek suture with ophiolites of the Late Cambrian-Early Ordovician, the Baidaulet-Akbastau arc terrane of the Ordovician and the Tekturmas - North Balkhash suture with ophiolites of the Middle Ordovician are marked out. All these ophiolites contain ashy jaspers as interlava layers. Petrochemistry shows evolution of magmas from calc-alkaline (early stages) to arc tholeiite (late stages) that proves supra-subduction zone origin of them. Middle Devonian ophiolites exist in the North Junggar suture, being synchronous with Devonian orogenic volcanic belt that probably means their back-arc setting with regard to their arc tholeiite and calc-alkaline affinity.

The mafic-ultramafic cumulates belong mainly to orthopyroxene-type except Tekturmas-North-Balkhash ophiolites belonging to clinopyroxene-type. Residual mantle peridotites are presented in all cases by harzburgites and dunites except of those in Tekturmas and North Balkhash sutures, where harzburgites contain rare lherzolitic bodies. Dyke swarms occur regularly, but sill swarms are widely developed instead of dykes in Early Cambrian - Middle Ordovician ophiolites.

Key words: Kazakhstan, Caledonian and Variscan ophiolites.

INTRODUCTION

Ophiolites are widely distributed in Central Kazakhstan having total length of sutures over 2500 km. Various questions on ophiolite geology in Central Kazakhstan have been discussed during last 45 years [3, 4, 5, 25, 33, 39, 49, 56]. However, very complex tectonic structure and poor outcropping conditions have complicated investigation of ophiolites in Central Kazakhstan. Ophiolites in Central Kazakhstan were not systematically regarded till recent time. Absence of macrofossils and unsatisfied geochronological data bore different and contrasting speculations about ophiolite setting and evolution. Firstly, ophiolites were regarded as Early Paleozoic [49], then as Late Precambrian [8, 10] and Late Precambrian (Vendian) - Early Cambrian [4] complexes. But ideas of Early Paleozoic age were developed since 1970 by N. Pupyshev and B. Nazarov [35] on the basis of radiolarian biostratigraphy. However, these fossils were poorly preserved. Since 1977 this situation was changed, when Early Ordovician conodonts were found in cherts and jaspers of Atasu region regarded before as the Late Precambrian [19]. Today, our knowledge about the age of basalts and cherts was completely changed after the discovery of these microfossils in all ophiolite sutures [30]. Petrochemical study was not promoted till 1970s [26]. Today, there are enough data to judge about petrochemical affinity of all ophiolites.

Before these changes, there were different points of view on the evolution, origin and tectonic setting of ophiolites. First group of scientists [4, 5, 6, 40] regarded all ophiolites as allochthonous bodies or remnants of one large nappe [6] of Vendian-Early Cambrian oceanic lithosphere generated before the development of Early Paleozoic island arcs.

Yu. Zaitsev [56] and E. Patalakha [39] interpreted ophiolites as autochthonous bodies in geoanticlinal zones, where serpentinized mantle rocks were protruding polychronously through the whole Paleozoic. All modern structural style was regarded as stable since the Early Paleozoic (according to conodonts), and ophiolites were regarded as a crust of former narrow basins formed synchronously with basalt-andesite volcanics in adjacent structures [30, 57].

Basing on these age data, T.Kheraskova [25] suggested the presence of a single back-arc (Maikain-Balkhash) basin, which has successively dismembered by new island arcs during their evolution. However, there are remnants of some independent back-arc basins, not of single basin, as will be shown below.

TECTONIC SETTING OF OPHIOLITES

According to the modern paleontological data [17, 30, 53, 55], basalt lavas and cherts in Central Kazakhstan belong to wide age spectrum ranging from the Vendian (Late Proterozoic) to the Middle Ordovician and in one case to the Middle Devonian (Fig.1).

Age data permit to separate ophiolites into three major groups such as: Vendian-Early Cambrian Caledonian ophiolites of pre-island arc origin (their age affinity is not always confident, but certainly Pre-Middle Cambrian); Early Cambrian-Middle Ordovician Caledonian ophiolites belonging to separate sutures (these complexes are synchronous with adjacent island arc terranes); Middle Devonian ophiolites of the Variscan Junggar area (they occur synchronously with Devonian orogenic volcanic belt in Caledonian part of Central Kazakhstan [42, 59] surrounding Variscan area in modern structure).

The structure of Kazakhstan has formed during Caledonian-Variscan successive accretion. Variscan collisional belts limit now triangular form of the Kazakhstan superdomain. The main structural line within superdomain is a border of the Western and the Eastern domains. This line determines face-to-face junction of differently orientated structures of the Western and the Eastern domains and probably represents the trace of pre-Late Llandoverian transform fault. Two domains were regarded firstly [4] as Western microcontinental (1100 Ma) and Eastern Vendian-Early Paleozoic oceanic domains, but indeed, the Western domain is also composed of some ophiolite sutures, which divide numerous continental terranes and the Stepnyak arc terrane. Eastern domain does not contain large continental terranes. The Boshekul-Chinghiz and Baidaulet-Akbastau arc terranes, which are separated by ophiolite sutures, form a backbone of the structure of domain.

So, there are no principal tectonic differences among Western and Eastern domains. Ophiolitic sutures mark boundaries among continental and island arc terranes [50]. Some sutures contain only basalts and chert covers (East Erementau zone).

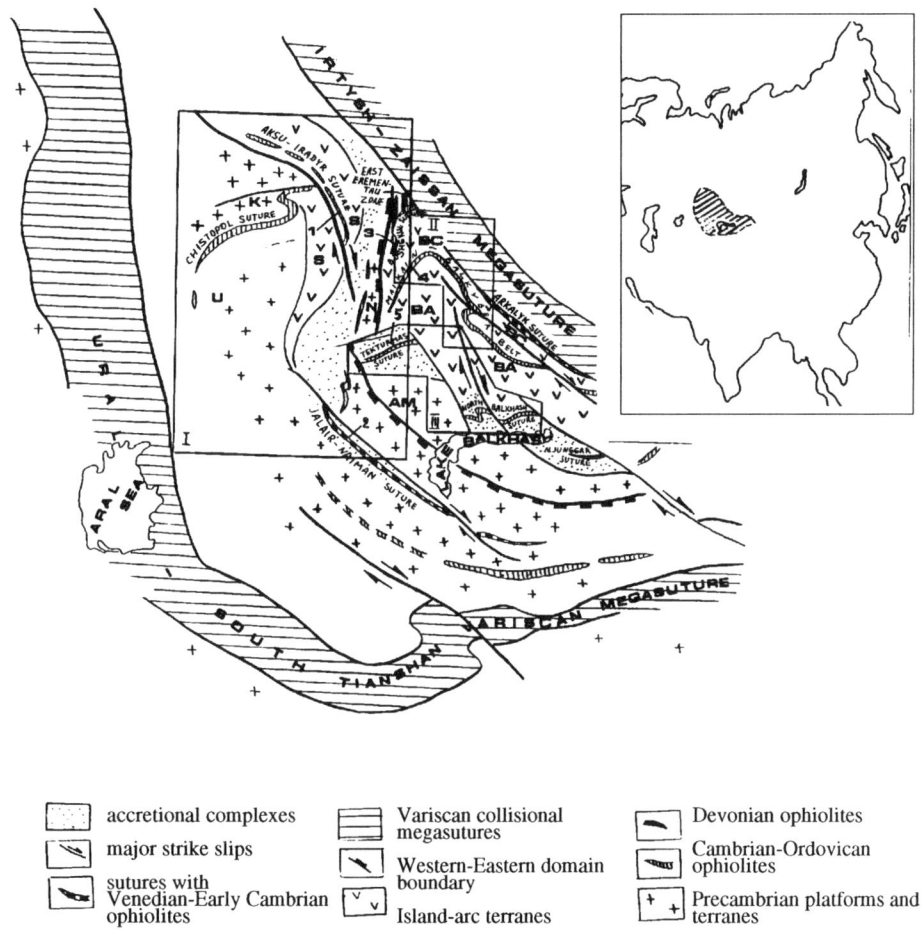

Figure 1. Distribution of ophiolite sutures in Kazakhstan. Numbers show locatios of well-preserved ophiolite massifs:1 - Aksu-Iradyr suture, 2 - Andasai, 3 - Semizbugu, 4 - Karaulcheku, 5 - Tekturmas and Basarbai; letters are terranes: K - Kokchetav, U - Ulutau, AM - Aktau-Mointy, N - Niyaz, BC - Boshekul-Chinghiz, BA - Baidaulet-Akbastau, S - Stepnyak.

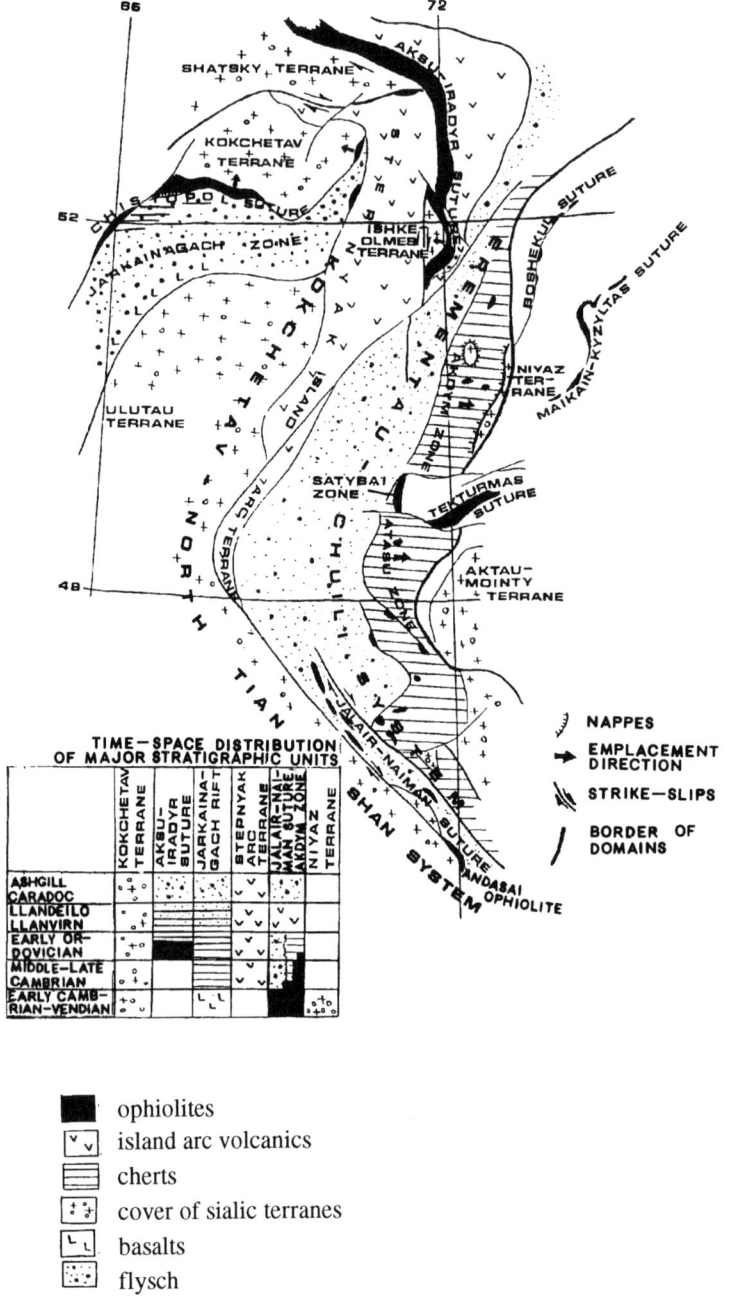

- ophiolites
- island arc volcanics
- cherts
- cover of sialic terranes
- basalts
- flysch

Figure 2. Geological setting of ophiolites in the Western domain.

WESTERN DOMAIN

Ophiolites of Western domain are not as well studied as in the Eastern domain. There are Kokchetav-North Tianshan and Erementau-Chu-Ili accretional systems within Western domain (Fig.2) [40, 57].

KOKCHETAV-NORTH TIANSHAN SYSTEM

Ophiolites were studied here by O.V.Minervin, N.P.Mikhailov, V.N.Moskaleva [33], E.M.Spiridonov [47] and some other authors. There are Chistopol and Aksu-Iradyr sutures, which probably represent fragments of former single suture separated by a dextral strike-slip fault (Fig.2).

Chistopol suture

The Chistopol suture bounds eastern and southern margins of Kokchetav continental terrane and extends for 140 km. Chistopol suture is situated among Kokchetav and Ulutau continental terranes. The northern rim of the Ulutau terrane is reworked by Late Precambrian-Early Paleozoic Jarkainagach rift [40]. Structural position of the suture is not clear now, but presence of numerous tectonites on northern periphery of ultramafic bodies suggests northward thrusting of ophiolites onto the Kokchetav continental terrane during Taconian collisional event [57].
Chistopol suture consists of serpentinite melange, fragments of cumulate sequences represented by websterites, pyroxenites and gabbro [33]. Relationships among these rocks are not yet described. Mafic-ultramafic bodies are associated with basalts and cherts, where Arenigian conodonts are known [23]. All authors suffer very poor exposure of the ophiolites, which are studied in many cases only from borehole samples [33].
Petrochemical data [23, 33] indicate arc tholeiite affinity of gabbro and basalts.

Aksu-Iradyr suture

Aksu-Iradyr suture extends in a north-south direction for 200-250 km inside of Stepnyak island arc terrane along dextral strike-slip fault. Remnants of the Kokchetav and Shatsky Precambrian terranes (such as Ishkeolmes terrane) are set now within Stepnyak terrane. All geologists are confident about allochthonous setting of ophiolites, but one group [40] accounts for westward direction of thrusting and the other group [9] is confident about eastward direction of emplacement.
An almost full ophiolite sequence is reported by Spiridonov [47] from little ophiolite massif near eastern rim of Ishkeolmes terrane. There are residual peridotites of dunite-harzburgite type at the base, which are covered by lherzolite-websterite-pyroxenite-gabbro cumulates and dolerites of sheeted dyke swarm (Fig.3). Spilitic pillow-basalts are related to primitive tholeiites. There are no full accordance in the age of basalts [47]. Recent interpretation of Spiridonov (personal commun.) suggests the presence of both Early Cambrian and Arenigian basalts. However, only Arenigian age is supported by microfossils, and presence of the Early Cambrian in this structure is not clear.

EREMENTAU-CHU-ILI SYSTEM

Ophiolites occur in the Jalair-Naiman suture [11]. Some little massifs of Atasu [16] and Akdym zones [14] and of Satybai suture, where serpentinite melange with blueschists of undetermined age is described [31], are regarded as northward extension of the Jalair-Naiman suture. These ophiolites are included into the 1500 km long accretionary system of Central Kazakhstan collided among island arc and continental terranes on the north and among two continental terranes on the south (Fig.2). Eastward emplaced ophiolitic allochthones are underlain by Middle Ordovician (Llandeillan or Caradocian) olistostrome bodies [9, 41].

Allochthones consist of slices composed of serpentinite melange, basalts and Late Cambrian-Llanvirnian cherts, which cover pillow-lavas that defines conventionally the Vendian-Early Cambrian age of ophiolites of the Jalair-Naiman suture [28]. There are also some geochronological data of unconfident quality [24] describing 680-830 Ma of gabbro-amphibolites and 1500 Ma of residual peridotites using thermoluminiscent method.
The archaeocytes from limestones are described in basalts in Akdym zone [40] that proves at least Early Cambrian age of ophiolites. Middle-Late Cambrian basaltic suites are preserved at the bases of some chert slices [28].
The remnants of Arenigian basalts in association with serpentinite melange are described near the western rim of the Aktau-Mointy continental terrane [16].
These facts may indicate that long-lasting spreading could really take place within such giant suture.

Andasai ophiolite
Full ophiolite sequences are described in the Jalair-Naiman suture. The massifs of the Jalair-Naiman suture have complex and irregular composition. The best one is Andasai massif (Fig.3). There are serpentinized harzburgites and dunites with small chromite lenses and rodingitized gabbro dykes. Cumulates are lying conformably over residual peridotites, being composed of successive sequence of lherzolites, wehrlites, clinopyroxenites, hornblendites, troctolites, olivine gabbro, gabbro and gabbro-norites with total sequence of about 0.5-0.7 km thick [11]. Described rocks occur in tectonic slices and in many cases their rims are sheared and brecciated. Basalt bodies have tectonic relationships with cumulates, but these complexes are always found together. Basalt pillow lavas contain quartzite, jasper, limestone and dolomite layers. Basalts are overlain by sandstones and cherts with brachiopods of the Late Cambrian [28].
Petrochemical data show intermediate composition of cumulative rocks lying between continental and oceanic tholeiites [11]. Basalts belong to low-Ti normal tholeiites and E-MORB rocks [48].

EASTERN DOMAIN

There are Boshekul-Chinghiz system, which undergone first major tectonic accretion in Middle Llandoverian [58], and Junggar-Balkhash system, where multiple major deformations were successively migrated in south-eastward direction ranging from Middle Devonian to Early and Middle Carboniferous [57, 58].

BOSHEKUL-CHINGHIZ SYSTEM

Boshekul-Chinghiz system stretches for 750 km, having north-east extension in the west and south-east extension in the east (Fig.4). Accretional events are recorded in this structure in the Middle Ordovician, Middle-Late Ordovician, Middle Llandoverian [40], but culminative event corresponding to collision of the Boshekul-Chinghiz and Baidaulet-Akbastau island arc terranes belongs to the Middle Llandoverian [50, 53].
Ophiolites of various ages occur in Arkalyk, Boshekul and Maikain-Balkybek sutures. In many cases ophiolites occur as melange-type bodies, but sometimes well preserved or slightly dismembered massifs are known (see for names and locations of these massifs on Figure 4).

Arkalyk suture
The Arkalyk suture is a part of accretionary complex in the northern frontal side of the Boshekul-Chinghiz arc terrane. This complex contains only rare fragments of serpentinite melange containing basalts ranging from Early to Late Cambrian [53] and Ordovician. Petrochemical affinity of lavas is not clear.

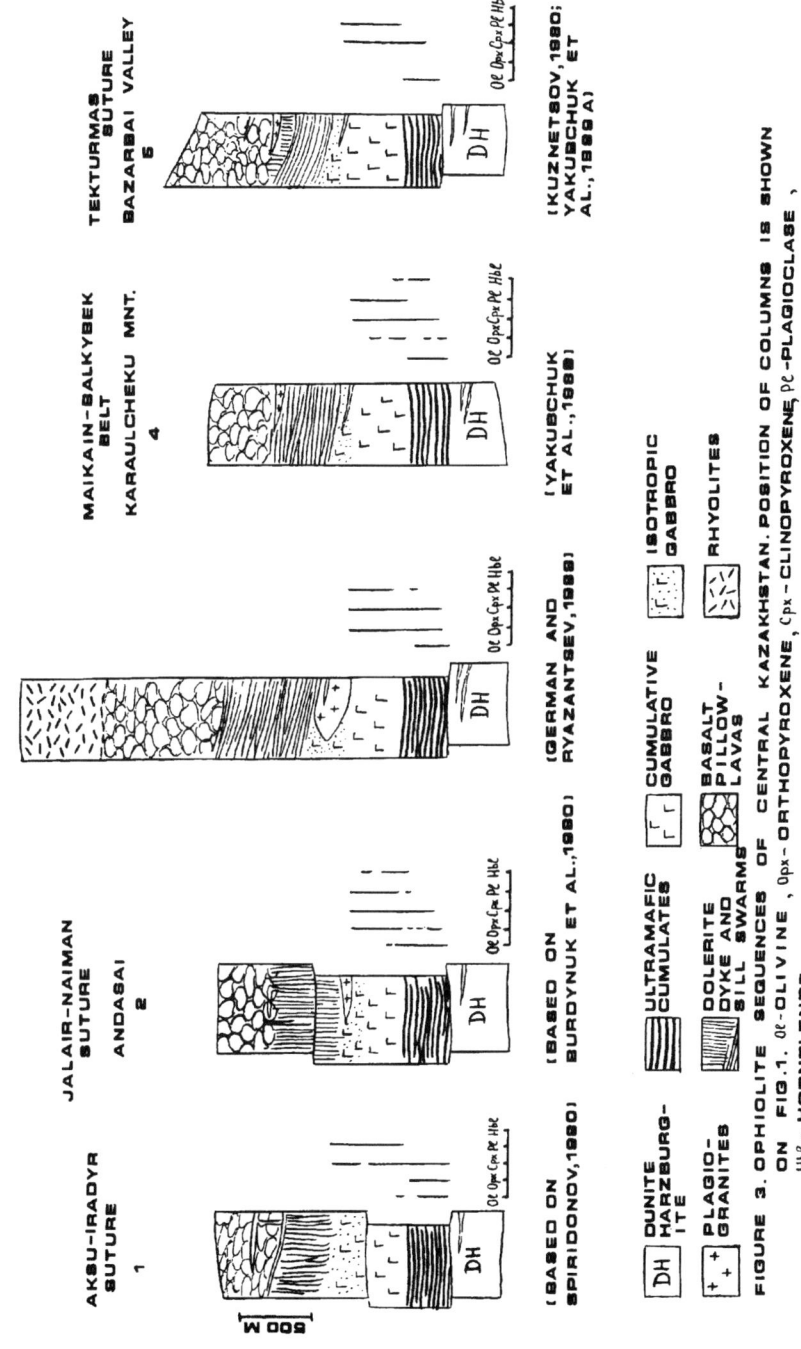

FIGURE 3. OPHIOLITE SEQUENCES OF CENTRAL KAZAKHSTAN. POSITION OF COLUMNS IS SHOWN ON FIG.1. Ol - OLIVINE, Opx - ORTHOPYROXENE, Cpx - CLINOPYROXENE, Pl - PLAGIOCLASE, Hbl - HORNBLENDE

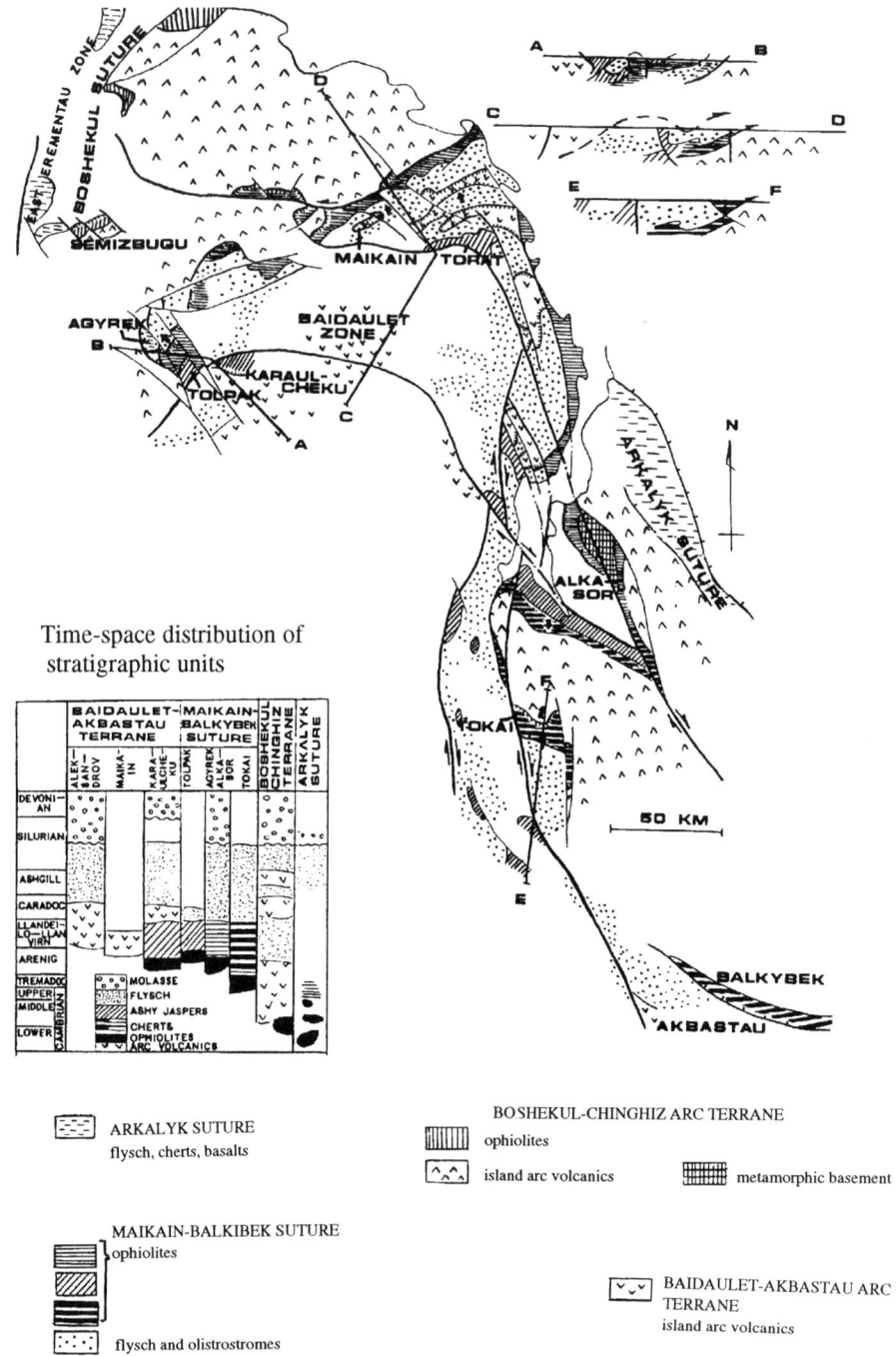

Figure 4. Geological seting of ophiolites in the Boshekul-Chinghiz system.

Boshekul suture

Frontal nappes of the Boshekul suture underlying by Middle Ordovician olistostromes are overthrust upon complexes of the East-Erementau zone. The East-Erementau zone contains only Vendian-Early Cambrian (according to oncolites [9]) ocean-island and N-MORB basalts with siliceous and limestone layers. There are no evidences of presence of lower parts of oceanic lithosphere in this zone. Boshekul suture contains ophiolites, which volcanics form the base of the stratigraphic column of Boshekul-Chinghiz island-arc terrane.

Semizbugu ophiolite

Dismembered ophiolite sequence of Boshekul suture is observed near the Semizbugu Mountain (Fig.3) [9]. Details of ophiolite composition are described by L.German and A.Ryazantsev [18]. Ophiolite sequence begins from serpentinized dunites and harzburgites. Cumulative sequence includes orthopyroxenites, gabbro-norites, gabbro, troctolites, granodiorites, tonalites, trondhjemites and plagiogranites. Sheeted sill swarm (800-1000 m thick), 1800 m of basalts and 1250 m of rhyolites and rhyo-dacites are on the top of reconstructed sequence. Acid and mafic volcanics reveal the Na-type alkalinity of the evolved tholeiite melt. The progressive low-pressure metamorphism up to amphibolite facies manifests itself locally in ophiolites. According to B.Khromykh [27], ophiolites overlain by a thick series of Early Cambrian calc-alkaline volcanics of the Boshekul-Chinghiz island arc terrane dated as Botomian (Early Cambrian) on macrofauna from limestone lenses; the gabbroids of this complex in the north of suture are radiometrically dated [27] at 525-540 Ma, and the plagiogranites are dated by Re-Os method on molybdenum at 568 ± 60 Ma, indicating Early Cambrian age.

Maikain-Balkybek suture

The Maikain-Balkybek suture has complex configuration in plan (Fig.4). Suture is situated between the Boshekul-Chinghiz and Baidaulet-Akbastau island arc terranes [50, 53]. The ages of ophiolites range from the Late Cambrian-Early Ordovician in the eastern part [44, 53] and mainly within the Early Ordovician in the western part [55]. However, there are heavily altered calc-alkaline Early Cambrian basalts and andesite-basalts in olistostromes [29, 40], but their affinity to ophiolites is not clear. Some authors interpret these bodies as remnants of former oceanic crust [25, 40], but their petrochemistry suggests island arc affinity.

True ophiolites form northwestward-displaced synsedimentary (Taconian) and co-folded (Middle of Llandoverian) nappes with melange-type or dismembered ophiolites (Tokai, Agyrek, Tolpak Mtns); well-preserved paraautochthonous ophiolites lying at he base of Caradocian island-arc formation occur along northwestern rim of the Baidaulet-Akbastau arc terrane (Karaulcheku and Torat Mountains).

Karaulcheku ophiolite

A weakly dislocated ophiolite sequence is in the Karaulcheku paraautochthone (Fig.3) studied in details recently by V.Stepanets and L.German [54]. It consists of tectonized, multiply folded serpentinized harzburgites and dunites at the base, followed with unconformity(?) by rhythmically stratified sequence of dunites, lherzolites, wehrlites, bipyroxene hornblende rocks, pyroxenites, cumulative gabbro-norites (total thickness up to 700 m) and thin isotropic gabbro at the top. The stratified bodies of acid rocks composed of tonalites to trondhjemites up to 50 m thick are also present. There is a continuous swarm of up to 500 m thick of subsills instead of traditional dyke swarm. This sill swarm is formed by dolerites (95%) and keratophyre (5%) bodies penetrated sometimes in cumulates and residual peridotites. Sill swarm is gradually replaced by the pre-Middle Arenigian 500 m thick basalts overlain by the Middle Arenigian - Llanvirnian (and, perhaps, including Llandeilian) ashy jaspers [55]. These ashy jaspers are uncoformably(?) overlain by the Caradocian-Ashgillian andesite-basalts and trachyandesites.

Petrochemical characteristics

Petrochemical data show tholeiite and arc-tholeiite composition of basalts from different ophiolite sequences regardless their age except for some E-MORB basalts from Agyrek pile (Fig. 5). Gabbro cumulates are of low-Ti type except for some high-Ti samples from Agyrek nappe pile (Fig. 6). There are discrepancies between the data of Figures 5 and 6 for Alkasor ophiolite. In Figure 5, the basalts show TiO_2 decreasing trend as FeO/MgO increases, but the gabbros of the same ophiolites show TiO_2 increasing trend as FeO/(FeO+MgO) increases. Alkasor gabbros are the poorest in TiO_2, but Alkasor basalts are the richest in TiO_2. This suggests the absence of genetical connection between the basalts and gabbros of Alkasor ophiolite. However, this fact is not typical, because other massifs do not reveal such phenomenon. Residual peridotite chromian spinel composition, which indicates SSZ origin of these ophiolites, does not differ from place to place that shows relative homogeneity of mantle rocks (Fig.7).

JUNGGAR-BALKHASH SYSTEM

Ophiolites of Junggar-Balkhash accretional-collisional system occur in Tekturmas, North-Balkhash and North-Junggar sutures. Tekturmas and North-Balkhash sutures are separated in modern structure by Late Paleozoic volcanic belt (Fig. 8), but they have the same age and composition of plutonic and stratigraphic units in both cases [2, 50]. The Tekturmas and North Balkhash sutures are located between Baidaulet-Akbastau Ordovician arc terrane to the north and hypothetical continental terranes, which are interpreted as northern buried part of the Aktau-Mointy continental terrane, to the south [50]. The former relationships of these two sutures are interpreted as possible transform junction in the Ordovician or as strike-slip disruption of former single structure before origin of Late Paleozoic volcanic belt [50].

Tekturmas suture

The ophiolites of Tekturmas suture occur as nappe piles, where northernmost paraautochtonous Bazarbai ophiolite lies at the base of Middle Paleozoic terrigeneous continuous sequence, but southernmost nappe piles, where Tekturmas ophiolite occurs, are thrust northward over each other and Bazarbai ophiolite at the Middle Devonian [52]. The Tekturmas ophiolite is overthrust from south by Middle Paleozoic terrigenous formations (Fig. 8). Main collisional event occurred at the Middle Devonian, although the Caradocian-Ashgillian synsedimentary thrusting is also known [17].

Tekturmas and Bazarbai ophiolites
Tekturmas and Bazarbai ophiolites have different ages of origin. The Late Arenigian(?)-Llanvirnian Tekturmas ophiolite of melange type includes serpentinized harzburgites and dunites. Occasional lherzolitic bodies are found [31]. Blocks of disintegrated stratified complex (clinopyroxenites, gabbro) and plagiogranites are plunged into serpentinized mass. Geochronological age obtained by K-Ar method reveals whole-rock 444 Ma of gabbro and 534 Ma of pyroxene from gabbro. Thin sequences of non-metamorphosed isotropic gabbro are also known [18]. Amphibolites with clear layering were formed back at the magmatic stage, since these metamorphic rocks are pierced with small non-metamorphosed diabase dykes. I.Kuznetsov [31] determined 600°C of metamorphism. The geochronological age of amphibolites from different blocks in melange was obtained by K-Ar method on amphibole at 365 Ma, on former(?) pyroxene at 538 Ma [31] and in one case 771 Ma [3]. Such various data show influence of later alterations. Basalts of Tekturmas ophiolite (Sr^{87}/Sr^{86} ratio is 0.7053 ± 0.0021) occur in the basalt slices (up to 300 m thick). The volcanics are of Late Arenigian(?)-Llanvirnian age on conodonts that is not at all in accordance with geochronological age of amphibolites and gabbro. This fact probably means the presence of metamorphic and magmatic rocks of different ages mixed together during thrusting, or geochronological data are not enough satisfactory.

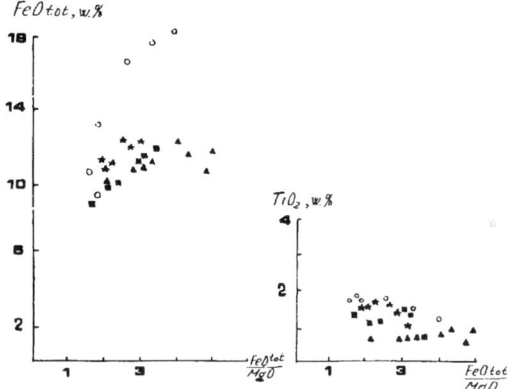

Figure 5. Basalt plots on FeO^{tot} and TiO$_2$ versus FeO^{tot}/MgO diagrams (Miyashiro 1975). Stars - Agyrek mnts; solid triangles - Tolpak Mtn; solid squares - Karaulcheku Mtn; open circles - Alkasor. Sourced from Yakubchuk (1991).

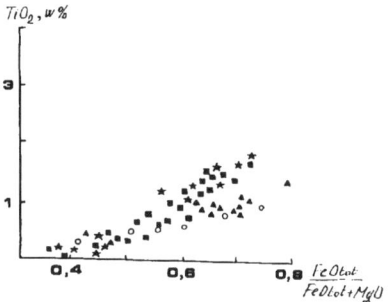

Figure 6. TiO$_2$ versus FeO^{tot}/FeO^{tot}+MgO for gabbroic rocks (Serri and Saitta 1980). See for symbols fig. 5. Graph sourced from Yakubchuk, 1991.

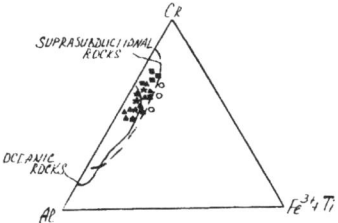

Figure 7. Al-Cr-(Fe^{3+}+Ti) diagram for spinels of residual peridotites. The fields are from Bogatikov *et al.* (1988). See for symbols fig. 5. Graph sourced from Kuznetsov *et al.* (1990 a).

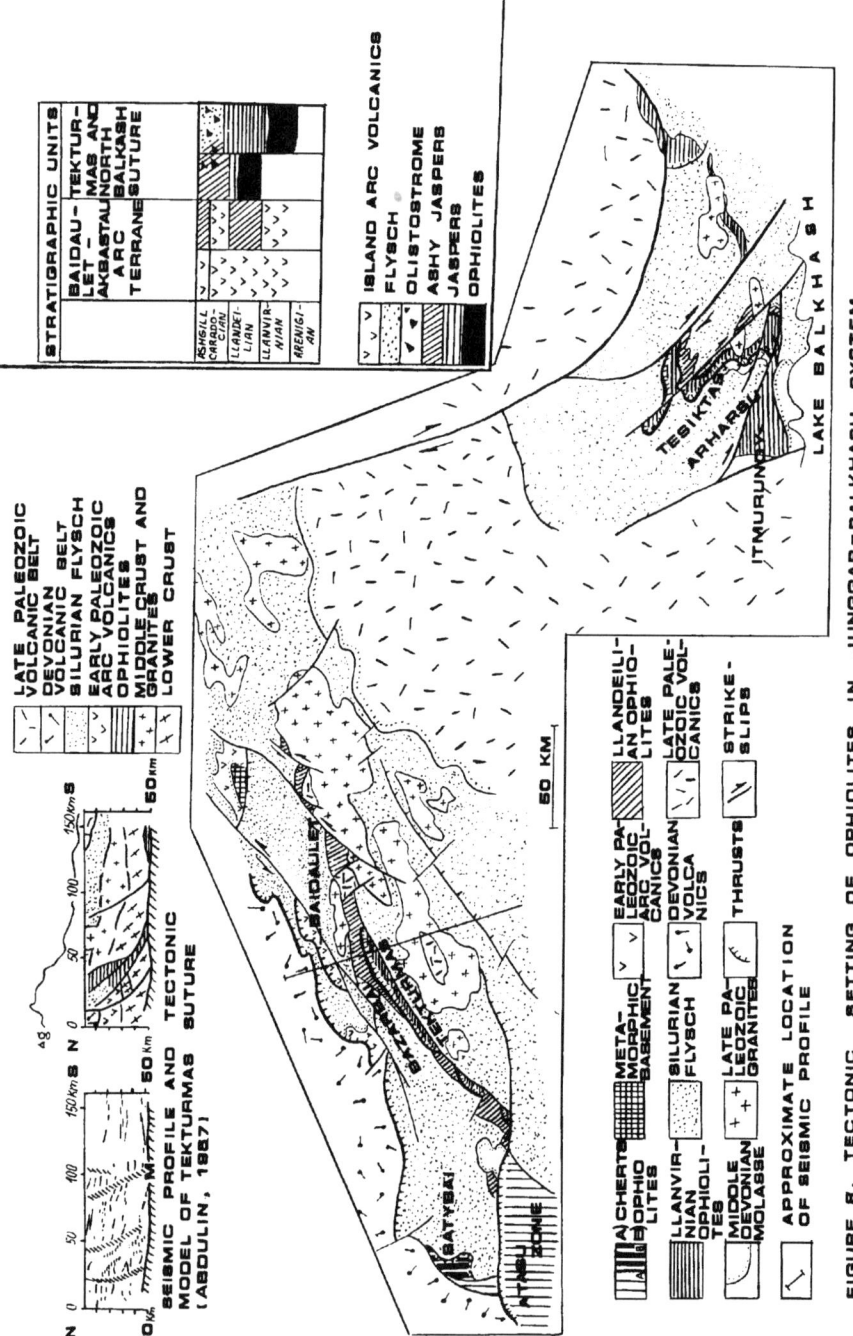

FIGURE 8. TECTONIC SETTING OF OPHIOLITES IN JUNGGAR-BALKHASH SYSTEM.

The Bazarbai dismembered ophiolite (Fig.3) includes serpentinized peridotites in the northern part of the suture, stratified and isotropic gabbro (up to 200 m), plagiogranites, a dyke-sill swarm (600-700 m) of diabases (95%) and keratophyres (5%), basalts with Llandeilian conodonts from interlava chert beds. Some relics of ultramafic cumulates (pyroxenites and olivine gabbro) are known from serpentinite melange. So, it is possible to reconstruct an almost full sequence.

North-Balkhash suture
Practically all workers recognize the lithological and temporal similarity of ophiolites between the Tekturmas and North Balkhash sutures. The North Balkhash suture contains completely dismembered ophiolites (Fig.8). There are no examples of full sequence, however, all members of ophiolites are represented in melange [51]. Representative cumulative sequence in the North Balkhash ophiolites occurs in the Tesiktas and Arkharsu massifs, which probably represent evidences of two different magmatic events. Both massifs, however, have similar sequences of cumulative rocks, i.e., dunites, wehrlites, clinopyroxenites and gabbro.

Petrochemistry of ophiolites of Tekturmas and North-Balkhash sutures
The petrochemistry is regarded here together bacause of noted similarity. The earlier basalts in every suture have generally higher alkalinity, TiO_2 and P_2O_5 contents like in cumulative gabbroids [37]. Earlier melts correspond to abyssal tholeiites (Fig. 9), whereas later magmas correspond already to island-arc tholeiites. At the same time, the relatively high alkalinity of early basalts does not allow to consider them as tholeiites because of high K_2O content, and they should rather be interpreted as E-MORB or calc-alkaline basalts [26]. Younger basalts undoubtedly correspond to arc tholeiites [37].
Cumulates of the Tekturmas ophiolite display higher titanium content than cumulates of the Bazarbai ophiolite (Fig.10). Amphibolites of the melange-type Tekturmas ophiolite are enriched by titanium too. Perhaps, this regularity is also valid in the North Balkhash suture, where cumulates of the Tesiktas massif ought to be regarded as products of earlier magmatic events than cumulates of the Arkharsu massif. The belonging of both cumulates to the low-titanium ophiolites evidences a back-arc setting of their origin [43], and successive changes from relatively earlier high-titanium magmas to relatively low-titanium magmas show possible ensialic origin of former back-arc basin according to G.Serri and M.Saitta [43].
The ultrabasic restites are 75 to 80 per cent serpentinized, so, the only informative mineral is often chromian spinel, which is uniform enough within residual peridotites of two sutures (Fig.11) [32], corresponding to ultramafics from supra-subduction zones [7]. However, these data are not enough representative, because harzbugites and dunites of the Tekturmas ophiolite do not reveal former chromian spinel, which is emplaced into magnetite in all analyzed samples. Only lherzolites of residual peridotites of Tekturmas ophiolite contain former spinels, which indicate contrast differences to those from harzburgites of Bazarbai ophiolite.

North-Junggar suture
The North Junggar suture is situated in the southeasternmost part of the Junggar-Balkhash system. Nobody described this melange as ophiolite, but as tectonized layered massifs [33]. However, there are all lithologic members such as serpentinized residual peridotites, gabbro, plagiogranites, basalts of the Eifelian-Givetian that permits to say about presence of Middle Devonian ophiolites. This assumption is supported by a presence of such ophiolites in the Darbut system of Western Xinjang, China [15].
Petrochemical data, which are compiled by A.F.Stupak (personal commun.) suggest island-arc tholeiite affinity of basalts. Spinel chemistry of serpentinized peridotites corresponds to supra-subduction zone of these mantle rocks. Ophiolites of the North-Balkhash suture are not enough studied, and reported data have preliminary character.

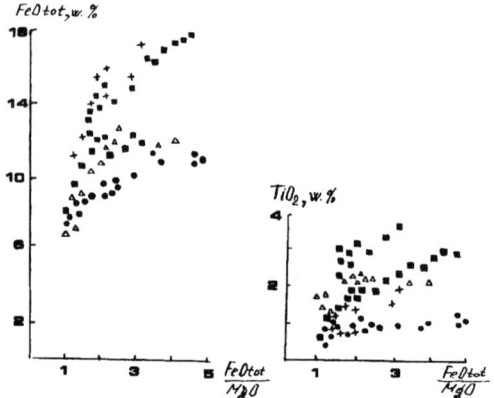

Figure 9. Basalt plots of Tekturmas and North Balkhash sutures on FeO^{tot} and TiO_2 versus FeO^{tot}/MgO diagrams (Miyashiro 1975). Llanvirnian: solid squares - Tekturmas Mtn, open triangles - Itmurundi Mtn; Llandeilian: solid circles - Bazarbai valley, crosses - Turetai, Kazyk, Tesiktas Mountains. Graph sourced from Novikova *et al.* (1991)

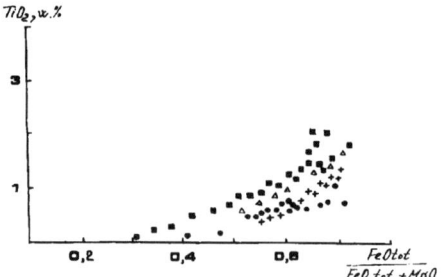

Figure 10. TiO_2 versus $FeO^{tot}/FeO^{tot}+MgO$ for gabboro (Serri and Sattia 1980). See for symbols fig. 9. Graph sourced from Kuznetsov *et al.* (1990 a).

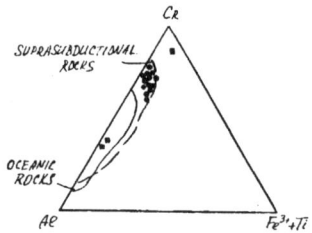

Figure 11. Al-Cr-(Fe^{3+}+Ti) diagram for spinels of residual peridotites. The fields are from Bogatikov et al. (1988). See for symbols fig. 9. Graph sourced from Kuznetsov *et al.* (1990 a).

DISCUSSION

Composition
Ophiolite sequences of Central Kazakhstan are equipped with all members [13], but instead of dyke swarms, sill or dyke-sill swarms are often developed (Fig.3). Relationships among dyke and sill generations show earlier dyke intrusion and later sill formation. Probable explanation of these facts is successive injection of sills in decollement zone among lavas and cumulates or inside of dyke swarm as in zone of inhomogeneity within ophiolite column on periphery of magma chamber during spreading [54]. Some sill(?) bodies are found in residual peridotites that determines two levels of decollement, i.e., at the Moho level and within inner crustal level, which have, perhaps, originated because of different velocity of movements in mantle and crust [36].

Time-space distribution
All regarded ophiolites may be divided into three generations according to their time distribution and age relationships with volcanics of adjacent island-arc terranes.
Ophiolites of Caledonian epoch reveal pre-island arc origin of Vendian - Early Cambrian ophiolites of the Jalair-Naiman suture and basalts of the East Erementau zone; syn-island-arc Early Cambrian - Middle Ordovician ophiolites are developed within Maikain-Balkybek, Tekturmas, North Balkhash, Arkalyk, Aksu-Iradyr and Chistopol sutures.
The sutures, which contain Early Cambrian-Middle Ordovician ophiolites, forming subsynchronous pairs with arc volcanic structures, are interpreted as the remnants of three different back-arc basins [50, 51], i.e., Boshekul-Chinghiz island arc terrane and Maikain-Balkybek ophiolite suture, Baidaulet-Akbastau island arc terrane and Tekturmas-North Balkhash sutures, Stepnyak island arc terrane and Aksu-Iradyr - Chistopol sutures.
Variscan epoch is characterized by presence of Middle Devonian ophiolites within North-Junggar suture being synchronous with the Devonian orogenic volcanic belt of Central Kazakhstan.

Petrochemistry and geodynamic environments
Vendian - Early Cambrian ophiolites are interpreted as the remnants of two different basins originated after destruction of 1100 Ma continental crust [57]. Perhaps, basalts and ophiolites of East Erementau zone and Arkalyk suture are the remnants of large basin (Paleo-Asian ocean), because these structures are faced to the Irtysh-Zaissan Variscan megasuture, where ophiolites of the Ordovician and Devonian are known [40]. Ophiolites of Erementau - Chu-Ili system (in Jalair-Naiman and Satybai sutures, Atasu and Akdym zones) may be interpreted as the complexes of autonomous spreading basin (like Western Philippine sea or Western Tasman sea) separating Precambrian continental massifs of former single (?) continent [57]. Alkaline to tholeiite character of former melts support an idea of little sizes of paleobasin.
Late Cambrian - Early Ordovician ophiolites of Maikain- Balkybek suture do not reveal such contrast petrological differences among earlier and later complexes, and their origin as a result of back-arc spreading behind the former Boshekul-Chinghiz arc is not so disputable. An absence of remarkable chemical variations through the time may indicate ensimatic origin of Maikain-Balkybek back-arc basin complexes, as it was shown by Serri and Saitta for modern back-arc basins [43].
Early Ordovician ophiolites of Aksu-Iradyr and Chistopol sutures do not yet reveal data on successive changes of magmatism (Fig.13), but they belong to abyssal and arc tholeiites [23, 47]. Regional tectonic setting of these ophiolites is very specific, and possible environments of their origin may not be regarded out of composition and age of formation of adjacent Kokchetav continental terrane, Stepnyak arc terrane and ophiolite sutures included in the giant Erementau - Chu-Ili accretional system. The most remarkable specifics is a position of Chistopol-Aksu-Iradyr suture between the Kokchetav metamorphic terrane, where 530 Ma (i.e., the Middle-Late Cambrian) diamond-bearing gneisses with coesite, which protolith rocks are nevertheless of 2000 Ma, have been recently discovered [12, 45, 46], on the one hand, and Stepnyak island arc terrane with manifestation of

island-arc magmatism since the Middle Cambrian till the Late Ordovician (with short interruption and rearrangement during Arenigian), on the other hand. In addition, 530-540 Ma age is revealed for layered ultramafic-mafic plutons, which are located within Kokchetav metamorphic terrane [33]. A very wide temporal activity (since the Vendian till the Early Ordovician (Arenigian)) of spreading processes is revealed for the Erementau - Chu-Ili system (Fig.2). An original interpretation of these facts supposes the presence of subduction events in this region since the Middle Cambrian,when oceanic crust of Erementau - Chu-Ili system began to underthrust westward beneath Kokchetav continental terrane, which was involved in a very deep subduction up to mantle environments (900-1000°C, 30 - 40 kbar) [45]. This event is characterized by gradual changes of shallow-water sediments into deep-water facies of the cover of Kokchetav terrane that also reflects the deeping of metamorphic basement. However, this process was interrupted in the Arenigian by manifestation of extension and spreading with formation of ophiolites (Chistopol and Aksu-Iradyr sutures) along the boundary of Kokchetav and Stepnyak terranes. These changes were reflected immediately in subsequent rapid uplift that is fixed in the shallowing of sedimentary environments of the cover accumulation of the Kokchetav terrane. However, it was episodic event of spreading, and ophiolites should be rather interpreted as SSZ complexes. Compressional influence from Erementau - Chu-Ili basin was, perhaps, too great, and a short-term divergent process was changed into convergent at the beginning of the Middle Ordovician, when reorigin of calc-alkaline volcanism in Stepnyak terrane and flysch accumulation took place.

The most contrast petrochemical composition is revealed for Middle Ordovician complexes of the Tekturmas and North Balkhash sutures. These data show enrichment of early magmas by alkaline components (especially in Tekturmas and North Balkhash sutures) corresponding to former alkaline melt; later ophiolites reveal arc-tholeiite and abyssal tholeiite (rare) affinity. Basalts and gabbros of ophiolites of Tekturmas and North-Balkhash sutures contain very high TiO_2. Residual peridotites include rare lherzolitic bodies. Cumulates are orthopyroxene-free (Fig.3). The basalts and gabbros may be originated in alkali basalt magma formed by a low degree of partial melting of mantle rocks. These ophiolites are apparently different from other subduction-zone ophiolites of older age in nearby areas. They resemble partly (Llanvirnian Tekturmas ophiolite) in petrologic nature the Alpine-Ligurian plagioclase-type ophiolites originated in a mid-ocean ridge cut by many transform fault zones [21]. However, the regarded ophiolites, being of clinopyroxene-type, should be rather regarded as back-arc complexes formed during earliest stages (Llanvirnian or even latest Arenigian) of a back-arc basin openning behind Baidaulet-Akbastau island arc, being synchronous in origin with its calc-alkaline volcanics. On the other hand, Bazarbai ophiolite, which is younger in age (Llandeilian) and synchronous with calc-alkaline volcanics of the Baidaulet-Akbastau island arc, reveals arc tholeiite characteristics that does not permit to regard older and younger complexes as a single arc system with changes of types of magmatism across island arc as in the present Northeast Japan. Hence, an interpretation of origin of these two ophiolites, as a result of successive openning of new ensialic back-arc basin during almost 20 Ma (i.e., during Llanvirnian-Llandeilian according to the Geological Time Scale [20]) behind the Baidaulet-Akbastau arc, is preferable, as it was shown by Serri and Saitta [43] for modern back-arc basins.

Middle Devonian ophiolites are also of SSZ-origin representing the latest episodes of mafic magmatism within already little southeastern part of Junggar-Balkhash system.

Mineralogy of cumulates

Comparison of cumulative sequences of Caledonian ophiolites belonging to different generations reveals interesting results showing an absence of principal differences in mineralogy of cumulates between pre- and syn-island arc generations (Fig.3). However, the variations, which are reflected in the olivine-orthopyroxene-clinopyroxene-(hornblende)-plagioclase composition of cumulates, on the one hand, and in the very simple olivine-clinopyroxene-plagioclase composition of cumulates, on the other hand, are observed within Early Cambrian - Middle Ordovician generation that was discussed above as possible reflection of various degree of mantle partial melting that is determined by the homogeneous dunite-harzburgite composition of residual peridotites of different structures except of lherzolitic harzburgites in the Tekturmas and North Balkhash sutures, where simple mineralogical association of cumulates is observed. Both groups, however, may be attributed to the supra subduction zone ophiolites according to data of A.Ishiwatari [22]. Presence of magmatic hornblende in some cumulates support an idea of SSZ origin of them, because this fact demonstrates presence of hydrous fluids in magma melt, which could be emanated from subducting plates, as it is proposed by Ozawa [38] for Japanese ophiolites. In this case, the composition of cumulates of Jalair-Naiman suture also reflects an influence of subducting lithosphere into a former crust of autonomous spreading basin.

CONCLUSION

1. Ophiolites in the Central Kazakhstan occur in the numerous sutures among accreted island arcs and continental terranes. Melange-type and dismembered ophiolites dominate. Intact, full-membered ophiolites are very rare, and occur only in the Karaulcheku massif. Specific feature of these ophiolites is a frequent presence of sill or dyke-sill swarms instead of traditional sheeted dyke swarms.
2. Time-space distribution of ophiolites and island arc volcanics reveal Vendian-Early Cambrian pre-island-arc ophiolites (rare) and Early Cambrian-Middle Ordovician syn-island-arc ophiolites (dominant) of Caledonian epoch and Middle Devonian ophiolites of Variscan epoch (only in Junggar).
3. Petrochemical data and tectonic interpretation show similarity of Vendian-Early Cambrian, Early Cambrian-Middle Ordovician and Middle Devonian ophiolites. They are mostly SSZ ophiolites. However, the first generation was formed before origin of firsts island-arc complexes in Kazakhstan [50], and two latter generations are synchronous with island-arc and orogenic volcanic complexes.
4. Paleontological data indicate that early magmatic events had alkaline or calc-alkaline specifics and later - arc-tholeiite or MOR tholeiite character in the Maikain-Balkybek and Tekturmas-North Balkhash sutures.

Acknowledgements
This paper was presented during ophiolite symposium of the 29th International Geological Congress taken place in Kyoto. I thank sincerely Professor R.G.Coleman (Stanford University, California), Professor R.Laurent (Universite LAVAL, Quebec) and Dr. A. Ishiwatari (Kanazawa University, Japan) for fruitful discussion and unformal interest. I thank Dr. Akira Ishiwatari for improving the English of this paper. I thank Council of Young Scientists of Moscow State University and Andrew Gidaspov, Michael Katz and Larisa Kurkovskaya from Start Market Ltd (Moscow) for essential financial support.

REFERENCES

1. A.A. Abdulin (Ed.). *Geology of the Uspensk tectonic zone*. vol.5. Nauka, Alma-Ata (1967). (in Russian).
2. N.A. Afonichev, V.Ya. Koshkin, A.E.Mikhailov and N.A.Pupyshev. About age of the Urtynzhal series of central Kazakhstan. *Izvestia Akademii Nauk SSSR*, ser. geol. N 7, 90-93 (1976). (in Russian).
3. R.M. Antoniuk. Oceanic crust of eugeosyncline area of central Kazakhstan. In: *Tectonics of the Ural-Mongolian fold belt*. A.L. Yanshin (Ed.). pp. 67-74. Nauka, Moscow (1974). (in Russian).
4. R.M. Antoniuk, G.F. Lyapichev, N.G. Markova, O.M. Rozen and S.G. Samygin. Structures and evolution of the Earth crust of central Kazakhstan. *Geotektonika* N 5, 71-82 (1977). (in Russian).
5. A.V. Avdeev and A.A. Kovalev. *Ophiolites and evolution of the southwestern part of the Ural-Mongolian fold belt*. Moscow University Press, Moscow (1989). (in Russian).
6. V.F. Bespalov. The tectonic nappe system of Kazakhstan. *Geotektonika* N 2, 78-94 (1980). (in Russian).
7. O.A. Bogatikov et al. (Eds). *Ultramafic rocks*. Nauka, Moscow (1988). (in Russian).

8. A.A. Bogdanov. Tectonic features of the Paleozoids of central Kazakhstan and Tienshan. *Biul. Moskovskogo Obshchestva Ispytateley Pryrody*, otd. geol. N 6, 8-42 (1965). (in Russian).
9. V.I. Borisenok, A.V. Ryazantsev, K.E. Degtyarev, A.S. Yakubchuk et al. Paleozoic geodynamics of central Kazakhstan. In: *Tectonic research and middle-large scale geomapping*. Yu.M. Pusharovskii (Ed.). pp. 81-95. Nauka, Moscow (1989). (in Russian).
10. R.A. Borukaev. *The pre-Paleozoic and Lower Paleozoic of northeastern Central Kazakhstan (Sary-Arka)*. Nedra, Moscow (1955). (in Russian).
11. A.G. Burdyniuk, E.S. Kichman, Yu.A. Gabov and V.N. Kokurkin. Basite-ultrabasite formation. In: *Chu-Ili ore belt. Geology of Chu-Ili region*. A.A. Abdulin et al. (Eds). pp. 174-186. Nauka, Alma-Ata (1980). (in Russian).
12. J.S. Claoue-Long, N.V. Sobolev, V.S. Shatsky and A.V. Sobolev. Coesite from diamond-bearing gneisses of the Kokchetav terrane. *Geology* 19, 710-713 (1991).
13. R.G. Coleman. *Ophiolites: an ancient oceanic lithosphere?* Springer Verlag, Berlin (1977).
14. K.E. Degtyarev. Ultramafics of southern Erementau (central Kazakhstan). *Vestnik Moskovskogo Universiteta*, ser. geol. N 3, 74-78 (1992). (in Russian).
15. Y. Feng. Characteristics of ancient plate tectonics in Western Junggar. *Bull. Xi'an Inst. Geol. Min. Res. Chinese Acad. Geol. Sci.*, N 18. (in Chinese).
16. N.A. Gerasimova. Stratigraphy of Ordovician of the Atasu anticlinorium. In: *Geology of early geosynclinal complexes of central Kazakhstan*. Yu. Zaitsev (Ed.). pp. 53-96. Moscow University Press, Moscow (1985). (in Russian).
17. N.A. Gerasimova, M.Z. Novikova, L.A. Kurkovskaya and A.S. Yakubchuk. New data on the Lower Paleozoic stratigraphy of the Tekturmas ophiolite belt (Central Kazakhstan). *Bull. Moskovskogo Obshestva Ispytatelei Pryrody* 67, N 3, 61-76 (1992). (in Russian).
18. L.L. German and A.V. Ryazantsev. A microgabbro zone in ophiolitic massifs and the problem of parental magma. *Vestnik Moskovskogo Universiteta*, ser.geol. N5, 71-74 (1988). (in Russian).
19. N.M. Gridina and T.V. Mashkova. Conodonts from siliceous-terrigenous suites of the Atasu anticlinorium. *Izvestia Akademii Nauk Kazakhskoi SSR*, ser. geol. N 6, 48-55 (1977). (in Russian).
20. W.B. Harland, R.L. Armstrong, A.V. Cox, L.E. Carig, A.G. Smith and D.G. Smith. *A geologic time scale*. Cambridge Univ. Press (1989).
21. A. Ishiwatari. Alpine ophiolites: product of low-degree mantle melting in a Mesozoic transcurrent rift zone. *Earth and Planet. Sci. Lett.* 76, 93-108 (1985/86).
22. A. Ishiwatari. Time-space distribution and petrologic diversity of Japanese ophiolites. In: *Ophiolite genesis and evolution of the oceanic lithosphere*. Tj. Peters et al. (Eds). pp. 723-743. Kluwer Academic Publishers, Dordrecht (1991).
23. K.S. Ivanov, V.A. Sakharova et al. New data on age of volcanic-siliceous units of the Kokchetav massif rim (northern Kazakhstan). *Doklady Akademii Nauk SSSR* 301, 158-163 (1988). (in Russian).
24. E.I. Ivanova. Utilization of the thermoluminiscent analysis for specification of relative age of geological formations. *Izvestia Akademii Nauk Kazakhskoi SSR*, ser. geol. N 9, 35-45 (1973). (in Russian).
25. T.N. Kheraskova. *Vendian-Cambrian formations of the Caledonides of Asia*. Nauka, Moscow (1986). (in Russian).
26. T.N. Kheraskova, M.Z. Novikova and N.I. Zardiashvili. Specification of composition of early geosynclinal volcanic formations of central Kazakhstan. *Izvestia Akademii Nauk SSSR*, ser. geol. N 6, 47-61 (1979). (in Russian).
27. B.F. Khromykh. New data on Vendian-Paleozoic evolution and metalogenesis of the Boshekul ore region. *Izvestia Akademii Nauk Kazakhskoi SSR*, ser. geol. N 6, 20-34 (1986). (in Russian).
28. E.S. Kichman, E.V. Alperovich, A.F. Kovalevskii and A.V. Kozhev. Riphean. Cambrian. In: *Chu-Ili ore belt. Geology of Chu-Ili region*. A.A. Abdulin et al. (Eds). pp. 30-41. Nauka, Alma-Ata (1980). (in Russian).
29. S.P. Koneva. *Stenotecoids and unarticuate brachiopods of the Lower and lower part of the Middle Cambrian of central Kazakhstan*. Nauka, Alma-Ata (1979). (in Russian).
30. L.A. Kurkovskaya. Complexes of conodonts from siliceous and volcanic-siliceous units. In: *Geology of early geosynclinal complexes of central Kazakhstan*. Yu. Zaitsev (Ed.). pp. 164-176. Moscow University Press, Moscow (1985). (in Russian).
31. I.E. Kuznetsov. Ultrabasites of the Tekturmas anticlinorium. In: *Problems of geology of central Kazakhstan*. Yu. Zaitsev (Ed.). vol. 1. pp. 122-139. Moscow University Press, Moscow (1980). (in Russian).

32. I.E. Kuznetsov, M.Z. Novikova and A.S. Yakubchuk. Evolution of magmatism of ophiolitic zones of Central Kazakhstan. In: *Geodynamic environments of origin, geochemical aspects of genesis of basites and ultrabasites*. O.M.Glazunov et al. (Eds). pp. 6-10. Inst. Geochemistry, Irkutsk (1990). (in Russian).
33. N.P. Mikhailov, V.N. Moskaleva et al. *Petrography of central Kazakhstan*. vol.2. Nedra, Moscow (1971). (in Russian).
34. A. Miyashiro. Classification, characteristics and origin of ophiolites. *J. Geology* 83, 249-281 (1975).
35. B.B. Nazarov. *Radiolaria of the Lower-Middle Paleozoic of Kazakhstan*. Nauka, Moscow (1975). (in Russian).
36. A. Nicolas. Dynamic magma chamber at fast spreading ridges: evidence from the Oman ophiolite. *Abstract, 29th IGC*. vol.1. p. 134. Kyoto (1992).
37. M.Z. Novikova, L.L. German, I.E. Kuznetsov and A.S. Yakubchuk. Ophiolites of the Tekturmas zone. In: *Metalogenesis and ore deposits of Kazakhstan*. A. Abdulin et al. (Eds). pp. 92-102. Gylym, Alma-Ata (1991). (in Russian).
38. K. Ozawa. Ultramafic tectonite of the Miyamori ultramafic complex in the Kitakami Mountains, northeast Japan: hydrous upper mantle in an island arc. *Contrib. Miner. Petrol.* 99, 159-175 (1988).
39. E.I. Patalakha and V.A. Belyi. *Ophiolites. Itmurundy-Kazyk anticlinorium*. Nauka, Alma-Ata (1980). (in Russian).
40. A.V. Peive and A.A. Mossakovskii (Eds). *Tectonics of Kazakhstan*. An explanatory note to the tectonic map of Eastern Kazakhstan (scale 1:2 500 000). Nauka, Moscow (1982). (in Russian).
41. A.V. Ryazantsev, L.L. German, K.E. Degtyarev, A.L. Kotlyar and E.V. Fedorov. Lower Paleozoic chaotic complexes of Eastern Erementau (central Kazakhstan). *Doklady Akademii Nauk SSSR* 296, 406-409 (1987). (in Russian).
42. A.M.C. Sengor. The Paleo-Tethyan suture: A line of demor- cation between two fundamentally different architectural styles in the structure of Asia. *Island Arc* 1, 78-91 (1992).
43. G. Serri and M. Saitta. Fractionation trends of the gabbroic complexes from high-Ti and low-Ti ophiolites and the crust of the major oceanic basins: a comparison. *Ofioliti* 5, 241-264 (1980).
44. N.N. Sigacheva, S.G. Samygin, D.I. Musatov, O.E. Belyaev and S.M. Liberman. Large-scale maps of some regions of Kazakhstan (northwest Chinghiz). *XXI All-Union tectonic conference*. Abstracts. pp. 69-71. Moscow (1988). (in Russian).
45. V.S. Shatskii, N.V. Sobolev, A.A. Zayachkovskii, Yu.N. Zorin and M.A. Vavilov. A new manifestation of micro-diamond in metamorphic rocks as an evidence of the regional character of ultra-high-pressure metamorphism. *Doklady RAN* 321, 189-193 (1991). (in Russian).
46. N.V. Sobolev, V.S. Shatskii, M.A. Vavilov and S.V. Goryainov. Coesite inclusion in zircon from diamond-bearing gneisses of Kokchetav massif. *Doklady RAN* 321, 184-188 (1991). (in Russian).
47. E.M. Spiridonov. Geosynclinal basite complexes of northeast Central Kazakhstan. In: *Problems of geology of Central Kazakhstan*. Yu. Zaitsev (Ed.). vol. 1. pp. 102-121. Moscow University Press, Moscow (1980). (in Russian).
48. S.G. Tokmacheva, N.M. Zhandaev and V.P. Fet'ko. Spilite-diabase formation. In: *Chu-Ili ore belt. Geology of Chu-Ili region*. A. Abdulin et al. (Eds). pp. 186-189. Nauka, Alma-Ata (1980). (in Russian).
49. T.N. Trusova. *Lower Paleozoic ultrabasic and basic rocks of central Kazakhstan*. Nauka, Moscow (1948). (in Russian).
50. A.S. Yakubchuk. Tectonic setting of ophiolitic zones in the structure of Paleozoids of Central Kazakhstan. *Geotektonika* N 5, 55-68 (1990). (in Russian).
51. A.S. Yakubchuk. *Tectonic position and mineral deposits of ophiolites (Central Kazakhstan as example)*. Ministry of Geology of the USSR, Moscow (1991). (in Russian).
52. A.S. Yakubchuk, A.F. Chitalin and E.Yu. Baraboshkin. Variscan tectonics of Tekturmas ophiolitic zone (central Kazakhstan). *Geotektonika* 5, 61-74 (1989). (in Russian).
53. A.S. Yakubchuk and K.E. Degtyarev. The character of conjunction among Chinghiz and Boshekul directions in Caledonides of northeastern Central Kazakhstan. *Doklady Akademii Nauk SSSR* 317, 957-962 (1991). (in Russian).
54. A.S. Yakubchuk, V.G. Stepanets and L.L. German. Swarms of sheeted dykes, subparallel to the layering in ophiolitic massifs, - evidences of spreading. *Doklady Akademii Nauk SSSR* 298, 1193-1198 (1988). (in Russian).
55. A.S. Yakubchuk, V.G. Stepanets, M.Z. Novikova, N.A. Gerasimova and L.A. Kurkovskaya. Specification of axial paleospreading zone in Ordovician ophiolites of Central Kazakhstan. *Doklady Akademii Nauk SSSR* 307, 1198-1202 (1989). (in Russian).

56. Yu.A. Zaitsev. Mantle ultramafic protrusions - a specific type of geosynclinal deep structures in Paleozoic eugeosyncline of Central Kazakhstan. In: *Problems of geology of Central Kazakhstan*. Yu. Zaitsev (Ed.). vol. 1. pp. 140-182. Moscow University Press, Moscow (1980). (in Russian).
57. Yu. A. Zaitsev. *Evolution of geosynclines*. Nedra, Moscow (1984). (in Russian).
58. Yu.A. Zaitsev. Areageosynclines and their role in geotectogenesis. *Acta Univ. Carolina. Geologica* 1, 55-73 (1990).
59. L.P. Zonenshain, M.I. Kuzmin and L.M. Natapov. *Plate tectonics of the territory of USSR*. vol.1. Nedra, Moscow (1990). (in Russian).

Geological and Structural Conditions Localizing Ornamental Stone Occurrences in the Ophiolites of the Itmurunda Zone, Kazakhstan

I. KOVALENKO[1], G. AEROV[2] and Z. BAGROVA[3]

[1]*VNIISIMS, Alexandrov 601600, Russia*
[2]*Kazzoloto Association, Balkhash 472210, Kazakhstan*
[3]*Sevzapgeologia SGE, PGE, St. Petersburg 199155, Russia*

Abstract: Early Paleozoic ophiolite massifs of the Itmurunda zone form a complex fold-block structure. They comprise gabbro, pyroxenite, peridotite, dunite, and plagiogranite. These rocks suffered multiple deformation and metamorphism throughout the Paleozoic time. Various ornamental stones such as jadeite, "icy" quartz, blue and variegated jaspers, corundum, etc. were formed through the complex evolutional history of the ophiolite. Localization of these ornamental stones is controlled by particular structural features, or is limited to particular metamorphic units.

Keywords: ultramafic rocks, mélange, jadeite, icy quartz, decorative serpentinite, blue jasper, corundum, metasomatite

INTRODUCTION

This paper is based on the original data that the authors have obtained through an integrated study of the Itmurunda ophiolite belt. The study includes geological mapping using remote sensing data (airborne and satellite imagery) and analysis of regularity in geological distribution patterns of ornamental stone occurrences.

GEOLOGIC OUTLINE

The Itmurunda ophiolite zone is an integral part of northern Balkhash anticlinorium and sits just in a junction of Kyzyk and Itmurunda anticlines. In fact, the ophiolite occurs as a complex of intricately folded thrust sheets that has been partly transformed into a polymictic mélange [1]. The ophiolites are composed largely of dunite and peridotite with subordinate pyroxenites, gabbroids, and basic volcanic rocks containing lenses of siliceous rocks. The northern margin of the Itmurunda ophiolite is in thrust contact with an overlying, folded, flyschoid graywacke complex of the early Silurian, Devonian and early Carboniferous ages (Figs. 1 and 2).

Polymictic mélange has formed along the northern part of the ophiolite's outcrop area. In this mélange, relatively massive blocks of ultramafic rocks, diabase porphyrite, spilitic-siliceous rocks, granitic rocks, jasper, garnet hornblendite, garnet–epidote–muscovite schist and glaucophane schist occur as inclusions in the serpentinite groundmass, or "cement", which exhibits both boudinage structure and schistosity. The granitic rocks include the plagiogranite associated with the ultrabasic rocks and the Lower Visean microcline granite which was also subject to boudinage deformation. This fact and some other evidences indicate that the mélange was formed in the post-Visean time. As the ophiolite mélange contains high-pressure metamorphic rocks such as garnet hornblendite and glaucophane schist, the northern plate may originally have formed a basement of the mélange.

The central and southern bodies of schistose serpentinite bear massive ultramafic blocks with primary structure, and are regarded as a monomictic serpentinite mélange. The northern polymictic serpentine melange is separated from the central monomictic serpentine melange by a strip of basic volcanic rocks, which are also present in the southern part of the ophiolite (Fig. 1).

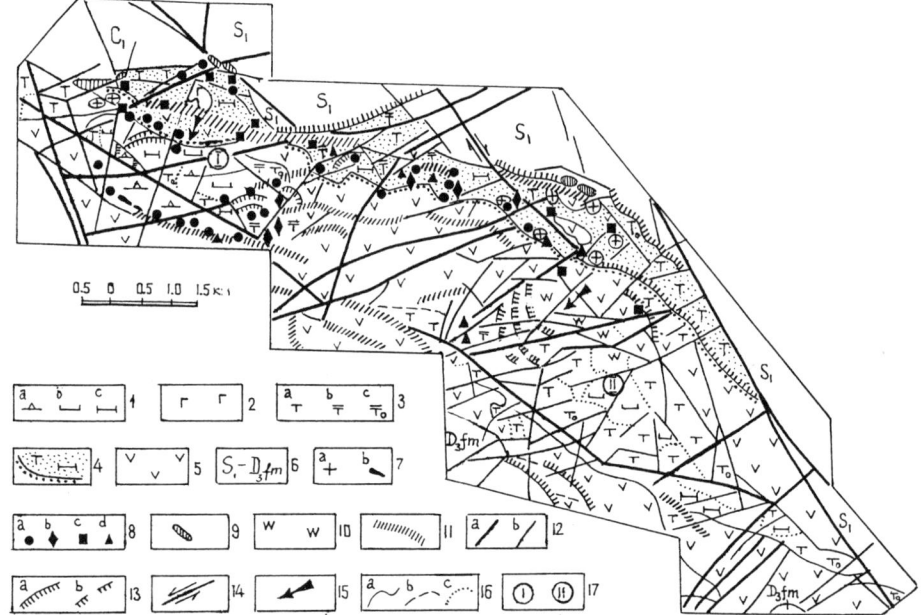

Figure 1. Geological-structural sketch map of the Kenterlau and Itmurunda massifs of the Itmurunda ophiolite with ornamental stone occurrences, compiled by I. Kovalenko, Z. Bagrova, and L. Akhmetova (1992).
1 – (a) undivided ultramafic rocks, (b) dunites, (c) undivided peridotites; 2 – (a) undivided pyroxenites, (b) gabbro; 3 – (a) undivided serpentinite, (b) microantigorite, (c) recrystallized antigorite; 4 – polymictic melange; 5 – spilites, diabase porphyrites, siliceous rocks (Itmurundian suite, O_2); 6 – undivided volcano-sedimentary country rocks (S-D); 7 – (a) plagiogranites, (b) albitite bodies, dikes, and veins; 8 – metasomatites: (a) jadeitites, (b) "icy" quartz bodies, (c) decorative jasper, (d) decorative serpentinite; 9 – listwanite; 10 – relics of silicified waste mantle; 11 – zones of tectonites, blastomylonites, and intense schistosity; 12 – (a) faults and zones of long faults, (b) minor fractures: 13 – (a) thrust faults and frontal zones of large thrust sheets, (b) assumed thrust faults; 14 – strike-slip faults; 15 – direction of large-scale block displacement; 16 – geological boundaries: (a) proved, (b) assumed, (c) boundary between rock facies; 17 – massifs: (I) Kenterlau, (II) Itmurunda.

The Itmurunda ophiolite is separated into three massifs. The western massif (Kenterlau) is composed of apoharzburgite serpentinites, the central massif (Itmurunda) is characterized by a larger share of dunite, while the eastern massif (Arkharsu) is nothing else than a layered intrusion whose basal portion composed of ultramafic rocks (serpentinite melange) grades upward into wehrlite, pyroxenite, and gabbro (Fig. 2).

The plagiogranites and sodium-metasomatic rocks related to the ultramafic rocks prevail the margins of the ophiolite, whereas its central part bears jadeitite and jadeite-albitite.

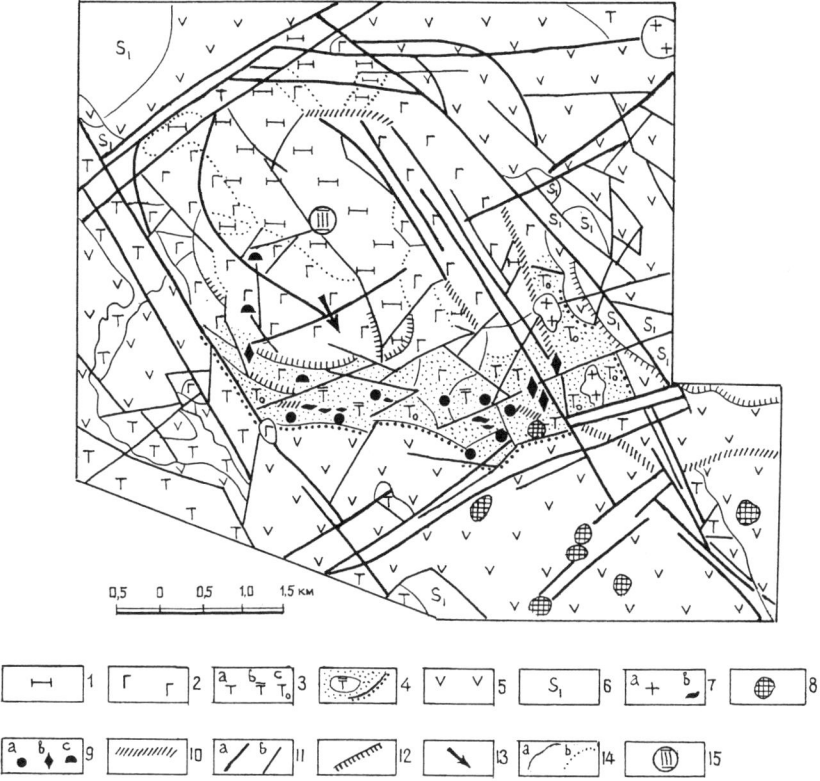

Figure 2. Geological–structural sketch map of the Arkharsu massif of the Itmurunda ophiolite with ornamental stone occurrences, compiled by I. Kovalenko, Z. Bagrova, and L. Akhmetova (1992).
1 – undivided peridotites; 2 – gabbro; 3 – (a) undivided serpentinites; (b) microantigorite, (c) recrystallized; 4 – polymictic serpentinite melange; 5 – spilites, diabase porphyrites, siliceous rocks (Itmurundian suite, O_2); 6 – undivided volcanogenic–sedimentary host rock mass (S_1); 7 – (a) plagiogranites, (b) albitites; 8 – central parts of assumed unexposed minor intrusions; 9 – metasomatites: (a) jadeites, (b) "icy" quartz, (c) corundum mineralization; 10 – zones of tectonites, blastomylonites, and intense schistosity; 11 – (a) faults and long fault zones, (b) minor faults and fractures; 12 – thrust faults and frontal zones of large thrust sheets; 13 – direction of large block displacement; 14 – geological boundaries: (a) proved, (b) between rock facies varieties; 15 – (III) center of the Arkhrsu massif.

GEOPHYSICAL DATA

The gravity field geophysical data indicate that the ophiolite is situated in a zone of steep gravitational gradient. The Kenterlau and Itmurunda massifs show higher positive gravity anomaly (up to +1.5 to 3 mGal), while the Arkharsu massif is marked by a positive gravity anomaly of nearly circular shape with the intensity up to +7.5 mGal.

In the anomalous magnetic field $\Delta T_{(a)}$, the ophiolites are marked by a complex mosaic field's strip that runs along the border of relatively quiet negative and positive fields. Rocks of the Itmurunda massif turn out to be most highly magnetized (up to 2,500 mOe); rock magnetization level for the Kenterlau and Arkharsu massifs is lower (up to 2,000–2,150 mOe).

According to the deep seismic sounding (DSS) data, a structural pattern of the field is influenced by an elevated portion in the configuration of the basaltic layer's top surface (10–12 km depth), which produces an anomaly with the north–south elongation. A transverse pinch with a sinistral dislocation is identified within this anomaly (Fig. 3). Running as a narrow linear strip in a WNW direction, the ophiolite belt coincides well with the elongated pattern of 8-km thick contour of the gabbro–diorite layer of the lower crust (Fig. 4). The anomalous geological and geophysical features of the southeastern edge of the ophiolite (Arkharsu massif) appear to be attributed to a larger thickness (9 km) of the gabbro–diorite layer which occurs here closer to the surface at the depth of about 2 km (Fig. 5).

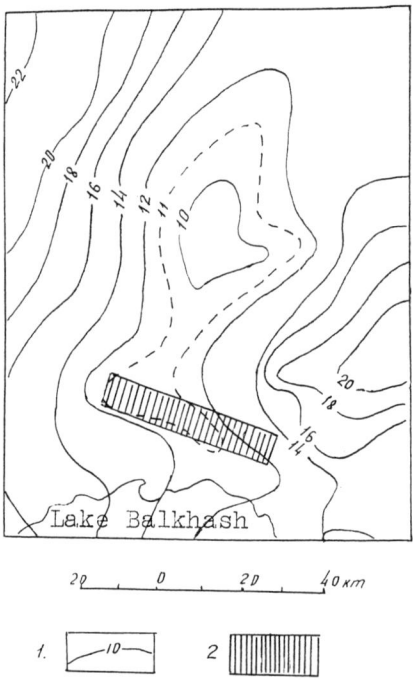

Figure 3. Contour map of the basaltic layer's roof topography (after F. Moiseenko and A. Gromov, 1985). 1 – Isodepth contour in kilometer; 2 – Area of geological mapping (Figs. 1 and 2).

FAULT SYSTEMS

Tectonic framework of the belt is controlled by two major conjugate regional fault systems, one of which is orthogonal (east–west and north–south) and the other is diagonal, and both are deep–seated and have long evolutional history. These fault systems predetermined the large-scale block structure of the northern Balkhash anticlinorium in which the Itmurunda anticline and the adjacent Itmurunda fault are located.

In general, the structural position of the ophiolite is influenced in the first place by the faults striking in nearly E–W and WNW–ESE directions, or in 270–290° and 300–310°, respectively. These faults were active during the formation of the ultramafic bodies, thrust faults, flow structures, shear zones, and boudinaged belts. In terms of kinematics, these faults are classified as sinistral strike-slip faults and thrust faults wherein the long–term compression prevailed.

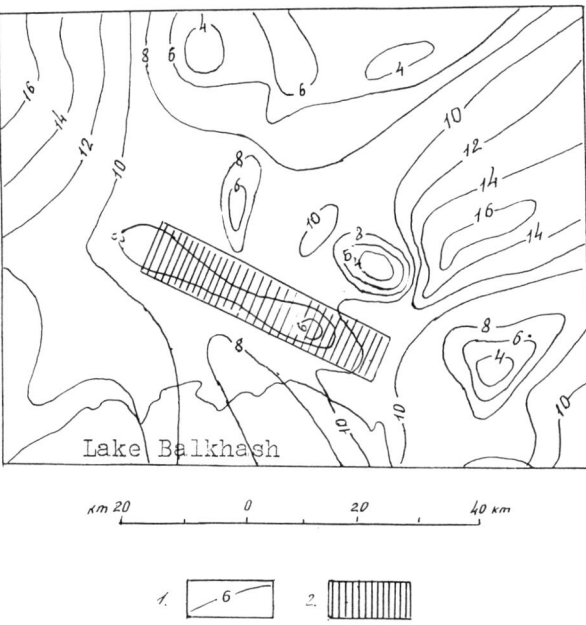

Figure 4. Contour map of the gabbro–diorite layer's thickness (after F. Moiseenko and A. Gromov, 1985). 1 – Thickness contour in kilometer; 2 – Area of geological mapping (Figs. 1 and 2).

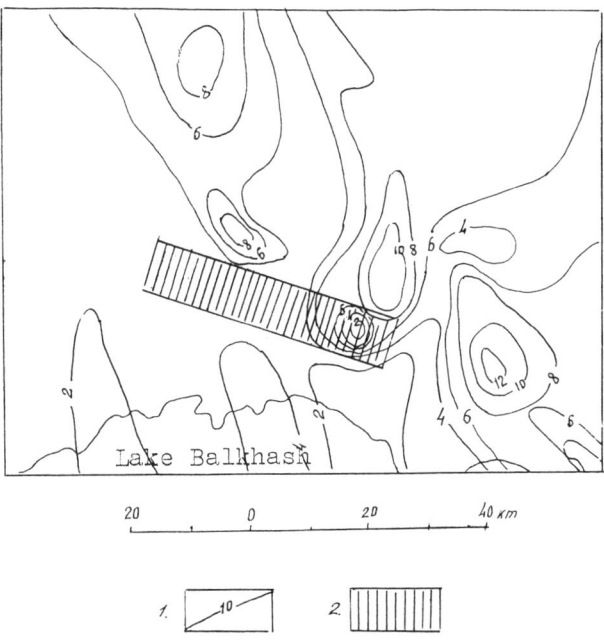

Figure 5. Contour map of the gabbro–diorite layer's roof (after F. Moiseenko and A. Gromov, 1985). 1 – Isodepth contours in kilometer; 2 – Area of geological mapping (Figs. 1 and 2).

The faults striking N–S and NE–SW (350–10° and 60–70°) are identified as the strike-slip faults and dip-slip faults, the latter being in a subordinate number. Unlike the E–W oriented faults wherein compression prevailed, the faults striking N–S and NE–SW were induced largely by the stretching forces.

The Itmurunda ophiolite has undergone complex tectonic evolution in which two major stages are distinguished. During the first, Caledonian stage, the ophiolite was formed and emplaced by thrusting, and in the second, Hercynian stage, fold-block deformation occurred and serpentinite was formed with the development of schistosity and boudinage structure.

METAMORPHISM

The ophiolite has been subject to polymetamorphism, and is dominated by the regional greenschist facies rocks, while the glaucophane schist facies rocks are present along the faults. The amphibolite facies rocks appear sporadically, and are regarded as relics. Metamorphic changes in the ultramafic rocks resulted in the formation of microantigorite serpentinite (macroflaky antigorite serpentinite in the fault zone). Diabase porphyrite and basalt have been changed into spilites, whereas plagiogranite has been changed into albitites and other sodium metasomatites [2].

ORNAMENTAL STONES

The ornamental stones such as jadeite, "icy" quartz, decorative serpentinite, blue and variegated jaspers, corundum, etc. were formed through the complex evolutionary history of the Itmurunda ophiolite as described above. Each of these stones has its own place in the evolutionary history of the ophiolite and possesses specific causes for its localization (Table 1).

Jadeite and "icy" quartz are localized in apoharzburgite serpentinites, and their position is marked by massive blocks of ultramafic rocks that are confined by the E–W trending zones of mylonitization (Figs. 1 and 2). Jadeite formations are always accompanied by albitite and other sodium metasomatites with a complex composition. "Icy" quartz bodies which accompany the jadeite tend to occur at the endogenous contacts with the ultramafic rocks. The mineralized zones often appear together with a glaucophane–greenschist metamorphism. Jadeite and "icy" quartz bodies concentrate in the massifs of Kenterlau and Arkharsu.

Decorative serpentinites are localized in the blocks of microantigorite apoperidotite serpentinites which are mechanically rigid. Such rocks are characterized by a well-developed thermal metamorphism and low intensity of gravity field (according to geophysical data), where the rocks seem to have been loosened and heated. Most extensive outcrops of the decorative serpentinites are found within the confines of the central block.

Corundum is to be found exclusively within the outcrop areas of finely fissile gabbro-amphibolites of the eastern block. Corundum is confined to the gabbro–serpentinite contacts and fault zones striking NW–SE.

Blue jaspers may be near-fault metasomatites which in addition to cryptocrystalline quartz contain up to 10–20 wt.% of thinly acicular blue amphibole (glaucophane). These jaspers are confined to the polymictic areas and zones of glaucophane–greenschist metamorphism. Their distribution pattern is structurally controlled by the E–W striking thrust faults.

Table 1. Localization of ornamental stone occurrences in the Itmurunda ophiolite.

Ornamental stone	Tectonic units	Lithologies			Metamorphic facies
		Country rocks	Wallrock formation	Accompanying metasomatites	
Jadeite	Serpentinite Combination of mylonitization zones and stiff blocks of ultramafic rocks	Apoharzburgite microantigorite serpentinite	serpentinite–mylonites with antigorite composition	Albitite, chlorite–phlogopite formations	Glaucophane–greenschist facies
"Icy" quartz	Serpentinite Narrow linear zones striking E–W.	Apoharzburgite microantigorite serpentinites	Serpentinites, macroflaky antigoritic, chloritized	Amphibole–quartz–albite–rocks with szechenyite, glaucophane	Glaucophane–greenschist facies
Decorative serpentinite	Tectonically stable blocks apart from the active zones.	Apoperidotite serpentinites with micro–antigorite			Greenschist facies local heating
Blue jasper	Zones of endogenous contacts with ultra–mafic rocks. Poly–mictic . E–W faults.	Antigorite schists	Plagioclasites	Glaucophane bearing metasomatites	Glaucophane–greenschist facies with relics of amphibolite facies
Corundum	Prolonged formation of faults striking NW–SE.	Interlayers of serpentinized pyroxenites in gabbro	Finely banded gabbro–amphibolites, albitized	Prehnite–zoisite–margarite–formations, rodingites	Amphibolite plus superposed greenschist facies

CONCLUSION

The ornamental stones described above may have been formed in particular stages of the tectonic and metamorphic evolution of the Itmurunda ophiolite. The decorative serpentinite and corundum may be a product of the early endogenic alteration of the ophiolite itself, whereas jadeite, "icy" quartz and blue jaspers may have formed later due to the action of both endomorphism and exomorphism, the latter affected the country rocks as well.

The most remarkable feature of the Itmurunda ophiolite zone is that the ophiolite has been subject to tectonic activity far more than once. The activity included, in particular, folding, faulting and metamorphism, and these resulted in the ophiolite's complex, folded thrust–sheet attitudes, subsequent formation of , essential transformations of the ophiolite's substance, and eventually in the formation of a wide range of ornamental stone varieties.

The Itmurunda ophiolite shows the rock assemblage (dominant serpentinized peridotite, gabbro, diabase, basalt, and plagiogranite in association with blueschist) and the occurrence (thrust sheets) in common with the well-known ophiolites in the world [3], and its multistage deformation and metamorphism through a long time span in Paleozoic suggest the evolutional history characteristic of the circum-Pacific multiple ophiolite belts [4].

Acknowledgements: The authors are grateful to Dr. A. Ishiwatari for the advice that he kindly offered while this paper was in writing.

References:
1. P.V. Ermolov, V.G. Stepanets and N. Seitov. *Kazakhstan ophiolites.* AN KazSSR, Karaganda (1990).
2. I.V. Kovalenko and A.V. Sviridenko. A regularity in a distribution pattern of jadeites in the Itmurunda deposit area. In: *On the mineralization in ultramafites.* I. Romanovich (Ed.). pp. 125-133, Nauka, Moscow (1985).
3. R.G. Coleman. *Ophiolites.* Mir, Moscow (1979). (in Russian).
4. A. Ishiwatari. Ophiolites in the Japanese Islands: Typical segment of the circum-Pacific multiple ophiolite belts. *Episodes,* **14**, 274-279.

Riphean ophiolites of the Northern Baikal Region (East Siberia)

M.I. GRUDININ and I.A. DEMIN
Institute of the Earth's Crust, Russian Academy of Sciences, Lermontov St. 128, Irkutsk, RUSSIA

Abstract. The complete sequence of ophiolites in the middle course of the Vitim River has been studied. It is represented by: a) ultrabasic complex including a group of ultrabasic massifs and a number of small serpentenite bodies; b) gabbroid complex represented by a series of cumulative gabbroids with peridotites and pyroxenites, and undifferentiated intrusives of the basic composition; c) sheeted-dyke complex composing the apical part of the undifferentiated Sunuekit intrusive; d) volcanic complex, made up of basic and acid volcanites of the Kelyana series. Compositions of the all the members of ophiolites correspond to typical series of this kind, developed in similar structures. The age of all the above rock associations being part of the known Baikal-Muya ophiolite belt, based on the microfossil finds in volcanics, is determined as Riphean. All ultrabasic and gabbroid massifs as well as other members of the ophiolite sequence formed in spreading conditions. At present they are allochthonous being thrusted over younger Vend-Cambrian carbonate-terrigenous deposits and intruded by younger Paleozoic granitoids.

INTRODUCTION

Ophiolites are widely developed in the folded frame of the Siberian platform with the age of emplacement varying in the wide range as well. Over the southern coast of Lake Baikal individual members of Lower Proterozoic ophiolites occur locally [1]. Occurrences of Proterozoic ophiolites are reported to be present in China [2]. New evidence on Riphean ophiolites in the East Sayan Range and North Baikal Region has been obtained recently [5, 7, 10].

The emplacement of ophiolite sequences and formation of ophiolite belts is commonly predetermined by durable restructurings in the Earth crust evolution and associated with the intensive tectonic and magmatic activity. The history of the Baikal-Muya ophiolite belt emplacement was complicated and durable as well. To reconstruct the history of geological development in the region, it would be interesting and important: a) to assess the extent of completeness of the ophiolite sequence; b) to consider its geological and structural position with respect to enclosing rocks, and c) to study mineral and chemical compositions of all the members of the ophiolite sequence.

This paper represents new data on ophiolites of the Middle-Vitim mountain terrain (Baikal-Muya ophiolite belt) where all their principal varieties have been discovered.

SUBSTANTIATION OF THE OPHIOLITE SEQUENCE

At the Middle-Vitim mountain terrain within the known Baikal-Muya ophiolite belt, all members of ophiolites occur, represented, according to R.G.Coleman [3], by a complete

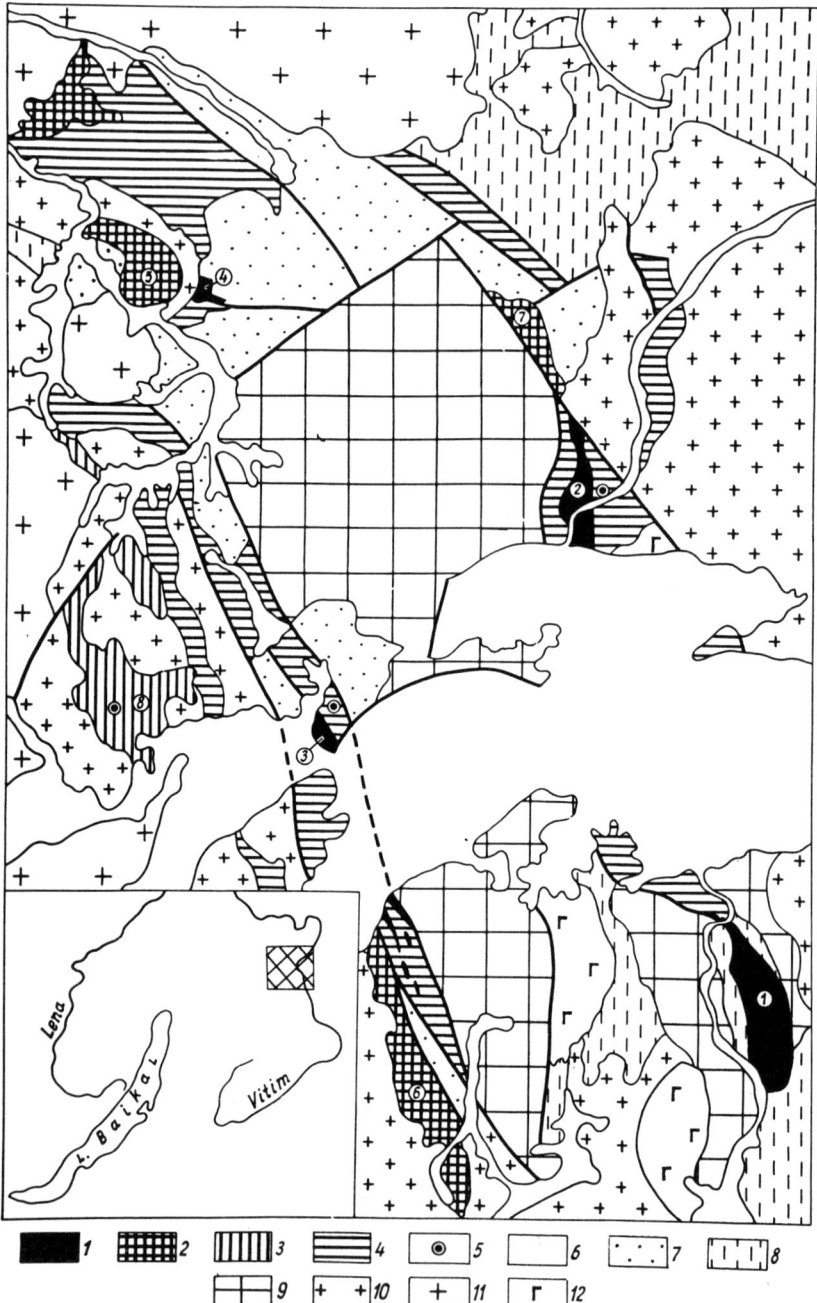

Figure 1. Geological sketch map of the Middle-Vitim mountainous terrain (Northern Baikal Region). 1-5 represent members of ophiolite sequence: 1 - dunite-harzburgite complex, 2 - massifs of differentiated gabbro, 3 - undifferentiated gabbroids (upper gabbro), 4 - volcanic complex, 5 - finds of the complex of parallel dykes; 6 - Quaternary deposits, 7 - Vend-Cambrian carbonate-terrigenous formations, Lower Proterozoic terrigenous-carbonate formations, 9 - exposures of the crystalline basement of the Siberian platform, 10 - Upper Proterozoic plagiogranites, 11 - Paleozoic granites, 12 - massifs of anorthosite-gabbro formation.

The following massifs are indicated by figures: 1 - Shaman, 2 - Param, 3 - Kelyana, 4 - Kaalu, 5 - Middle-Mamakan, 6 - Irokinda, 7 - Yanguda, 8 - Sunuekit.

sequence. Massifs of Alpine-type hyperbasites were formed in conditions of paleoocean in situation of intensive spreading of the previously produced continental crust [10]. Similar conditions were typical of intrusives of differentiated basites that corresponded to cumulative layered ocean-type gabbro, as well as of rather uniform massifs of gabbro and gabbro-diabases with a series of parallel dykes associated with volcanics of the Kelyana series (Fig.1). On the whole the Riphean paleoocean of the Northern Baikal region is characterized by juxtaposition of various rock types of ophiolitic origin. In the upper reaches of the Mamakan and Kelyana rivers the formation of ophiolite sequence are thrusted over Vend-Cambrian carbonate-terrigenous deposits. The Kaalu massif is an example of such tectonic juxtaposition in the central segment of the thrust, where different rock types (hyperbasites, gabbroid and volcanic rocks) in form of various slabs and blocks are juxtaposed (Fig.2). Such strongly altered ultrabasic and basic rocks contain, apart from common types, rodingitized gabbroids and rodingites in composition of which serpentine, amphibole, saussurite, sericite, chlorite along with

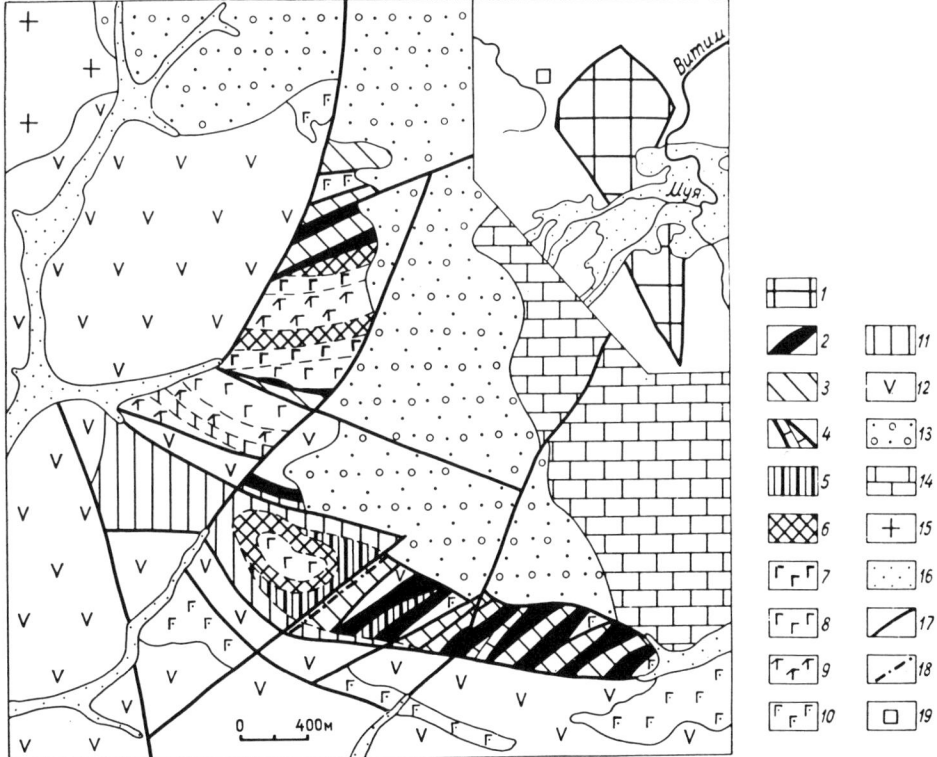

Figure 2. Geological sketch map of the Kaalu massif. 1 - strongly metamorphosed formations of the Muya block; 2-12 ophiolite complex: 2 - apoharzburgite serpentinites, 3 - diallagites, 4 - carbonate deposits, 5 - peridotites, 6 - pyroxenites, 7 - melanocratic gabbro, 8 - mezocratic gabbro, 9 - leucocratic gabbro, 10 - upper gabbro, 11 - talc-listwanite rocks, 12 - effusive-terrigenous formation of the Kelyana suite (R); 13 - conglomerates and gritstones of the Padrokan suite (V), 14 - dolomites and limestones of the Sidelta and Yanguda suites (C), 15 - granites, 16 - Quaternary deposits, 17 - dislocations with a break in continuity, 18 - zone of serpentinite melange, 19 - mapped section.

vesuvianite and garnet, occur. Analogous tectonic zones occur also Kelyana ultrabasic massif where they are represented by tectonites (up to 10 m thick) which are composed by rocks both of ultrabasic and, in lesser degree, basic composition.

Ultrabasic complex includes the Shaman, Param and Kelyna nonfeldspathic hyperbasite massifs and a series of small bodies of serpentinites. All large massifs (of which the Kelyna massif is the smallest - 7 km^2 and the Shaman massif the largest - 110 km^2) are largely tectonized, i.e. broken down into separate blocks; in some massifs there are thick zones of tectonites. Geophysical observations reveal that the lower boundary of such massifs as the Shaman and Param ones have the configuration of a rather gentle concave-downward curve with gravimetry fixing no incurrent channels [16].

All rocks of the mentioned complex are, as a rule, intensely serpentinized, talcized and carbonatized. In the Shaman massif serpentinites in marginal parts are transformed into olivinites. Relicts of primary olivine are still preserved only in the central parts of the massif. Intensely serpentinized and talcized dunites and, much less commonly, serpentinized harzburgites and lherzolites along with pyroxenites are also present in the ultrabasic complex. Major rockforming minerals are represented mainly by olivine (up to 94% Fo), enstatite (6-8% Fs) and diopside. Composition of all these minerals in all varieties is very similar [6]. Accessory minerals of dunites, harzburgites and, to a lesser degree, pyroxenites, are chromspinellids, the average content of which amounts to 1-2% reaching at times 5 and even 8%. In all the rocks there are traces of magnetite, rarely, sulphides.

Gabbroid complex consists of both a series of cumulative (with peridotites and pyroxenites) gabbroids and of less differentiated intrusives of basic composition. The former are represented by the Middle-Mamakan, Irokinda and Yanguda massifs. The most representative of these is the Middle-Mamakan massif, located in the interfluve of the Middle and Right Mamakan Rivers (Middle-Vitim mountainous terrain). It is made up of rocks of ultrabasic and predominantly basic composition. Metavolcanics of the Kelyana series are enclosing rocks of the massif that is overlain on the south-west by the Vend-Cambrian deposits of the Padrokan and Yanguda suites [14].

All the rocks of the massif, similar to the rocks of the previously discussed ultrabasic complex are largely metamorphosed (amphibolized, saussuritized, serpentinized, etc.). Marginal parts especially the ones confining to contacts with granitoid intrusions, are, as a rule, quartzized; and in zones of crushing and schist-forming processes they are metamorphosed into amphibole schists. According to petrochemical composition, ultrabasic and basic rocks of the massif refer to peridotites and pyroxenites and also to gabbroids and thin horizons of anorthosites.

Lherzolites, websterites and diallagites occur in ultrabasic rocks. Lherzolites are not widely distributed in the massif and are represented mainly by serpentinized differences an d only in separate cases a considerable amount (up to 30%) of olivine is observed in them, ortho- and clinopyroxene amount to 20%. The major part of such rocks (about 50%) is represented by serpentine and tremolite. Olivine and rhombic pyroxene were found in form of individual idiomorphic or elongated accumulations; the composition of the former refers to forsterite-chrysolite (10-14% Fa) and of the latter - to enstatite-bronzite (12-15% Fs). Monoclinic pyroxene is commonly represented by tabular colourless crystals of diopside. Websterites are a transitional difference from lherzolites to non-olivine pyroxeites and melanocratic gabbroids. These rocks contain about 20% of orthopyroxene (bronzite - 12-15% Fs), over 30% of clinopyroxene (diopside) and about 5% of olivine (chrysolite with 15% Fa). Saussuritized plagioclase occurs locally. In ultrabasic rocks of this massif there are diallagites, serpentinites and horblendites. Fe-content in orthopyroxene from non-olivine websterites is much higher (up to 25% Fs).

Basic rocks (mesocratic and leucocratic gabbroids, largely amphibolized and saussuritized, along with altered anorthosites) make the large part of the massif (over 80% of the total area). Unaltered gabbroids contain up to 10-15% of rhombic pyroxene (Fe-content ranges between 16 and 26% Fs) up to 40-50% of augite and over 50% of plagioclase in composition of which from 45 to 70% of anorthite component is found. Insignificant quantity (less then 1%) of biotite, apatite and ore mineral is present. Anorthosites are represented by entirely altered plagioclasites, composed by epidote-zoisite aggregate with carbonate and by finest neocrystallization of albite.

The distinctive feature of rocks from the described massif is almost ubiquitous serpentinization, amphibolization, saussuritization, etc. During these processes that must have had many stages, not only the mineral composition of rocks changed but also redistribution of ore-forming components took place. This conclusion is based upon the study both of badly altered differences, where ore minerals are either absent or concentrate in shape of newly formed veinlets or nest-like separations, and slightly altered or unaltered rocks where ore minerals are just weakly resorbed or have fully preserved their primary shape.

The Irokinda massif is similar to the just described one, though it is even more granitized and, moreover, is a subject to a significant cataclasm. Gabbro, gabbronorites, olivine gabbro, troctolites, pyroxenites and peridotites along with serpentenites occur in the massif.

Undifferentiated intrusives of basic composition from gabbroid complex (upper gabbro after N.L.Dobretsov [4]) are represented by the Sunuekit massif (about 140 km^2 in area) also located in volcanics of the Kelyana suite and intruded by later granitoids.

In contrast to the differentiated Middle-Mamakan massif, the Sunuekit massif is made up of fairly uniform fine- and medium-grained gabbro and gabbro-diabases which are devoid of ultabasic differentiates. The rocks of the massif are characterized by prevailing mesocratic outlook and ophite texture; plagioclase is intensely saussuritized or sericitized. Variations in its composition appear to be insignificant (all measurements in relicts of mineral are limited by 45-50% An). Pyroxene is, as a rule, almost completely amphibolized and only in rare cases relicts of poorly coloured augite are observed. Such rocks contain a significant amount of magnetite, titanomagnetite, often apatite.

Complex of parallel dykes of basic composition was discovered in the bedrock at the right bank of the middle course of the Sunuekit river. Here the complex appears to be the apical part of the Sunuekit massif. Similar outcrops of parallel dykes have also been encountered in other exposures of the same river but they are not expressed so well being subjected to granitization. These outcrops of parallel dykes are about 40 m wide and from 0,2 to 5 m thick, with well-defined zones of chilling. Near contacts these zones are composed by fine-grained differences with gradual transition into small- and medium-grained diabases (Fig.3). All dykes of diabases dip very gently at an angle of 20° to 335-340° NW.

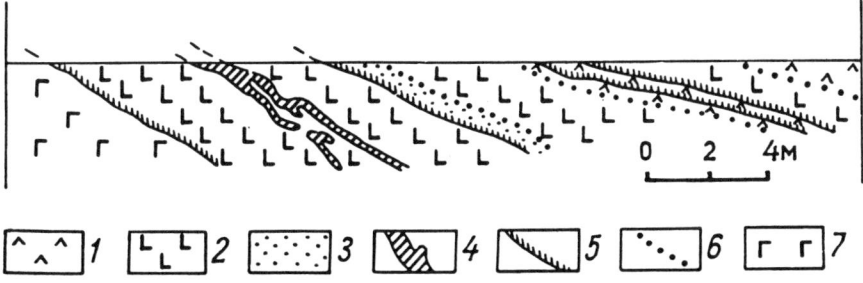

Figure 3. Cross-section of the complex of parallel dykes (middle course of the Sunuekit river). 1-3 - diabases of dykes: 1 - medium-grained, 2 - small-grained, 3 - fine-grained; 4 - thin non-crystalline dykes of diabases, 5 - zones of chilling; 6 - boundaries of gradual transition between diabases with different grain-size, 7 - gabbroids of the Sunuekit massif.

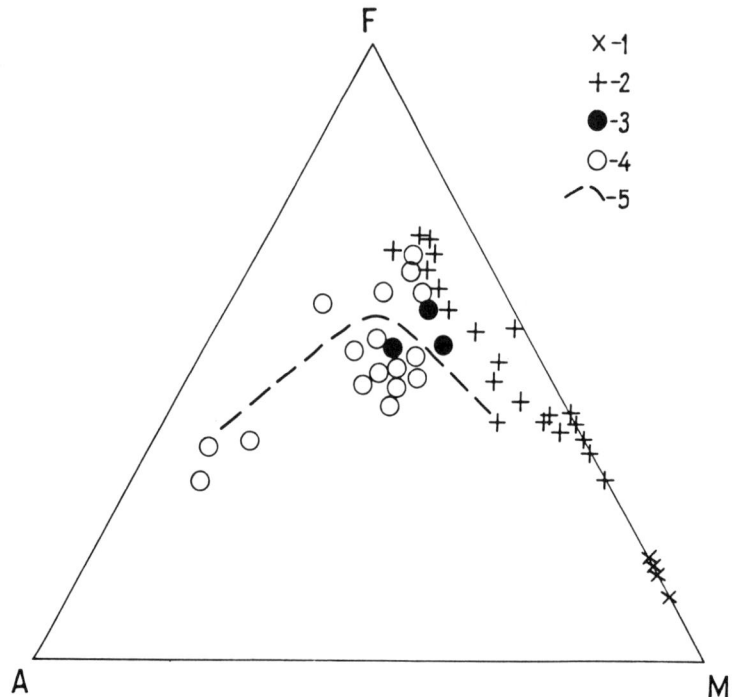

Figure 4. AFM diagramm. 1 - hyperbasites of the Kelyana massif; 2 - gabbroids of the Irokinda and Sunuekit massifs; 3 - diabases of the parallel dykes complex (Sunuekit river); 4 - metavolcanics of the Kelyana series (the Kelyana river basin); 5 - line separating tholeiitic and calc-alkali series.

Petrographical composition of dykes is close to that of gabbroids enclosing them. Similar to gabbroids of the massif dykes consist of saussuritized and sericitized deoxidized plagioclase (close to andesine) and totally amphibolized clinopyroxene. From other minerals epidote occurs everywhere; crystals of apatite, quartz and albite are not uncommon. Mineralogical composition is characterized by high (up to 10%) content of ore mineral - magnetite (titanomagnetite). Dykes of diabases, analogous to the rocks of the massif, are metamorphosed in amphibolite facies.

High content of ore mineral has naturally told on chemical properties of diabases. All of them are characterized by high Fe-content and higher content of Ti-oxides (Table 1). Such rocks with high Fe- and Ti-content are typical of ophiolite sequences of the Mediterranean Region [3], Mugodjar dyke complex [9], ophiolites of the Polish Sudetes [12] and other oceanic sequences of spreading zones. The most important petrochemical parameters of the dyke complex under study are, therefore, quite comparable with basalts of mid-oceanic ridges and in the AFM diagram fit well the usual tholeiitic trend (Fig.4). The same is confirmed by their trace element composition: almost equal quantity of Sr, Ni, Co in both. At the same time differing Ti-, V- and especially Cr-content is quite remarkable. The content of the latter in dykes is considerably lower than that in enclosing gabbroids, whereas dykes of diabases contain more Ti and V. These variation point directly at the order of emplacement of ophiolite sequences.

Volcanic complex is represented in this region by formations of the Kelyana suite [11, 13], composed by transformed sedimentary-volcanic formations. According to G.L. Mitrofanov [11], this series is dominated by volcanics of andesite-basaltic composition. The development

Table 1.
Chemical composition of the ophiolite complex (Northern Baikal region).

	1	2	3	4	5	6	7	8	9	10	11	12	13	14	15	16	17	18	19
SiO_2	34.35	37.69	38.20	38.50	41.10	35.22	40.58	48.66	47.06	44.02	53.59	51.96	54.11	48.33	48.98	47.77	47.42	47.4	47.8
TiO_2	0.02	0.09	0.01	0.01	0.01	0.11	0.21	0.96	1.25	0.10	0.88	0.94	0.75	1.96	2.16	2.33	1.95	1.59	1.9
Al_2O_3	0.65	1.00	0.95	0.75	0.35	0.00	3.30	10.55	7.10	27.60	16.10	15.10	16.00	15.10	14.30	14.20	15.45	15.51	16.34
Fe_2O_3	2.63	6.15	2.18	2.19	1.58	13.64	7.34	3.25	4.27	1.04	5.58	4.32	4.22	5.24	7.84	8.04	5.91	3.1	3.98
FeO	1.95	1.74	8.48	7.58	7.49	5.67	5.65	8.49	9.87	1.11	5.03	6.11	5.30	10.06	8.89	9.34	8.89	9	7.77
MnO	0.11	0.10	0.15	0.15	0.11	0.03	0.13	0.22	0.29	0.07	0.16	0.17	0.17	0.25	0.28	0.26	0.25	-	-
MgO	40.89	38.33	46.47	48.27	46.46	33.28	24.50	13.18	15.00	2.68	5.30	5.50	6.20	5.60	5.20	5.20	3.74	6.8	5.29
CaO	0.20	0.00	0.14	0.42	0.00	0.31	8.04	10.07	10.68	14.05	8.40	8.90	8.60	10.20	9.30	9.20	10.31	10.1	9.82
Na_2O	0.09	0.01	0.01	0.05	0.12	0.09	0.86	1.59	0.30	1.94	1.62	2.90	1.73	1.55	1.55	1.46	2.77	2.41	2.76
K_2O	0.03	0.00	0.05	0.05	0.10	0.00	0.23	0.62	0.04	0.82	0.58	1.20	0.59	0.34	0.28	0.43	0.44	0.28	0.45
P_2O_5	0.01	0.01	0.01	0.01	0.01	0.01	0.02	0.02	0.01	0.02	0.15	0.19	0.15	0.38	0.48	0.42	0.34	0.29	0.41
CO_2	2.38	0.55	-	-	-	0.66	0.00	0.11	1.87	0.00	0.22	0.77	0.22	0.00	0.44	0.00	0.22	-	-
H_2O^-	1.79	1.15	0.32	0.04	0.40	0.43	0.67	0.40	0.38	0.44	0.50	0.44	0.27	0.22	0.20	0.21	0.24	-	-
H_2O^+	14.28	12.96	2.38	1.42	2.21	10.43	9.03	1.69	2.03	6.30	2.10	1.10	1.20	0.90	0.28	1.00	2.37	-	-
Total	99.38	99.78	99.35	99.44	99.94	99.88	100.56	99.81	100.15	100.19	100.21	99.60	99.51	100.13	100.18	99.86	100.30		
Ni	1600	1800	2500	2400	3300	720	770	230	220	38	62	57	71	61	57	62	61	120	100
Cr	2400	1800	2400	8160	3100	1040	1200	180	220	210	130	140	190	83	59	50	93	290	240
Co	130	11	130	140	150	36	95	55	62	10	46	37	59	62	70	62	56	48	41
V	12	44	28	34	20	24	16	80	270	32	220	230	210	430	420	510	400	450	430
Sc	5	11	-	-	-	13	71	6.1	120	13	26	0	0	-	-	0	-	50	39
Rb	0	1	-	-	-	5	4	3.4	8	14	9	24	24	10	8	15	7	6.3	10.9
Li	0	1	-	-	-	7	3	390	6	0	430	6	8	5	2	4	0	0	0
Ba	0	0	-	-	-	93	2	590	0	990	250	370	460	210	240	230	180	64	140
Sr	0	0	-	-	-	38	-	350	990	310	270	270	240	270	260	270	340	220	320

Ultrabasic complex: 1,2 - serpentinites (Kelyana massif); 3-5 - dunites (Shaman massif). Gabbroid complex (Middle-Mamakan massif): 6 - peridotite; 7 - pyroxenite; 8,9 - gabbro-norites; 10 - anorthosite. Upper gabbro: 11-13 - gabbro-diabases (Sunuekit massif). Complex parallel dykes: 14-17 - gabbro-diabases (Sunuekit River). Volcanic complex: 18-19 - middle chemical composition of metabasalts [13]. Oxides wt.%, trace elements ppm.

of acid differences in volcanics is very limited, they occur mainly in marginal parts of troughs. Sedimentary rocks are neither widely distributed in this complex and occur in form of limestones and flinty slates. The lowermost part of the Kelyana cross-section is represented (in the region of the Param hyperbasic massif) by various schists, corresponding in composition to tholeiitic basalts - 1500 m. Schists give way upwards to thin-laminated calcareous-flinty slates with a sheet of porhyrites - 750-800 m. And finally, pelitomorphic limestones intercalated with green tuffs, crystal tuffs, sand tuffs and porphyritoids - 500 m, occur. All the rocks of this complex are transformed to epidote-amphibolite and amphibolite facies of metamorphism.

Chemical characteristics of the principal members of the ophiolite association in the Northern Baikal Region (Table 1) indicate that their composition is rather typical of ophiolite complexes. Rocks of this complex are characterized by very high MgO-content and low content of $FeO+Fe_2O_3$ and TiO_2 which is a direct evidence of its restite nature. The gabbroid complex in general corresponds to typical members of cumulative basites from ophiolite sequences in other regions. Analyses of these rocks placed on the AFM diagram produce trends corresponding to calc-alkali and tholeiitic rocks series. Chemical features of undifferentiated gabbroids and a complex of parallel dykes described above, are also rather typical of that sort of formations. Volcanic complex represented by rather variable magmatic formation, is the least studied member of the ophiolite association, though metabasalts are known for high Mg-content. Besides, as was mentioned before, undifferentiated gabbroids and the related complex of parallel dykes are characterized by high Fe- and Ti-content. Here tendency is observed, to accumulation of TiO_2 and Fe_2O_3 in the upper parts of the ophiolite sequence.

DISCUSSION

The data obtained in recent years on position and composition of ultrabasic and basic rocks in the region suggest that they are sufficiently diverse, formed in different geodynamical settings and suffered significant metamorphic transformation [7]. At the same time, part of ultrabasites and basites in the Middle-Vitim mountain terrain which suffered the greatest transformations and closely associate with each other, is consistent to typical ophiolites and constitute a single sequence. Recent finds of A.M.Stanevich [15] of Riphean microfossils between volcanic rocks of this association put an end to doubts concerning the age of the present ophiolites. To be fair we should note that the same viewpoint had been previously expressed [4, 11] by some workers.

It is also worth while mentioning that the recently obtained data on the composition of diabase dykes [7] indicate that they correlate well with complexes of parallel dykes of the

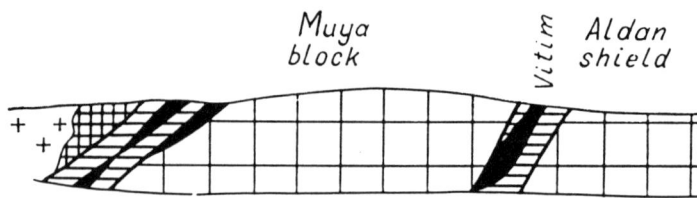

Figure 5. Ophiolite sequence of the eastern branch of the Baikal-Muya ophiolite belt. The legend is the same as in Fig.1.

known ophiolite sequences. Unusual, very gentle dip of the complex of parallel dykes is, to our mind, quite explicable. As was mentioned before [16], such hyperbasic massifs of the Baikal-Muya ophiolite belt are allochthonous. Moreover, the position of ophiolites in this belt with respect to Muya block in the North Baikal Region (Fig.5), is very similar to position of ophiolites around the Gargan block in the East Sayan Range (East-Sayan ophiolite belt) [5]. In both cases ophiolites are overthrusted upon these blocks as a series of tectonic slices. Thus, we have every reason to infer that ophiolites both of the Baikal-Muya and East-Sayan belts are allochthonous and that their present position is a result of later displacements of separate crustal segments.

CONCLUSIONS

1. Judging from geological structural position, mineral compositions and chemical features, Alpine-type hyperbasites, cumulative and undifferentiated gabbroids with sheeted-dyke complex, and metavolcanics closely associated with the above rocks types, form a single ophiolite association and are represented in the study area by complete (according to R.G.Coleman) sequence.

2. The position of hyperbasic and gabbroid bodies and distribution of a series of parallel dykes chilled on one side, enclosed in gabbroids is a direct evidence of spreading conditions during their emplacement.

REFERENCES

1. S.B. Branded, M.I. Grudinin, V.S. Lepin and Yu.V. Menshagin. On the ultrabasic-basic association in the southern Baikal Region, *Doklady Akademii Nauk SSSR.* **292**, N2, 422-425 (1987) *(in Russian).*
2. Chuan Min Sun. Ophiolites richesen Ti du Proterozoique de Shimian et de Yanbian dans la Province du Sichuan en Chine. *C.R. Acad. Sci., Ser.2.* **310**, 1673-1679 (1990).
3. R.G. Coleman. *Ophiolites.* Springer, Heidelberg. (1977).
4. N.L. Dobretsov. Ophiolites and the problem of the Baikal-Muya ophiolite belt. In: *Magmatism and metamorphism of Baikal-Amur Railway Region and their role in formation of useful minerals.* pp. 11-19. Nauka, Novosibirsk (1983) *(in Russian).*
5. N.L. Dobretsov, E.G. Konnikov, V.M. Medvedev and E.V. Sklyarov. Ophiolites and olistostromes of the East Sayan Range. In: *Riphean-Lower Paleozoic ophiolites of North Eurasia.* pp.34-58. Nauka, Novosibirsk (1985) *(in Russian).*
6. M.I. Grudinin. *Basite-hyperbasite magmatism of the Baikal mountainous terrain.* Nauka, Novosibirsk (1979) *(in Russian).*
7. M.I. Grudinin, I.A. Demin and B.G. Mitrofanov. Petrogeochemical types of gabbroid massifs of the Middle-Vitim mountainous terrain, *Geologiya i Geofizika.* N9, 15-23. (1991) *(in Russian).*
8. M.I. Grudinin, I.A. Demin and S.N. Kovalenko. The dyke complex of the Baikal-Muya ophiolitic belt, *Doklady Akademii Nauk SSSR.* **320**, N1, 165-168. (1991) *(in Russian).*
9. S.N. Ivanov, V.G. Korinevsky and G.P. Belyanina. Relicts of the rift oceanic valley in the Urals, *Doklady Akademii Nauk SSSR.* **221**, N4, 939-942 (1973) *(in Russian).*
10. E.G. Konnikov. On the problem of the Baikal-Muya belt ophiolites, *Geologiya i Geofizika.* N3, 119-129. (1992) *(in Russian).*
11. G.L. Mitrofanov. *The Precambrian and Early Paleozoic in the central part of the Baikal mountainous terrain.* Dr.Sc. Thesis. Irkutsk (1978) *(in Russian).*
12. W. Narembsky and A. Majerowicz. Ophiolites of the surroundings of the Gory Sowie mts. block and initialites of the polish part of Sudetes. In: *Riphean-Lower Paleozoic ophiolites of the North Eurasia.* pp. 86-106. Nauka, Novosibirsk (1983) *(in Russian).*
13. A.I. Peskov. *Island-arc and oceanic complex of the Muya segment of the Precambrian Baikal-Vitim belt.* Dr.Sc. Thesis. Moscow (1990) *(in Russian).*

14. L.I. Salop. *Geology of the Baikal mountainous terrain. Stratigraphy. vol.1.* Nedra, Moskow (1964) *(in Russian).*
15. A.M. Stanevich and V.A. Zheleznyakov. Discovery of acritarch microfossils in the Kelyana series in the Middle Vitim. In: *The Late Precambrian and Early Paleozoic in Siberia.* pp. 135-146. Nauka, Novosibirsk (1990) *(in Russian).*
16. Yu.A. Zorin, M.I. Grudinin and E.Kh. Turutanov,E.Kh. Morphology of the Shaman and Param hyperbasic massifs, *Doklady Akademii Nauk SSSR.* **238**, 181-184 (1978) *(in Russian).*

A Late Proterozoic Ophiolite Pulse

A.S. YAKUBCHUK[1], A.M. NIKISHIN[1] and A. ISHIWATARI[2]

[1] *Historical and Regional Geology Department, Moscow State University, Leninhills, Moscow 119899 RUSSIA*
[2] *Department of Earth Sciences, Faculty of Science, Kanazawa University, Kanazawa 920-11 JAPAN*

Abstract
The world-wide distribution of Late Proterozoic ophiolites (younger than 1000 Ma) are thoroughly reviewed for the first time to deduce Late Proterozoic ophiolite pulse. Numerous radiometric, paleontological and stratigraphic data are used for age definition. The age data of the Late Proterozoic ophiolites spread over the full time range (1000–570 Ma) with a pronounced concentration at 750 Ma and a lesser pulse at 600 Ma. Although the Late Proterozoic ophiolites are distributed among all continents, the 750 Ma and older ophiolites dominate Arabia, Africa and South America, and the 600 Ma ophiolites are abundant in Central Eurasia. The magnitude of the 750 Ma pulse may have been as great as the major Phanerozoic pulses in Early Paleozoic (450 Ma) and Late Mesozoic (150 Ma). The 750 Ma pulse established here suggests Earth's major periodic igneous pulses with 300 m.y. interval. The pulses may be related to the collapses of supercontinents caused by superplumes.

Key words: ophiolite pulses, Late Proterozoic, superplumes

INTRODUCTION

Phanerozoic ophiolite pulses were well established by Abbate et al. [1] and displayed in histograms by some authors [37, 43, 44, 65]. Some other authors briefly discussed the problem of Precambrian ophiolites [12, 61]. However, no one extended the histogram back to the Proterozoic. The first attempt to do so was made by Yakubchuk and Nikishin [92]. The aim of this paper is to define more precisely the time-space distribution of the Late Proterozoic ophiolites, to check the magnitude of the Late Proterozoic ophiolite pulses by the number of ophiolites and the length of ophiolite sutures, and to discuss their role in the Earth evolution. The "ophiolitic" complexes older than 1000 Ma are not regarded as ophiolites here, because they do not adequately correspond to the definition of "ophiolite" [61]. In this paper, the terms "Late Riphean" (1000–680 Ma) and "Vendian" (680–570 Ma) are often used to denote earlier and later parts of the Late Proterozoic, respectively. The expression "1 Ma" stands for the moment one million years ago from now, and 1 m.y. represents the duration of time one million years long. Serious evaluation of the age data is beyond the scope of this paper, and published data are mostly accepted in their face values.

LATE PROTEROZOIC OPHIOLITES IN EURASIA

The Late Proterozoic ophiolites are most abundant in the Central Asian segment of the Ural–Mongolian fold belt such as in Kazakhstan, Altai-Sayan, Mongolia and the Transbaikalian regions. Age data may be more reliable for European ophiolites, but is obscure in Central

Asia where the age is determined mostly on paleontological evidences and stratigraphic position of basalt formations. The ophiolite occurrences are noted below from west to east.

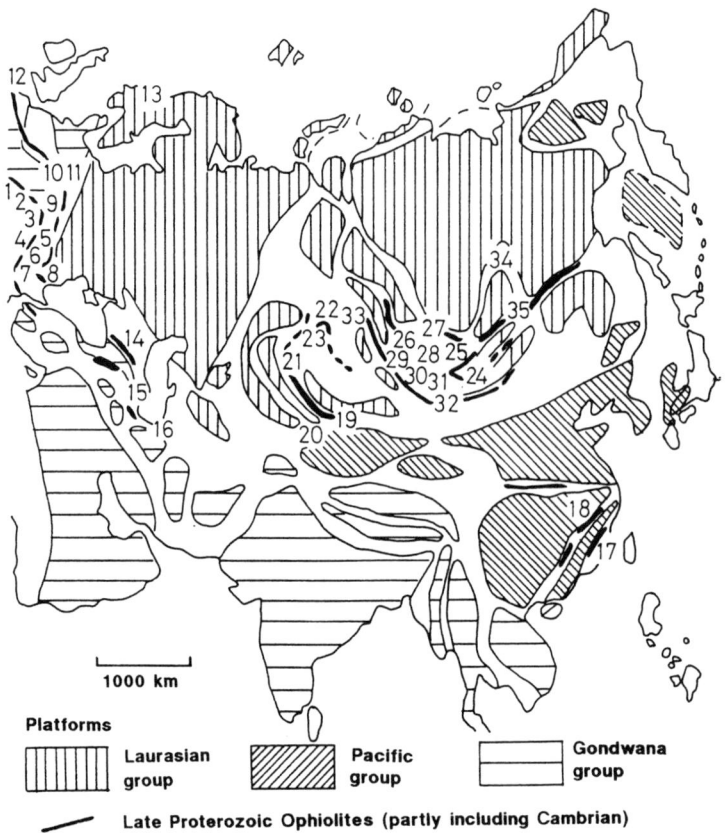

Figure 1. Late Proterozoic ophiolites of Eurasia. Platforms are shown according to Leonov and Khain [55]. Numbers indicate ophiolite localities: 1 - Chamrousse; 2 - E. Alps; 3 - Plankogel and Ritting Mtn.; 4 - Medvedica Mtn.; 5 - Mechek; 6 - Panonian basin basement; 7 - Stara Planina; 8 - South Carpathia; 9 - Marian-Lazensky; 10 - Central Saxonia; 11 - Sachavskie Gory; 12 - Cadomian; 13 - Seve Nappe; 14 - Dzirula; 15 - Miskhan; 16 - Central Iranian massif; 17 - Zhenghe-Jiande; 18 - Sibao; 19 - Aksu; 20 - N. Tarim; 21 - Jalair-Naiman; 22 - Erementau; 23 - Boshekul; 24 - Bayan-Chonggor; 25 - E. Chubsuguul; 26 - Dzabkhan; 27 - Kurtushibin; 28 - S. Tuwa; 29 - W. Sayan; 30 - Jiddin; 31 - Khan-Khukkei; 32 - Khan-Taishir; 33 - Kuznetsk-Alatau; 34 - Baikal-Muya; 35 - Mongol-Okhotsk suture. Compiled by the first author.

1. Alpine Europe

An early Paleozoic or older ophiolite is present at Chamrousse (Fig. 1, No. 1) in the Belledonne range, Western Alps. The ophiolite is dated as 496 Ma by the Sm–Nd method on amphibolites [73]. The age of another old ophiolite in the Eastern Alps is 500±45 Ma (Fig. 1, No. 2) [28]. The ophiolites at Plankogel and Ritting Mountains in the Eastern Alps (Fig. 1, No. 3) are of 700 Ma age dated by Sm–Nd method [64]. In Yugoslavia, Late Proterozoic ophiolites are described at the Medvedica Mountain near Zagreb (Fig. 1, No. 4) and Frushka-Gora to the northwest of Belgrade, where the amphibolites are dated by Rb–Sr method as 1010 Ma and 750–790 Ma, respectively [36]. Metaharzburgite and eclogites of 592 Ma age

[5] are described from Mechek region in southwestern Hungary (Fig. 1, No. 5). The basement of the Panonian basin (Fig. 1, No. 6) also contains ophiolites, whose ages are thought to be between 1000 and 700 Ma by geological position [42]. Geologically inferred Late Precambrian ophiolites are also known from Stara-Planina, Bulgaria (Fig. 1, No. 7) [32]. The Romanian South-Carpathian Mountains contain 850±50 Ma ophiolites (Fig. 1, No. 8), which consist of basic schist, gabbro, and ultramafic rocks, as well as gabbro-ultrabasic rock complex of 550-570 Ma age [81]. The Marian-Lazensky ophiolite of Czechoslovakia (Fig. 1, No. 9) with 575 Ma amphibolite is also cited [60].

2. Variscan and Caledonian Europe

Central Saxonian ophiolites (Fig. 1, No. 10) are dated by K-Ar method on gabbro as 730 Ma. The Rb-Sr data (705 Ma) confirm a Late Proterozoic age of the ophiolites, which are covered with Vendian sediments [89,90]. Ophiolites in Sachavskie Gory (Fig. 1, No. 11) of Polish Sudets may be of Vendian age due to their geologic position, though some other Middle Cambrian ophiolites are identified by the fossil fauna in the sedimentary beds among basalts [63]. Late Riphean ophiolites are described also in northwestern France within Cadomian complexes (Fig. 1, No. 12) [15]. The presence of pillow lavas erupted in the Late Precambrian and Cambrian continental margin is inferred as protoliths of eclogites in Scandinavian Caledonides (Seve Nappe) (Fig. 1, No. 13) [49].

3. Caucasus and Iran

Some Late Proterozoic ophiolites are also reported from Asian part of the Alpine orogen. The Late Riphean or Vendian(?) ophiolites are mentioned from the Dzirula block (Fig. 1, No. 14) in the Caucasus (Georgia) [42]. Another such ophiolite is found in the Miskhan massif, Armenia (Fig. 1, No. 15), where serpentinites, orthoamphibolites and albitites are characterized by a 620 Ma Rb-Sr isochron [2]. Similarly in the central Iranian massif (Fig. 1, No. 16), gneiss, schist, and serpentinite metamorphosed at 660 Ma are covered by Vendian sediments [40].

4. China

Zhang et al. [93] cited some ophiolite occurrences such as Zhenghe-Jiande ophiolite (Fig. 1, No. 17) on the eastern rim of Yantze craton, where it is dated as Vendian(?) according to geological position. Ophiolite of Sibao Group (Fig. 1, No. 18) in the same area gives 837 Ma age by U-Pb method [93]. Vendian(?) ophiolites are also reported from Qinling (Fig. 1, No. 19) [93]. Nakajima et al. [62] have recently found Late Proterozoic blueschist (698 to 718 Ma) from the Aksu region in the northwestern margin of the Tarim craton (Fig. 1, No. 20). This suggests the presence of older oceanic crust, which suffered subduction at that time.

5. Kazakhstan

Some locations with Late Proterozoic ophiolites are described in the Jalair-Naiman suture (Fig. 1, No. 21), where Vendian-Lower Cambrian basalt lavas are identified according to their geological position at the base of the Paleozoic [46]. Thermoluminiscent method gave 680 to 830 Ma ages for the gabbros [38], but the ages are not quite confident. Vendian-Early Cambrian basalts with microphytoliths in limestone layers [10] are reported from eastern Erementau (Fig. 1, No. 22), but the plutonic part is completely absent and only some bodies of serpentine melange are described [45]. Ophiolites of the Boshekul suture (Fig. 1, No. 23) are well dated to be 525-540(±20) Ma with Re-Os method on gabbro [45].

6. Mongolia

Almost all Mongolian ophiolites are dated as Late Proterozoic to Early Cambrian, but these age estimations are based mainly on paleontological data of overlying sedimentary formations. The Bayan–Chonggor ophiolites (Fig. 1, No. 24) are overlain by sedimentary cover with oncolites, and a Late Riphean or Early Vendian age is estimated [39]. Eastern Chubsuguul area (Fig. 1, No. 25) is described as the location of Late Riphean–Early Vendian ophiolites [39]. Dzabkhan (Fig. 1, No. 26) and Jiddin (Fig. 1, No. 30) ophiolites are covered by sediments with archaeocytes that indicate Precambrian–Early Cambrian age of ophiolites [39]. Ophiolites of the Khan–Khukhei Range (Fig. 1, No.31) contain Early Cambrian archaeocytes in interlava limestone [39]. The Khan–Taishir ophiolite (Fig. 1, No. 32) is attributed to the Late Vendian, being covered by sediments with archaeocytes [95].

7. Altai–Sayan region

Late Proterozoic–Early Paleozoic ophiolites in this region are not well-dated. The ophiolite succession is reconstructed from some slices in olistostrome in South Tuwa (Fig. 1, No. 28), where a gabbro is dated as 760 Ma by K–Ar method [57]. Kurtushibin ophiolite (Fig. 1, No. 27) may be of Vendian age in view of its geologic position [19]. Ophiolites of Western Sayan (Fig. 1, No. 29) may also belong to the Vendian according to the regional geologic setting [19]. Early Cambrian ophiolites are distinguished according to their geologic position in Kuznetsk Alatau (Fig. 1, No. 33) and Salair Range [19].

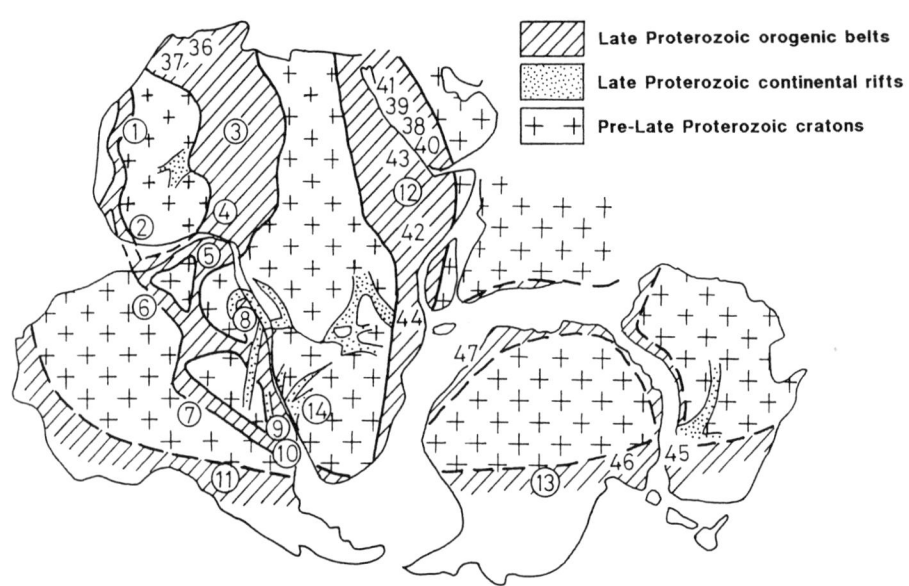

Figure 2. Late Proterozoic ophiolite belts of Gondwana. Numbers in circles indicate orogenic belts with ophiolitic complexes, and numbers without circles indicate ophiolite localities. 1 – Mauritanids; 2 – Rockellids; 3 – Ahaggar; 4 – Dahomey; 5 – Borborema, 6 – Araguaia; 7 – Paraguay; 8 – Mantiqeira; 9 – Ribeira; 10 – Gariep (Proto–South Atlantic); 11 – Sierras Pampeanas; 12 – Arabian–Mozambique; 13 – Transantarctic; 14 – Damara. 36 – Siroua; 37 – Bou Azzer; 38 – Darb Zubaydah, 39 – Yanbu; 40 – Nabitah; 41 – Jabal al Wask; 42 – Baragoi; 43 – Sudan Red Sea Hills; 44 – Morrua; 45 Dandas Trough; 46 – Sladgers; 47 – Lützow Holm Bay. Compiled by the second author using [41, 58, 75].

8. Transbaikalian region

This region bears abundant ophiolites, which are not well dated. The Baikal–Muya ophiolite (Fig. 1, No. 34) to the northeast of the Baikal Lake is attributed to Riphean by some authors [11,96] but to Early Proterozoic by others [27]. Early Cambrian ophiolites of the Mongol–Okhotsk suture (Fig. 1, No. 35) are also described [22].

LATE PROTEROZOIC OPHIOLITES IN ARABIA, AFRICA AND SOUTH AMERICA

Numerous Afro–Arabian and South American orogens, which formed during Pan–African orogenic events, contain ophiolitic remnants of Late Proterozoic age (Fig. 2). However, their ages are well established only in eastern Africa (Kenya) and Saudi Arabia.

1. Western Africa

A serpentinite belt is described from Mauritanids (Fig. 2, No. 1), which underwent 1000 Ma folding [41]. Similar rocks are found in Senegal, Sierra–Leone and Guinea (the Marampa serpentinite belt) (Fig. 2, No. 2). A possible continuation of this belt is probably the Araguaia orogen of South America (Fig.2, No. 6).

2. Anti–Atlas and Ahaggar–Togo–Ghana regions

Late Proterozoic ophiolites are reported by Boukhari et al. [24] from the Siroua massif (Anti–Atlas, Morocco) (Fig. 2, No. 36), where ophiolites are cut by 661±23 Ma granites, and are of pre–Vendian age. The Bou Azzer ophiolite is dated as 788±8 Ma on its gabbro (Fig. 2, No. 37) [54]. The Ahaggar suture (Fig. 2, No. 3) contains 800 Ma ophiolites dated by the Rb–Sr method as cited by Khain and Bozhko [41]. Its continuation is observed in Togo and Ghana within the Dahomey orogen (Fig. 2, No. 4). The age of this complex is Late Riphean, because it is intruded by 680±30 Ma granites [7].

3. Arabian–Mozambique orogen

The most famous occurrences of Late Proterozoic ophiolites are in the western margin of the Arabian shield, where the Darb Zubaydah suture (Fig. 2, No. 38) contains ophiolites with a gabbro dated as 830±20 Ma [77]. The Yanbu suture (Fig. 2, No. 39) has ophiolites of 724±24 Ma and 782±30 Ma [85], and the Nabitah suture (Fig. 2, No. 40) has those of 715 Ma (on gabbro) [85]. Engel et al. [26] reported Late Proterozoic ophiolites in the west Pokot area of northwestern Kenya, having 663±49 Ma Rb–Sr whole-rock isochron age of the related calc-alkaline volcanics and 584±25 Ma age of the associated metasediments. However, 982±40 Ma amphibolites and calc–silicates are also reported. Ophiolites of 728±28 Ma (Sm–Nd method) and 740±11 Ma (U–Pb method) are described from Jabal al Wask area (northern Saudi Arabia) by Pallister et al. [72] (Fig. 2, No. 41). Berhe [6] compiled numerous data on Late Proterozoic ophiolites from the whole Arabian-Mozambique orogenic belt. These data are concentrated in some intervals such as 820–870 Ma, near 750 Ma and 709–694 Ma. Some other data are reported from Kenya (796 Ma and 609 Ma ages for the Baragoi ophiolite (Fig. 2, No. 42), Sudan Red Sea Hills (Fig. 2, No. 43), 832 Ma Gebeit Mine volcanics, 920 Ma Haya terrane volcanics and 850 Ma Haya terrane granodiorites) and Mozambique (Fig. 2, No. 44) (950 Ma cherts associated with Morrua ophiolite) [6].

4. South America

Several Late Proterozoic orogens with ophiolites are known in South America. The Araguaia orogen (Fig. 2, No. 6), where serpentinite belt is described in its axial part, is a possible continuation of the Mauritanian orogen (Fig. 2, No. 1) in Africa. The belt extends southward into the Paraguay orogen (Fig. 2, No. 7) for more than 2500 km. Age data are not available from the serpentinite belt, which may have formed before the 1000 Ma tectonism [17].

The Ribeira orogen (Fig. 2, No. 9) contains ophiolites and serpentinite melange. The age of the complex is defined on the basis of stromatolites as 900 Ma [91]. Ophiolites of the same age may possibly be present in Mantiqeira (Fig. 2, No. 8) and Borborema (Fig. 2, No. 5). Ophiolites are not well studied in the Sierras-Pampeanas orogen (Fig. 2, No. 11).

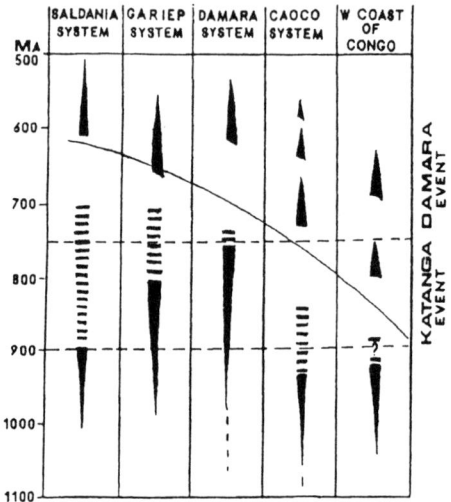

Figure 3. Time-space distribution of Late Proterozoic rifting (black cones below the curve), spreading (ophiolites) (black-white interchange) and closing (black cones above the curve) in Equatorial and South Africa and Eastern Brazil (Simplified after Porada [75]).

Porada [75] compiled data on younger ophiolites for Equatorial Africa and South America. His data show the contemporary nature of these Late Proterozoic ophiolites, whose ages are limited mainly within 950-700 Ma interval. The contemporary ophiolites spreads over Western Congolids(?), Gariep [48] (Fig. 2, No. 10) and Damara systems (Fig. 2, No. 14) of Africa on one hand, and in the Ribeira orogen on the other hand, though minor time lag may be present in some areas (Fig. 3).

LATE PROTEROZOIC OPHIOLITES IN THE OTHER CONTINENTS

Late Proterozoic ophiolites are described from the Dandas trough of Tasmania (Fig. 2, No. 45). Numerous geochronological data from these ophiolites make peaks at 800, 700 and 580 Ma [71], but it is difficult to estimate real formation age of the ophiolites due to Paleozoic reworking. Vendian age of these rocks may be attributed according to their position at the base of the Paleozoic section.

The Sladgers Group (Fig. 2, No. 46) in the Transantarctic mountains is regarded as a possible continuation of the Dandas trough. Basalt bodies are dated as Vendian on the basis of paleontological data (about 600 Ma) [25]. Hiroi et al. [33] reported a gabbroic-ultramafic complex from Lützow-Holm Bay of East Antarctica (Fig. 2, No. 47) as a fragment of oceanic lithosphere tectonically emplaced into the continental collision zone. Some Rb-Sr ages of the gabbroic rocks range from 1200 to 680 Ma, but most other K–Ar and Rb–Sr ages are concentrated at 500 Ma, suggesting a heating event coeval with the pervasive Early Paleozoic granite intrusion.

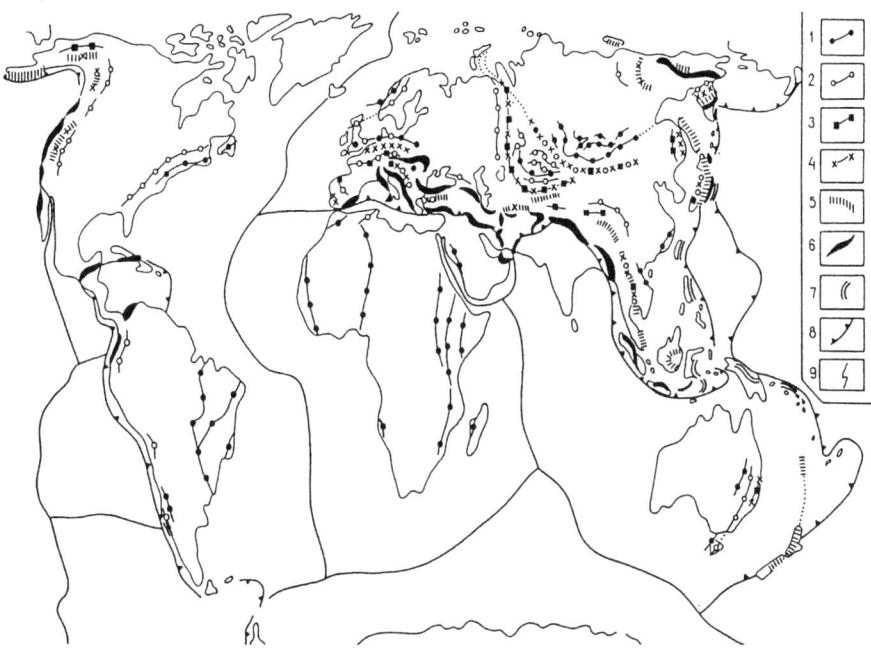

Figure 4. Sketch map showing distribution of Late Proterozoic and Phanerozoic ophiolites of the world. Ophiolites of different ages are shown by different symbols: 1 – Late Proterozoic and Early Cambrian, 2 – Middle Cambrian to Middle Ordovician, 3 – Late Ordovician and Silurian, 4 – Devonian, 5 – Carboniferous to Triassic, 6 – Jurassic and Cretaceous, and 7 – Tertiary. Convergent (8) and divergent (9) plate boundaries are also shown. Compiled by the first and second authors using [2–11, 13–16, 18–29, 31–42, 45–50, 54–57, 60, 62–64, 66, 70–75, 77–85, 88–95].

LATE PROTEROZOIC OPHIOLITE PULSE

A brief overview presented above shows that Late Proterozoic ophiolites are widely distributed on the Earth, and suggests the possibility to establish ophiolite pulses as those in Phanerozoic. In this paper, we suppose that ancient oceanic basins underwent the whole Wilson cycle. The only evidence that is left after a Wilson cycle is the length of the ophiolitic suture zone (Fig. 4), which we interpret to be more or less equal to the former length of the ancient ocean basin before its closure. The secular change in the intensity of ophiolite formation during the Late Proterozoic and Phanerozoic is demonstrated in Figure 5a using age data (mainly the age of basalt lavas) and the length of ophiolite suture on the basis of numerous data cited above [2–11, 13–16, 18–29, 31–42, 45–50, 54–57, 60, 62–64, 66, 70–75, 77–85, 88–95]. The histogram shows a clear, discrete pulse at about 750 Ma as well as another concentration in the Late Vendian (600 Ma) which apparently merges into Cambrian. The Phanerozoic part of this semi-quantitative histogram correlates well to the other diagrams based simply on the numbers of reported ophiolites [1, 37, 43, 44]. The 750 Ma pulse is also evident among the geochronologically well-studied ophiolites in Saudi Arabia [26, 72] (Fig. 5b) as well as among all ophiolite age data cited in this paper (Fig. 5c).

We take into account that the Late Proterozoic ophiolites are poorly preserved within Variscan and Alpine orogens, i.e. it is not possible to extend the Late Proterozoic sutures into these orogens, though it may be possible in Caledonian orogens. This difficulty, however, does not essentially change the quality of estimation, because those young orogens occupy only a minor part of the continents, and most of the Late Proterozoic ophiolites are distributed outside of the young orogens. We have to note the other possible errors in our method of account. (1) An underestimation of the length of ophiolite sutures because of water-covered areas such as Southwest Pacific or Kara Sea and because of platform covers such as West Siberian platform. (2) An overestimation of the role of ophiolites in fold belts, i.e. some of them might not be fragments of the ancient oceanic lithosphere. (3) A poor accuracy of basalt's age definition. So, there is no complete information, and real intensity of the peaks remains undefined. We can describe only the relative increase and decrease of ophiolite formation.

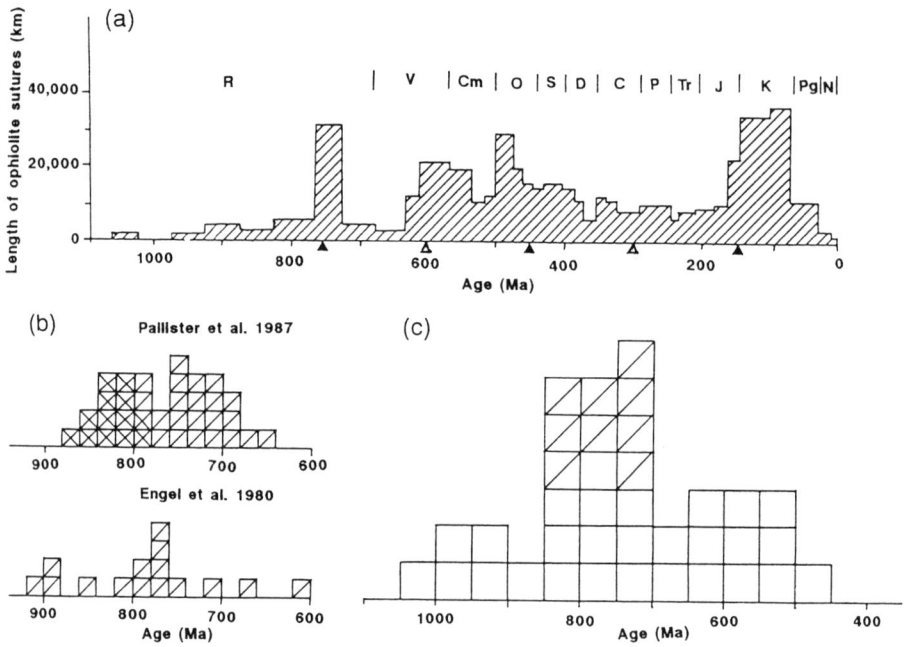

Figure 5. (a) Length of ophiolite sutures of each geologic age in Late Proterozoic (divided for each 50 Ma) and Phanerozoic (following major subdivisions of each period). Abbreviation of each geologic period is shown above, and nominal age of the major (750, 450 and 150 Ma) and minor (600 and 300 Ma) ophiolite pulses are marked by solid and open triangles beneath the time axis, respectively. The minor 300 Ma pulse is evident among the circum–Pacific orogenic belts [37], but is not visible in this diagram. Compilation by the first author. (b) Histogram of radiometric ages of the Arabian Late Proterozoic ophiolites [26, 72]. Each box represents one analysis. Crossed boxes are the data from the central area. (c) Histogram of radiometric ages of the Late Proterozoic ophiolites of the world cited in this paper. Barred boxes represent Arabian data. Compilation by the third author (b and c).

DISCUSSION

1. Pangean megacycles

As for the Late Proterozoic, we remark the major ophiolite pulses occurred at 750 Ma (represented by the ophiolites in Pan-African and South American orogens) and 600 Ma (represented by the ophiolites in Iapetus and Paleoasian oceans as well as in Pan-African orogens) (Fig. 5). An older pulse (900 Ma) may be present, but its magnitude is not clear. The earlier part of the Proterozoic (2500-1000 Ma) is rather characterized by the absence of ophiolites, suggesting prevalence of a thick basaltic crust (as thick as the continental crust) on the Earth in this period, the situation resembling the present Venus [61]. The Late Proterozoic ophiolites are distributed both in Eurasia and Africa-South America as well as in some other areas, and the Late Proterozoic ophiolite pulses were undoubtedly of global importance. Other ophiolite pulses of similar magnitude took place only in Early Paleozoic and Late Mesozoic.

Two megacycles are distinguished for the Late Proterozoic-Phanerozoic history of the Earth; the Late Riphean-Triassic and the Triassic-Quaternary (the latter is not finished yet) possibly corresponding with a break-up, reconstitution and new break-up of Pangea [59, 68, 87]. The biggest peaks of ophiolite pulses in the late Riphean and Jurassic-Cretaceous correspond to the epochs of quick break-up of the supercontinents, and the depressions in the histogram correspond to the epochs of reuniting of supercontinents (Fig. 5a).

Three types of global tectonic structures of the Earth in the Late Proterozoic-Phanerozoic are distinguished. (1) One supercontinent Pangea and one superocean Panthalassa. (2) Many separated continents of comparable sizes and many oceans divided by the continents. (3) One large continent with some small continents and one large ocean with some small oceans. The first type was typical for Middle Triassic, and possibly early Late Riphean (1000?-800? Ma) [68]. The second type was typical for the Late Mesozoic-Early Cenozoic (especially for Cretaceous) and, perhaps, for the Early Paleozoic (Ordovician) and late Late Riphean (800-700 Ma). The third type is typical for the modern situation, where one Eurasian-African-Australian supercontinent, giant Pacific ocean and some isolated continents (like Americas) and oceans (like Atlantic) exist, and, perhaps, for the Middle to Late Paleozoic and Late Vendian-Cambrian (630-500 Ma; the time of Gondwana accretion). The third type is similar to the first type in the global situation other than the ocean/continent distribution, but the third type persist for longer time than the first type.

The three types of Earth's ocean-continent distribution correspond to long-term changes of global magmatism, eustatic sea levels, and magnetic field as well demonstrated for the recent geohistory since 180 Ma. The Pangea-Panthalassa type corresponds to the low sea level, decrease of magmatic accretion of ophiolites and granites, high level of virtual dipole moment of magnetic field, and low frequency of its inversions. The structure with the maximum of the Pangea disintegration corresponds to high sea level, increase of magmatic accretion of ophiolites due to mantle plumes, increase of granitoid magmatism, low level of virtual dipole moment of magnetic field, almost full absence of magnetic inversion, and increase of "black shales" and petroleum in marine sediments.

2. Superplumes

An interpretation of such a long-term periodic changes of the Earth's global activity is developed by the second author on the basis of the model of Larson and Olson [30, 51-53]. They proposed a model of superplume. According to this model, many hot diapirs (or one big superplume, more than 1000 km in diameter) were separated from core/mantle boundary in the Cretaceous time, and caused changes in the outer core and its magnetic field. Such diapirs may arrive at the base of the lithosphere several m.y. after their take off from the core/mantle boundary.

The second author proposes "active" and "passive" models to explain how the superplumes control the Earth's magmatic and magnetic activity imprinted as geological records on the Earth's surface. The "active" model supposes that a voluminous detachment of a hot mass from the core/mantle boundary is the very reason of the Pangea disintegration. The "passive" model assumes that a steady regime of convective flows in the mantle was established during the epochs of the most intensive disintegration of Pangea. This regime started to collapse when thermal and convective stability of the outer core was provoked by the superplume detachment from the core/mantle boundary. Then, the mantle flows have been destabilized and reorganized to fit the new input of the hot mass, and numerous collisions of the separated continents have resulted in the course of reformation of the Eurasia-Africa-Australian supercontinent during the Cenozoic time. It is not possible at present, however, to give preference for either "active" or "passive" model.

Some scientists have shown that Proterozoic-Phanerozoic history of the Earth is characterized by megacyclicity [59, 68, 87]. Duration of megacycles may be about 600-800 m.y. with formation of a supercontinent at the end of every cycle. These cycles include shorter 300-400 m.y. cycles which are obvious on the Vail curve. The latter cycles also include shorter cycles. Nikishin and Khain [67] proposed that the closing of mid-ocean ridges took place during moments of continental collisions. These ridges were main zones where the Earth's interior heat is loosing. Closing of some of these ridges ought to be the reason of the sharp interruption of this process that could produce overheating and thermal extension. These short periodical phases correspond to fall ocean level, which was synchronous with reorganization of plate's kinematics, and, perhaps, with the phases of global compression.

The Earth is a giant heat-machine in which main "stove" is in the outer core. Nikishin accounts that working of this machine should include four factors: (1) The processes in the "stove" itself (in the outer core and its surface). (2) The processes taking place at the surface of the heat-machine (the continents play a role of "lids", and the mid-ocean ridges are "windows", through which the most intensive heat loosing occurs). (3) Heating mains among the "stove" and "windows" (upgoing mantle flows and plumes). (4) The "poker", which could influence on the surface of the "stove" (downgoing whole-mantle flows, which go through the whole mantle from main subduction zones toward the core visible in the recent seismograph [69]). The plate tectonic evolution during the Proterozoic-Phanerozoic time has been controlled by the thermo-magnetic pulsations of both periodical and irregular habit depending on both the processes in the "stove" itself and configuration of "windows" and "lids", which also affect the work of "stove" by "poker".

Finally, some hypotheses for the Earth's pulsating processes are formulated.
(1) Megacycles could be correlated to the autonomous cyclicity of the outer core, which produces superplumes at the core/mantle boundary in episodic manner.
(2) Interdependent movements of plates can influence on the character of mantle flows and Earth's heat loosing through the mid-ocean ridges. When plate collision become often, the subducted slabs influence remarkably on the core surface (the "poker" touch the "stove"), and the stable thermal regime near the core/mantle boundary is highly disturbed. This may be the reason of the frequent magnetic inversions. When collisions of continents become less frequent (e.g. in Cretaceous time), these fluctuations become less important, and the coalesced hot masses make a superplume to take off from the core/mantle boundary. After the superplume separation, thermal regime near the core/mantle boundary recovers stability, and a long-term stable magnetic field regime without inversion takes place [76, 86].

CONCLUSION

The data summarized in this paper show the presence of the Late Proterozoic ophiolite pulse. In view of the lesser possibility of ophiolite preservation with increasing age, this ancient pulse may be comparable with the Late Jurassic–Cretaceous pulse in magnitude. The Mesozoic is an epoch of intensive destruction of Pangea, and the same could have taken place during the Late Proterozoic., i.e. the destruction of former supercontinent like Late Paleozoic–Early Mesozoic Pangea. This idea was discussed in numerous international and Russian publications [41], and our results support these speculations.

The ages of the Late Proterozoic ophiolites range over the full time span (1000–570 Ma) with a marked concentration at 750 Ma and a less pronounced peak at 600 Ma (Fig. 5). The third author views periodicity with 300 m.y. interval for the Late Proterozoic–Phanerozoic ophiolite pulses at 750 Ma (late Late Riphean), 450 Ma (Ordovician), and 150 Ma (Jurassic–Cretaceous) with the intervening smaller pulses at 600 Ma (Vendian–Cambrian) and 300 Ma (Permo–Carboniferous). All these pulses may possibly be subordinate for the megacyclicity with 600 to 800 m.y. interval. Although the Late Proterozoic ophiolites are widely distributed among the continents, the 750 Ma and older ophiolites are especially abundant in Arabia, Africa and South America, and the 600 Ma and younger ophiolites are dominant in Central Eurasia.

The ophiolite pulses depend probably on the interaction of the processes such as autonomous, episodic instability of the outer core, formation and detachment of the superplumes at the core/mantle boundary, instability of the core/mantle boundary caused by the downgoing cold material from the upper mantle, change in the pattern of convection in the mantle, and rearrangement of the ocean–continent configuration on the surface.

Acknowledgements
The first two authors thank Organizing Committee of the 29th International Geological Congress for grants, which enabled their participation in the ophiolite symposium of IGC in Kyoto that stimulated the origin of this paper. All authors thank Prof. J. Malpas of Memorial University of Newfoundland (Canada) and Dr. C. Xenophontos of Geological Survey Department (Cyprus) for critical reading of the manuscript.

References
1. E. Abbate, V. Bortolotti, P. Passerini and G. Principi. The rhythm of Phanerozoic ophiolites. *Ofioliti*, **10**, 109–138 (1985).
2. V.A. Agamalian. Magmatic and metamorphic formations of metamorphic complexes of Armenian SSR. In: *Origin and evolution of metamorphic formations.* vol. 3, pp. 89–91. Nauka, Novosibirsk (1986) (in Russian).
3. T. Alabaster, J.B. Pearce and J. Malpas. The volcanic stratigraphy and petrogenesis of the Oman ophiolite complex. *Contrib. Mineral. Petrol.,* **81**, 168–183 (1982).
4. H. Bahlburg and C. Breitkreuz. Paleozoic evolution of active margin basins in the Southern Central Andes (northwestern Argentina and northern Chile). *J. of South America Earth Sci.* **4**, 171–188 (1991).
5. Z. Balla. Pre-Upper Carboniferous basites and ultrabasites of Hungary. In: *Riphean – Lower Paleozoic ophiolites of northern Eurasia.* N.L. Dobretsov and L.P. Zonenshain (Eds). pp. 136–148. Nauka, Novosibirsk (1985) (in Russian).
6. S.M. Berhe. Ophiolites in Northeast and East Africa: implications for Proterozoic crustal growth. *J. Geol. Soc. London,* **147**, 41–57 (1990).
7. J.M.L. Bertrand, R. Caby, J. Ducrot, J. Lancelot, A. Moussine-Pouchkine and A. Saadallah. The Late Pan-African intracontinental linear fold belt of the Eastern Hoggar (Central Sahara, Algeria): geology, structural development, U/Pb geochronology, tectonic implications for the Hoggar shield. *Precambrian Res.,* **7**, 349–376 (1978).
8. J.M. Bird and S.F. Dewey. Lithosphere plate–continental margin tectonics and the evolution of the Appalachian orogen. *Geol. Soc. Amer. Bull.,* **81**, 1031–1060 (1970).
9. N.A. Bogdanov and A.V. Fedorchuk. Geochemistry of Cretaceous oceanic basalts of the Olyutorsky Range (framework of the Bering Sea). *Ofioliti,* **13**, 113–124 (1987).

10. V.I. Borisenok, A.V. Ryazantsev, K.E. Degtyarev and A.S. Yakubchuk. Paleozoic geodynamics of Central Kazakhstan. In: *Tectonic research and middle-large scale geomapping*. Yu.M. Pusharovskii (Ed.). pp. 81-95. Nauka, Moscow (1989). (in Russian).
11. A.N. Bulgatov. *Tectonotype of Baikalids*. Nauka, Novosibirsk (1977). (in Russian).
12. K. Burke, J.F. Dewey and W.S.F. Kidd. World distribution of sutures - the sites of former oceans. *Tectonophysics* 40, 69-99 (1977).
13. A. Cina, A. Tashko and A. Tershana. The Bulqiza and Gosmiqe ultrabasic massifs, ophiolites of Albanides: a geochemical comparison. *Ofioliti* 13, 219-236 (1987).
14. B.M. Ciric. Complexe ophiolitique et formations comparable dans les Dinarides. *Memoires XXIII*, Belgrade (1984).
15. J. Cogne and A.E. Wright. L'orogen Cadomien. In: *Geologie de l'Europe, du Precambrien aux bassins sedimentaires post-Hercynieus*. pp. 29-55. Publications 26 CGI, Paris (1980).
16. R.G. Coleman. Ophiolites and accretion of the North American Cordillera. *Bull.Soc.Geol.France*, 6, 961-968 (1986).
17. U.G. Cordani, G. Amaral and K. Kawashita. The Precambrian evolution of South America. *Geol. Rundsch.* 62, 309-317 (1972).
18. R.D. Dallmeyer. Contrasting accreted terranes in the Southern Appalachian orogen, basement beneath the Atlantic and Gulf Coastal plains, and West African orogen. *Precambrian Res.* 42, 387-409 (1989).
19. A.B. Dergunov. *The Caledonides of the Central Asia*. Nauka, Moscow (1989). (in Russian).
20. Y. Dilek and E.M. Moores. Regional tectonics of the Eastern Mediterranean ophiolites. *Ofioliti* 13, 175-176 (1988).
21. G.R. Dunning and R.B. Pedersen. U/Pb ages of ophiolites and arc-related plutons of the Norwegian Caledonides: Implication for the development of Iapetus. *Contrib. Mineral. Petrol.* 98, 13-23 (1988).
22. S.I. Dril and S.S. Shudegova. A composition and nature of ultrabasic rocks of central zone of Mongol-Okhotsk suture. In: *Geodynamic conditions of formation, geochemical aspects of genesis of basic and ultrabasic rocks*. O.M. Glazunov (Ed.). pp. 11-15. Institute Geochem., Irkutsk (1990). (in Russian).
23. C.J. Ebinger. Tectonic development of the western branch of the East African rift system. *Geol. Soc. Amer. Bull.*, 101, 885-903 (1989).
24. A. El Boukhari, A. Chabane, G. Rocci and J.-L. Tane. Upper Proterozoic ophiolites of the Siroua Massif (Anti-Atlas, Morocco) a marginal sea and transform fault system. *Jour. Afr. Earth Sci.* 14, 67-80 (1992).
25. P.H. Elliot. Tectonics of Antarctica: a Review. *Amer.J. Sci.* 275A, 45-106 (1975).
26. A.E.J. Engel, T.H. Dixon and R.J. Stern. Late Precambrian evolution of Afro-Arabian crust from ocean arc to craton. *Geol. Soc. Amer. Bull.* 91, 699-706 (1980).
27. V.S. Fedorovskii. *Lower Proterozoic of the Baikal area*. Nauka, Moscow (1985). (in Russian).
28. H.W. Flugel and F.P. Sassi. In: *Pre-Variscan and Variscan events in the Alpine-Mediterranean mountain belts*. P. Grecula (Ed.).1 0pp. 38-49. Bratislava (1987).
29. W. Franke. Variscan plate tectonics in Central Europe - current ideas and open questions. *Tectonophysics* 169, 221-228 (1988).
30. M. Fuller and R. Weeks. Superplumes and superchrons. *Nature* 356, 16-17 (1992).
31. T. Grand, G. Mascle and M. Ohnenstetter. Structure syn-ophiolitique dans l'environnement des amas sulfures de Chypre et leur interpretation. *Doc. BRGM* 1, N 258, 285-305 (1988).
32. I. Haydoutov. Origin and evolution of the South Carpathian - Balkan Precambrian ophiolite. In: *Strukturny vyvoj Karpatsko-Balkanskeho orogennogo pasma*. p.27. Bratislava (1987).
33. Y. Hiroi, K. Shiraishi, Y. Motoyoshi, S. Kanisawa, K. Yanai, and K. Kizaki. Mode of occurrence, bulk chemical compositions, and mineral textures of ultramafic rocks in the Lutzow-Holm complex, East Antarctica. *Mem. Nat. Inst. Polar. Res.*, Spec. Issue 43, 62-84 (1986).
34. C.S. Hutchinson. *Geological evolution of South-East Asia*. Clarendon press, Oxford (1989).
35. D.H.W. Hutton. Strike-slip terranes and a model for the evolution of the British and irish Caledonides. *Geol. Mag.* 134 (5), 405-425 (1987).
36. M.Ilic. Ophiolite setting in geological evolution of Dinarides. *Acta Geol. Acad. Scient. Hungar.* XI, F. 1-3, 77-93 (1967).
37. A. Ishiwatari. Time-space distribution and petrologic diversity of Japanese ophiolites. In: *Ophiolite genesis and evolution of the oceanic lithosphere* (Proceedings of the Oman 90 symposium). pp. 731-751. Kluwer Academic Publishers, Dordrecht (1991).
38. E.I. Ivanova. Utilization of thermoluminiscent analysis for specification of relative age of geological formations. *Izvestia Akademii Nauk Kazakhskoi SSR*, ser. geol., N 9, 35-45 (1973). (in Russian).
39. K.B. Kepezhinskas, V.V. Kepezhinskas and N.S. Zaitsev. *Evolution of the Earth's crust of Mongolia during Precambrian-Cambrian*. Nauka, Moscow (1987). (in Russian).

40. V.E. Khain (Ed.). *Tectonics of Europe and adjacent areas. Variscides, epipaleozoic platforms, Alpids.* Nauka, Moscow (1978).
41. V.E. Khain and N.A. Bozhko. *Historical geotectonics. Precambrian.* Nedra, Moscow (1988). (in Russian).
42. V.E. Khain and S.G. Rudakov. About modern position of northern boundary of Gondwana in Europe and West Asia. *Geotektonika* N 4, 24-38 (1991). (in Russian).
43. V.E. Khain and K.B. Seslavinskii. Global changes of endogenous activity of the Earth during Paleozoic. *Vestnik Moskovskogo Universiteta,* ser. geol. N 6, 3-25 (1990). (in Russian).
44. V.E. Khain and K.B. Seslavinskii. Global changes of endogenous activity of the Earth during Mesozoic and Cenozoic. *Vestnik Moskovskogo Universiteta,* ser. geol. N 5, 3-24 (1990). (in Russian).
45. B.F. Khromykh. New data on Vendian-Paleozoic evolution and metalogenesis of the Boshekul ore region. *Izvestia Akademii Nauk Kazakhskoi SSR,* ser. geol. N 6, 20-34 (1986). (in Russian).
46. E.S. Kichman, E.V. Alperovich, A.F. Kovalevskii and A.V. Kozhev. Riphean. Jalair-Naiman synclinorium. In: *Chu-Ili ore belt. Geology of the Chu-Ili region.* A.A. Abdulin et al. (Eds). pp. 34-35. Nauka, Alma-Ata (1980). (in Russian).
47. R.J. Korsch, H.J. Harrison, C.G. Murray, C.L. Fergusson and P.G. Flood. Tectonics of the New England orogen. *BMR Bull.* 232, 35-52 (1990).
48. A. Kroener. *The Gariep group Late Precambrian formation in the Western Richtersveld, Northern Cape Province.* Precambrian Res. Unit., Tniv. Cape Town, Bull. 13 (1974).
49. K. Kullerud, M.B. Stephens and E. Zachrisson. Pillow lavas as protoliths for eclogites: evidence from a Late Precambrian - Cambrian continental margin, Seve Nappes, Scandinavian Caledonides. *Contrib. Mineral. Petrol.* 105, 1-10 (1990).
50. J.-Y. Labbe and P. St-Julien. Failles de chevauchement acadiennes dans la region de Weedon, Estrie, Quebec. *Can. J. Earth Sci.* 26, 2268-2277 (1989).
51. R.L. Larson. Geological consequences of superplumes. *Geology* 19, 963-966 (1991).
52. R.L. Larson. Latest pulse of Earth: Evidence for a mid-Cretaceous superplume. *Geology* 19, 547-550 (1991).
53. R.L. Larson and P. Olson. Mantle plumes control magnetic reversal frequency. *Earth and Planet. Sci. Lett.* 107, 437-447 (1991).
54. M. Leblanc. Proterozoic oceanic crust at Bou Azzer. *Nature* 261, N 5555, 34-35 (1976).
55. Yu.G. Leonov and V.E. Khain (Eds). *Tectonics of continents and oceans. An explanatory note to the International Tectonic Map of the World.* Scale 1 : 15 000 000, Nauka, Moscow, (1988). (in Russian).
56. J. Malpas and T. Calon. Shulaps ophiolite in the Canadian Cordillera. *Abstract, 29th IGC,* vol. 1, p.136, Kyoto (1992).
57. A.A. Meliakhovetskii and E.V. Sklyarov. Ophiolites and olistostromes of Western Sayan and Tuwa. In: *Riphean - Lower Paleozoic ophiolites of northern Eurasia.* N.L. Dobretsov and L.P. Zonenshain (Eds). pp. 58-71. Nauka, Novosibirsk (1985). (in Russian).
58. E.E. Milanovskii. *A riftogenesis during Earth's history (a riftogenesis of cratons).* Nedra, Moscow (1983). (in Russian).
59. E.E. Milanovskii and A.M. Nikishin. Character of megacycles of Earth, Mars and Moon evolution. *Doklady Akademii Nauk SSSR* 280, 1204-1209 (1985). (in Russian).
60. Z. Misar. Riphean and Paleozoic ophiolites and relative rocks of the Chekh massif. In: *Riphean - Lower Paleozoic ophiolites of northern Eurasia.* N.L. Dobretsov and L.P. Zonenshain (Eds). pp. 119-134. Nauka, Novosibirsk (1985). (in Russian).
61. E.M. Moores. The Proterozoic ophiolite problem, continental emergence, and the venus connection. *Science* 234, 65-68 (1980).
62. T. Nakajima, S. Maruyama, S. Uchiumi, J.G. Liou, X. Wang, X. Xiao and S.A. Graham. Evidence for Late Proterozoic subduction from 700-Myr-old blueschists in China. *Nature* 346, 263-265 (1990).
63. V. Narembski and A. Meierovich. Ophiolites of the Sov'i mountains rim and Early Paleozoic initial rocks of Polish Sudets. In: *Riphean - Lower Paleozoic ophiolites of northern Eurasia.* N.L. Dobretsov and L.P. Zonenshain (Eds). pp. 86-105. Nauka, Novosibirsk (1985). (in Russian).
64. F. Neubauer, W. Frisch, R. Schemerold and H. Schloser. Metamorphosed and dismembered ophiolite suites in the basement of Eastern Alps. *Tectonophysics* 64, 49-62 (1989).
65. A. Nicolas. *Structures of ophiolites and dynamics of oceanic litosphere.* Kluwer Academic Publishers, Dordrecht (1989).
66. A. Nicolas and E.D. Jackson. Repartition en deux provinces des peridotites des chaines alpines longement la Mediterrane: implications geotectonique. *Schweiz. Mineral. Petrogr. Mitt.* 52, 479-495 (1972).
67. A.M. Nikishin and V.E. Khain. About character of Mid-Oceanic Ridges total length changes during geological history of the Earth. *Doklady Akademii nauk SSSR* 320, 157-161 (1991). (in Russian).

68. A.M. Nikishin, V.E. Khain and L.P. Lobkovskii. Scheme of geological evolution of the Earth. *Doklady RAN* 323, 519-523 (1992). (in Russian).
69. P. Olson, P.G. Silver and R.W. Carlson. The large-scale structure of convention in the Earth's mantle. *Nature* 344, 209-215 (1990).
70. V.S. Oxman, L.M. Parfenov, A,V. Prokopiev, V.F. Timofeev and F.F. Tretyakov. The Cherskiy Range ophiolite (northeast Asia). *Abstract, 29th IGC*, vol.1, p. 138. Kyoto (1992).
71. P. Page, M. McCalloch and L. Black. Isotopic data on major events of the Precambrian of Australia. In: *Geology of Precambrian*, 27 IGC, Section C.05, Reports, vol. 5, pp. 14-35, Moscow (1984). (in Russian).
72. J.S. Pallister, J.S. Stacey, L.B. Fischer and W.R. Premo. Precambrian ophiolites of Arabia: geological settings, U-Pb geochronology, Pb-isotope characteristics, and implications for continental accretion. *Precambrian Res.* 38, 1-54 (1988).
73. C. Pin, F. Carme. A Sm-Nd isotopic study of 500 Ma old oceanic crust in the Variscan belt of Western Europe: the Chamrousse ophiolite complex, Western Alps (France). *Contrib. Mineral. Petrol.* 96, 406-413 (1987).
74. O. Piskin, M. Delaloye and P. Voldet. REE behaviour in the Kizil Dag ophiolitic rocks (Hatay, Turkey). *Ofioliti* 13, 193-201 (1987).
75. H. Porada. Pan-African rifting and orogenesis in Southern Equatorial Africa and Eastern Brazil. *Precambrian Res.* 44, 103-136 (1989).
76. M. Prevot, M.E.-M. Derder, M. McWilliams and J. Thompson. Intensity of the Earth's magnetic field: evidence for a Mesozoic dipole low. *Earth and Planet. Sci. Lett.* 97, 129-139 (1990).
77. J.E. Quick. Geology and origin the Late Proterozoic Darb Zubaydah ophiolite, Kingdom of Saudi Arabia. *Geol.Soc. Amer. Bull.* 102, 1007-1020 (1990).
78. H. Rai. Geochemical study of volcanics associated with Indus ophiolitic melange in Western Ladakh, Jammu and Kashmir, India. *Ofioliti* 13, 71-82 (1987).
79. Z. Reti. Comparison of the Mesozoic mafic and ultramafic complexes in northern Hungary. *Ofioliti* 13, 43-52 (1987).
80. M. Sandulescu. Structure and tectonic history of the northern margin of Tethys between the Alps and the Caucasus. *Mem. Soc. Geol. Fr.*, N 154 (2), 3-16 (1989).
81. H. Savu. Prealpine ophiolites and other basites and ultrabasites of Romanian Carpathians and Dobrodja. In: *Riphean-Lower Paleozoic ophiolites of northern Eurasia*. N.L. Dobretsov and L.P. Zonenshain (Eds). pp. 160-167. Nauka, Novosibirsk (1985). (in Russian).
82. A.M.C. Sengor, Y. Yilmaz and I. Ketlin. Remnants of a pre-Late Jurassic ocean in Northern Turkey: fragments of Permian-Triassic paleo-Tethys? *Geol. Soc. Amer. Bull.* 91, 599-609 (1980).
83. S. Sengapta, H.U. Ray, S.K. Acharya and J.B. de Smith. Nature of ophiolite occurrences along the eastern margin of the Indian plate and their tectonic significance. *Geology* 18, 439-442 (1990).
84. P. Spadea, M. Delaloye, A. Espinosa, A. Orrego and J.-J. Wagner. Mineralogy and chemistry of ophiolitic rocks from La Tetilla area, Southwestern Colombia (South America). *Ofioliti* 12, 258 (1987).
85. D.B. Stoeser and V.E. Camp. Panafrican microplate accretion of the Arabian Shield. *Geol.Soc.Amer.Bull.* 96, 817-826 (1985).
86. R.B. Stothers. Periodicity of the Earth's magnetic reversals. *Nature* 322, 444-446 (1986).
87. J. Sutton. Long-term cycles in the evolution of the continents. *Nature* 198, 731-735 (1963).
88. J.J. Veevers (Ed.). *Phanerozoic Earth history of Australia*. Oxford Sci. Publ. (1986).
89. C.-D. Werner. Proterozoische Metabasite in Sachsischen Grundgebirge (DDR) - Sammelhand. *Ophiolithe der PK IX, UK2.* Moskau (1979).
90. C.-D. Werner. Upper Riphean ophiolites of Saxonia (GDR). In: *Riphean - Lower Paleozoic ophiolites of northern Eurasia*. N.L. Dobretsov and L.P. Zonenshain (Eds). pp. 106-119. Nauka, Novosibirsk (1985). (in Russian).
91. E. Wernick. The Archean of Brazil. *Earth Sci. Rev.* 17, 31-48 (1981).
92. A.S. Yakubchuk and A.M. Nikishin. The ophiolites and oceanic crust pulses during Phanerozoic. *Abstract, 29th IGC*, vol.1, p. 142, Kyoto (1992).
93. Zh. M. Zhang, J.G. Liou and R.G. Coleman. An outline of plate tectonics of China. *Geol.Soc.Amer.Bull.* 95, 295-312 (1984).
94. P. Ziegler. *Evolution of Laurussia*. Kluwer Akad.Publ., AH Dordrecht (1989).
95. L.P. Zonenshain, M.I. Kuzmin, O. Tomurtogoo and V.V. Kopteva. Ophiolites of Western Mongolia. In: *Riphean - Lower Paleozoic ophiolites of northern Eurasia*. N.L.Dobretsov and L.P.Zonenshain (Eds). pp. 7-19. Nauka, Novosibirsk (1985). (in Russian).